Developments in Primatology: Progress and Prospects

Series Editor:

Russell H. Tuttle
Department of Anthropology
The University of Chicago, Il, USA

For other titles published in this series, go to
www.springer.com/series/5852

Kristiaan D'Août · Evie E. Vereecke
Editors

Primate Locomotion

Linking Field and Laboratory Research

Springer

Editors
Kristiaan D'Août
Laboratory for Functional Morphology
Department of Biology
University of Antwerp
Belgium
kristiaan.daout@ua.ac.be

Evie E. Vereecke
Anatomy
Faculty of Medicine
Catholic University Leuven, KULAK
Belgium
evie.vereecke@kuleuven-kortrijk.be

ISBN 978-1-4419-1419-4 e-ISBN 978-1-4419-1420-0
DOI 10.1007/978-1-4419-1420-0
Springer New York Dordrecht Heidelberg London

Printed on acid-free paper

Springer is part of Springer Science+Business Media (www.springer.com)

Foreword

Studies of primate locomotion in the field and in captivity spanned the entire 20th century and first decade of the current century, and, as highlighted in *Primate Locomotion: Linking Field and Laboratory Research*, they promise to continue for many more decades as newer generations of scientists devise and employ ever more refined tools and approaches. Major events in the evolutionary history of vertebrates such as the tetrapod shift from water to land, and befeathered reptiles taking to the air, have held special interest for scientists and laypersons alike. Standing prominently among these evolutionary puzzles, human bipedalism also generated great interest in how other primates are built and move and are motivated to do so, thereby stimulating research to test models of precedent positional behaviors and changes that might have occurred in the transition from quadruped to hominid biped. Although the anthropological bias has been strong, many scientists have also pursued topics on nonhuman primate species and a wide variety of other tetrapods simply for their own sake or to illuminate broad biomechanical principles that apply to them (Howell 1944; Young 1957; Hildebrand 1967; Alexander 1968, 2003; Biewener 2003).

Sir Arthur Keith must be counted among the earliest scientists to employ behavioral observations and laboratory experiments, in addition to comparative morphological studies on nonhuman primates to illuminate our peculiar mode of posture and locomotion. While a medical officer in Thailand (1889–1892), Keith set up a primitive dissection laboratory in the dense forest where he resided. His initial goal was to dissect gibbons (*Hylobates lar*) and sympatric colobine monkeys (*Trachypithecus germaini*: Groves 2001; Roos et al. 2008) to see whether they, like his patients, suffered from malaria (Keith 1940, 1950). Following earlier anatomists, he noted marked differences between their internal and external structures, with gibbons more closely resembling humans. He further observed distinct differences between how brachiating gibbons and quadrupedal monkeys negotiated the forest canopy.

When he returned to the United Kingdom he continued to dissect a greater variety of apes and monkeys and conducted experiments to understand possible selective effects of gravity on the human body in relation to obligate orthograde posture and locomotion. For instance, he inserted a mercury manometer into his stomach and rectum (one expects in that sequence) to measure pressures on the pelvic floor, abdominal wall, diaphragm, and viscera as he assumed a variety of postures (Keith 1923).

Basic research slowed during World Wars I and II, but during the latter, Elftman and Manter (1935a,b; Elftman 1944) published much-cited informative comparisons of human and chimpanzee footprints and feet as the subjects walked bipedally. Later researchers have supported many of their observations on the functional morphology of human and chimpanzee feet, but some of their generalizations from a single 5-year-old chimpanzee can be challenged. For instance, chimpanzees more commonly walk with extended lateral toes and an abducted hallux than with curled lateral toes and an adducted hallux (Tuttle 1970, 1987, 1990, 2008; Tuttle et al. 1990, 1991, 1992, 1998). I suspect the extent to which subjects are comfortable during experiments is a factor.

Studies, research papers, symposia, and books on primate locomotion and postcranial morphology in extant and fossil primates burgeoned from the 1960s onwards (Kinzey 1967; Kondo et al. 1975; Jenkins 1974; Morbeck et al 1979; Kondo 1985; Strasser et al. 1998; Ishida et al 2006; Stevens and Carlson 2008), and virtually all meetings of the American Association of Physical Anthropologists, International Primatological Society, and American Society of Primatologists have hosted symposia and podium and poster presentations on these topics.

Clearly, although we have learned a good deal in comparison with the level of pre-20th century knowledge, there are many more puzzles remaining to be solved and envisioned. As a pioneer in the adaptation and application of fine-wire electrode electromyography to apes (Tuttle et al. 1972, 1979, 1983, 1992; Tuttle and Basmajian 1973, 1974a,bc, 1977, 1978a,b; Tuttle 1974, 1994; Tuttle, Basmajian, and Ishida 1975, 1978, 1979; Ishida, Tuttle et al. 1978; Tuttle and Watts 1985; Tuttle, Hallgrímsson, and Basmajian 1994, 1999), I must warn that the return of useful information about the adaptive complexes of subject species, and especially the application of it to interpret fossil primates, is very limited. The same holds for new and refined technologies employed by researchers who report and reflect on their projects in *Primate Locomotion: Linking Field and Laboratory Research.*

As some of the authors remind us, the environments in which one must work are increasingly restricted by rules governing studies on primates, particularly great apes. The good news is that some researchers meet the challenge by creatively crafting protocols that limit or eliminate invasive techniques and physical restriction of their subjects. A further encouraging sign for future advances is that there are many more researchers, laboratories, field sites, and focal species than when I began collaborative research with John V. Basmajian in the United States and Hidemi Ishida, Tasuku Kimura, and Morihiko Okada in Japan. For instance, the 45 authors in *Primate Locomotion: Linking Field and Laboratory Research* are from 8 nations (Belgium, France, Germany, Greece, Japan, Madagascar, the United Kingdom, and the United States).

Finally, I urge all laboratory workers, especially ones who have spent their lives in urban settings, to venture into the field and spend notable spans watching primates and other animals moving on natural substrates. Films are fine, but they really are not the same as one's own direct observations to inform creative laboratory experiments and to bound evolutionary models based on them.

Russell H. Tuttle
Department of Anthropology
The University of Chicago, Il, USA

References

Alexander RMcN (1968) Animal Mechanics. Sidgwich & Jackson, London.

——— 2003. Principles of Animal Locomotion. Princeton University Press, Princeton, NJ.

Biewener AA (2003) Oxford University Press, Oxford.

Elftman H (1944) The bipedal walking of the chimpanzee. J Mammal 24:67–71.

Elftman H, Manter J (1935a) Chimpanzee and human feet in bipedal walking. Am J Phys Anthropol 20:69–79.

——— (1935b) The evolution of the human foot, with especial reference to the joints. J Anat 70:56–67.

Groves C (2001) Primate Taxonomy. Smithsonian Institution Press, Washington, DC.

Hildebrand M (1967) Symmetrical gaits of primates. Am J Phys Anthropol 26:119–130.

Howell AB (1944) Speed in Animals: Their Specialization for Running and Leaping. Hafner, New York.

Ishida H, Tuttle RH, Okada M, Kimura T (1978) Activities of hindlimb muscles in a bipedal gibbon. In: Chivers DJ, Joysey KA (eds), Recent Advances in Primatology, Vol. 3: Evolution. Academic Press, London, pp 459–462.

Ishida H, Tuttle R, Pickford M, Ohihara N, Nakatsukasa M (2006) Human Origins and Environmental Backgrounds. Springer, New York.

Jenkins FA Jr, ed (1974) Primate Locomotion. Academic Press, New York.

Keith A (1923) Man's posture: its evolution and disorders. Br Med J 1:451–454, 499–502, 545–548, 587–590, 624–626, 669–672.

——— (1940) Fifty years ago. Am J Phys Anthropol 26:251–267.

——— (1950) An Autobiography. Watts, London.

Kinzey WG, ed (1967) Symposium on primate locomotion. Am J Phys Anthropol 26:115–275.

Kondo S (1985) Primate Morphophysiology, Locomotor Analyses and Human Bipedalism. University of Tokyo Press, Tokyo.

Kondo S, Kawai M, Ehara A, Kawaura S (1975) Proceedings from the Symposia of the Fifth congress of the International Primatological Society. Japan Science Press, Tokyo.

Morbeck ME, Preuschoft H, Gomberg N (1979) Environment, Behavior, and Morphology: Dynamic Interactions in Primates. Gustav Fischer, Stuttgart and New York.

Roos C, Nadler T, Walter L (2008) Mitochondrial phylogeny, taxonomy and biogeography of the silvered langur species group (*Trachypithecus cristatus*). Mol Phylogenet Evol 47:629–636.

Stevens NJ, Carlson KJ, eds (2008) Bridging gaps between experimental and naturalistic approaches in the study of primate behavior. Int J Primatol 29:1395–1480.

Strasser E, Fleagle J, Rosenberger A, McHenry H (1998) Primate Locomotion: Recent Advances. Plenum Press, New York.

Tuttle RH, Watts DP (1985) The positional behavior and adaptive complexes of *Pan gorilla.* In: Kondo S (ed), Primate Morphophysiology, Locomotor Analyses and Human Bipedalism, pp. University of Tokyo Press, Tokyo, pp 261–28.

Tuttle RH (1970) Postural, propulsive, and prehensile capabilities in the cheiridia of chimpanzees and other great apes. In Bourne GH (ed), The Chimpanzee, Vol. 2. Karger, Basel, pp 167–253.

——— (1975) Electromyography of *Pan gorilla*: an experimental approach to hominization. In: Kondo S, Kawai M, Ehara A, Kawamura S (eds), Proceedings from the Symposia of the Fifth Congress of the International Primatological Society. Japan Science Press, Tokyo, pp 303–314.

——— (1987) Kinesiological inferences and evolutionary implications from Laetoli bipedal trails G-1, G-2/3, and A. In: Leakey MD, Harris JM (eds), Laetoli, a Pliocene Site in Northern Tanzania.Clarendon Press, Oxford, pp 503–523..

——— (1990) The pitted pattern of Laetoli feet. Nat Hist 3/90:60–65.

——— (1994) Up from electromyography: primate energetics and the evolution of human bipedalism. In: Corruccini RS, Ciochon RL (eds), Integrative Paths to the Past: Paleoanthropological Advances in Honor of F. Clark Howell. Prentice Hall, Englewood Cliffs, NJ, pp 269–284.

——— (2008) Footprint clues in hominid evolution and forensics: lessons and limitations. Ichnos 15:158–165.

Tuttle RH, Basmajian JV (1973) EMG of locomotion in gorilla and man. In: Stein R, Pearson KB, Smith RS, Redford JB (eds), Control of Posture and Locomotion. Plenum Press, London, pp 599–609.

——— (1974a) Electromyography of forearm musculature in gorilla and problems related to knuckle-walking. In: Jenkins FA Jr (ed), Primate Locomotion. Academic Press, New York, pp 293–347.

——— (1974b) Electromyography of the manual long digital flexor muscles in gorilla. In: Barnosell F (ed), Proceedings of the 6th Congreso Internacional de Medicina Fisica, Vol. II, 1972. Ministerio de Trabajo, Instituto Nacional de Prevision, Madrid, pp 311–315.

——— (1974c) Electromyography of brachial muscles in *Pan gorilla* and hominoid evolution. Am J Phys Anthropol 41:71–90.

——— (1977) Electromyography of pongid shoulder muscles and hominoid evolution I. Retractors of the humerus and "rotators" of the scapula. Yrbk Phys Anthropol 1976 20:491–497.

——— (1978a) Electromyography of pongid shoulder muscles. II. Deltoid, rhomboid and "rotator cuff." Am J Phys Anthropol 49:47–56.

——— (1978b) Electromyography of pongid shoulder muscles. III. Quadrupedal positional behavior. Am J Phys Anthropol 49:57–70.

Tuttle RH, Basmajian JV, Ishida H (1975) Electromyography of the gluteus maximus muscle in gorilla and the evolution of hominid bipedalism. In: Tuttle RH (ed), Primate Functional Morphology and Evolution. Mouton, The Hague, pp 253–269.

———(1978) Electromyography of pongid gluteal muscles and hominid evolution. In: Chivers DJ, Joysey KA (eds), Recent Advances in Primatology, Vol. 3, Evolution.Academic Press, London, pp 463–468.

——— (1979) Activities of pongid thigh muscles during bipedal behavior. Am J Phys Anthropol 50:123–136.

Tuttle RH, Basmajian JV, Regenos E, Shine G (1972) Electromyography of knuckle-walking: results of four experiments on the forearm of *Pan gorilla*. Am J Phys Anthropol 37:255–266.

Tuttle RH, Cortright G, Buxhoeveden D (1979) Anthropology on the move; progress in studies of non-human primate positional behavior. Yrbk Phys Anthropol 1979 22:187–214.

Tuttle RH, Hallgrímsson B, Basmajian JV (1994) Electromyography and elastic mechanisms in kunckle-walking *Pan gorilla* and *Pan troglodytes*. In: Thierry B, Anderson JR, Roeder JJ, Herrenschmidt N (eds), Current Primatology, Vol. 1: Ecology and Evolution. Université Louis Pasteur, Strasbourg, pp 215–222.

——— (1999) Electromyography, elastic energy and knuckle-walking: a lesson in experimental anthropology. In: Lindburg D, Strum SC (eds), The New Physical Anthropology. Prentice-Hall, Englewood Cliffs, NJ, pp 32–41..

Tuttle RH, Hallgrìmsson B, Stein T (1998) Heel, squat, stand, stride: function and evolution of hominoid feet. In: Strasser E, Fleagle J, Rosenberger A, McHenry H (eds), Primate Locomotion: Recent Advances. New York: Plenum Press, New York, pp 435–448.

Tuttle RH, Hollowed J, Basmajian JV (1992) Electromyography of pronators and supinators in great apes. Am J Phys Anthropol 87:215–226.

Tuttle RH, Velte M, Basmajian JV (1983) Electromyography of brachial muscles in *Pan troglodytes* and *Pongo pygmaeus*. Am J Phys Anthropol 61:75–83.

Tuttle RH, Webb DM, Baksh M (1991) Laetoli toes and *Australopithecus afarensis*. Hum Evol 6:193–100.

Tuttle RH, Webb DM, Tuttle NI (1991) Laetoli footprint trails and the evolution of hominid bipedalism. In : Coppens Y, Senut B (eds), Origine(s) de la Bipédie chez les Hominidés. C.N.R.S, Paris, pp 203–218.

Tuttle RH, Webb DM, Tuttle NI, Baksh M (1992) Footprints and gaits of bipedal apes, bears and barefoot people: perspectives on Pliocene tracks. In: Matano S, Tuttle RH, Ishida H, Goodman M (eds), Topics in Primatology, Vol. 3, Evolutionary Biology, Reproductive Endocrinology and Virology. University of Tokyo Press, Tokyo, pp 221–242.

Young JZ (1957) The Life of Mammals. Oxford University Press, Oxford.

Contents

Contributors

Peter Aerts
Laboratory for Functional Morphology, Department of Biology, University of Antwerp, Belgium; and Vakgroep Sport-en Bewegingswetenschappen, Ghent University, Belgium

Gilles Berillon
UPR 2147 CNRS, Dynamique de l'Evolution Humaine, Paris, France

Mary L. Blanchard
Primate Evolution and Morphology Group, School for Biomedical Sciences, The University of Liverpool, UK

Christophe Boesch
Max-Planck Institute for Evolutionary Anthropology, Germany

Kristian J. Carlson
Department of Anatomy, New York College of Osteopathic Medicine, USA

Andrew Chamberlain
University of Sheffield, UK

Susan M. Cheyne
Wildlife Conservation Research Unit, Department of Zoology, University of Oxford, UK

Robin H. Crompton
Primate Evolution and Morphology Group, School for Biomedical Sciences, The University of Liverpool, UK

Kristiaan D'Août
Laboratory for Functional Morphology, Department of Biology, University of Antwerp, Belgium, and Centre for Research and Conservation, Belgium

G. Daver
Department of Prehistory, Muséum national d'Histoire naturelle, Musée de l'Homme, France

G. Dubreuil
Station of Primatology, CNRS, France

Jean-Pierre Gasc
Laboratoire d'Anatomie Comparée, Muséum National d'Histoire Naturelle, France

Denise Guillot
University of Michigan, USA

Eishi Hirasaki
Osaka University, Japan

R. Jones
University of Salford, UK

Cornelia Krause
Institut für Spezielle Zoologie und Evolutionsbiologie, Friedrich Schiller Universität Jena, Germany

W. Scott McGraw
Department of Anthropology, The Ohio State University, USA

M.E. Morbeck
Department of Anthropology, University of Arizona, USA

Martin N. Muller
Department of Anthropology, University of New Mexico, USA

F. Multon
UFR-STAPS, University of Rennes 2, France

Masato Nakatsukasa
Kyoto University, Japan

G. Nicolas
UFR-STAPS, University of Rennes 2, France

Toshisada Nishida
Mahale Mountains Chimpanzee Research Project, Kyoto University, Graduate School of Science, Japan

Naomichi Ogihara
Kyoto University, Japan

Richel Payne
Royal Veterinary College, UK

Herman Pontzer
Department of Anthropology, Washington University, USA

David A. Raichlen
Department of Anthropology, University of Arizona, USA

Fidy Ralainasolo
University of Antananarivo, Madagascar

Jonah H. Ratsimbazafy
Durrell Wildlife Conservation Trust, Madagascar

Manuela Schmidt
Institut für Spezielle Zoologie und Evolutionsbiologie, Friedrich Schiller Universität Jena, Germany

D. Schmitt
Department of Evolutionary Anthropology, Duke University, USA

Paul W. Sciulli
Department of Anthropology, The Ohio State University, USA

Liza J. Shapiro
University of Texas at Austin, USA

Michael D. Sockol
Department of Anthropology, University of California at Davis, USA

Art Souther
Corvallis, Oregon, USA

Nancy J. Stevens
Department of Biomedical Sciences, Ohio University, Ohio, USA

D. Rick Sumner
Department of Anatomy and Cell Biology, Rush Medical College, USA

E.E. Vereecke
Anatomy, KULAK, Catholic University of Leuven, Belgium

B. de la Villetanet
LAPP, UMR 5199 PACEA University of Bordeaux 1, France

Ambigail E. Williams
Department of Biology, James Madison University, USA

Richard W. Wrangham
Department of Anthropology, Harvard University, USA

Roshna E. Wunderlich
Department of Biology, James Madison University, USA

Atsushi Yamanaka
Department of Oral Anatomy, Kagoshima University Dental School, Japan

Dionisios Youlatos
Department of Zoology, School of Biology, Aristotle University of Thessaloniki, Greece

Jesse W. Young
University of Texas at Austin, USA

Chapter 1
Introduction. Primate Locomotion: Toward a Synergy of Laboratory and Field Research

Evie Estelle Vereecke and Kristiaan D'Août

Researchers have studied primate locomotion over a considerable period of time, e.g., baboon locomotion by Muybridge (1899), and it continues to receive a great deal of attention from primatologists, anthropologists, and biomechanists worldwide. There are several good reasons for this, many boiling down to the primates possessing several "unique" features, which are thought to relate to their ancestral arboreal niche, and that presumably opened options for the evolution of hominins (as bipeds). In the past, primate locomotion had been tackled from a variety of perspectives. Field primatologists have collected quantitative data on locomotion and posture since the 1960s (e.g., Napier and Napier 1967; Richard 1970; Grand 1972; Rose 1973, 1976; Chivers 1974; Mittermeier and Fleagle 1976; Mittermeier 1978); in-depth biomechanics research on primate locomotion has been conducted since as early as 1935, with Elftman and Manter's study on chimpanzee bipedalism, and Fleagle and colleagues could be considered pioneers in combining both approaches (e.g., Fleagle 1974, 1976, 1992, 1999; Fleagle and Mittermeier 1980).

Whatever the specific research aims, ultimately any primatologist needs to understand the integrative story behind the species' locomotor behavior: how the individual is able (and has been able) to perform adequately in its natural habitat. The seminal paper by Arnold (1983) provides a good framework (see also Wunderlich et al.,Chapter 8). This framework can be expanded (e.g., Aerts et al. 2000), but in its basic form it links morphology, performance, and fitness. Usually, the performance gradient (linking morphology to performance) has been tackled by lab-based researchers in projects *ex situ*, whereas the fitness gradient (linking performance to fitness) has been dealt with mostly *in situ*, by field primatologists. Functional morphology, for instance, studying the relationship between skeletal structure and locomotion, is a good example of the performance gradient (see also

K. D'Août (✉)
Laboratory for Functional Morphology, Department of Biology, University of Antwerp, Belgium, and Centre for Research and Conservation, Belgium
e-mail: kristiaan.daout@ua.ac.be

K. D'Août and E.E. Vereecke (eds.), *Primate Locomotion: Linking Field and Laboratory Research*, Developments in Primatology: Progress and Prospects,
DOI 10.1007/978-1-4419-1420-0_1, © Springer Science+Business Media, LLC 2011

International Journal of Primatology, special volume, 2010). Insights enable the interpretation of fossils and, by doing so, the locomotor mode of extinct species.

Arnold's scheme is a full circle per generation, and does not allow for missing links if we are to understand fully the adaptive process. Therefore, it is essential that field and laboratory-based primatologists communicate or collaborate. The IPS symposium "Primate Locomotion: Linking *Ex Situ* and *In Situ* Research" in Edinburgh (August 2008) had exactly this purpose, and this book builds upon this initiative. It includes chapters by all symposium participants as well as chapters by invited authors who have contributed significantly to our understanding of primate locomotion and adaptation. As the aim of this volume is to bring together field and laboratory-based primatologists and stimulate future collaboration, we have attracted primatologists from a diversity of research backgrounds, each presenting their recent work and proposing opportunities and/or improvements that could be made by integrating both approaches.

This first chapter sets a general framework, illustrating how the various chapters support the idea of the book, and present issues that were raised during the general discussion at the conference.

There are various ways in which the lab and the field can approach each other, which can largely be attributed to four categories (all of which are illustrated in this volume).

In the first approach, field and lab stay basically separated, but both disciplines communicate by means of publications, meetings, and personal contacts. This approach is best established, as it is the option requiring the least effort. Often, the approach taken is dictated by the research question, leaving little room for a choice between laboratory or field research.

Schmitt (Chapter 2) provides a very good overview of the problems encountered in both field and laboratory studies. Not only does it map those problems in a transparent manner, but more importantly, it also proposes practical suggestions for solving these problems, for instance, by providing examples of how high-tech laboratory data can yield simplified proxies for ecologically crucial variables, e.g., energetic efficiency, and thus enable field workers to address easily the typical "laboratory" link.

This volume contains some clear examples of research that could not have been conducted in the wild, but is possible only in the laboratory, in zoos, or in other captive populations. Ogihara et al. (Chapter 4) use a combination of CT-scanning and high-resolution kinematic data to construct a dynamic model of Japanese monkey walking. Such advanced techniques are strictly limited to the laboratory, and are in fact a step beyond experimental laboratory studies (although field data can provide some input for modeling studies). Berillon et al. (Chapter 5) describe an integrated research project in which 3D-kinematics, dynamic measurements, and morphometrics of a large and well-documented baboon group, and for all ontogenetic stages, are combined. Again, studies from the wild would not provide the same level of detail in any of these topics. Interestingly, this study shows that typical quadrupeds may be adept bipedal walkers, and so the latter behavior, which has been observed in the wild, often considered atypical, may be an integral part of the baboon locomotor repertoire. Schmidt and Krause (Chapter 6) present data on the kinematics of the shoulder, which become fully visible only by using X-ray videography, limited to experimental setups in the laboratory. However, it is demonstrated how the resulting data can be brought to the field. Specifically, the laboratory data suggest that the invisible (for conventional, portable videography) aspects of shoulder function are quite similar within mammals. In this way, field data miss some information that can—with caution—be supplemented by knowledge from the laboratory, albeit of different individuals or different species). On the other hand, observational data from unconstrained wild individuals can indicate how large the proportion of locomotor behavior is in the total positional repertoire, often overlooked in experimental setups designed only for studying locomotion. Examples of neglected behaviors are sitting and sleeping, illustrated by McGraw and Sciulli (Chapter 12). In this study, detailed behavioral observations (such as posture and substrate use) in seven species of cercopithecids are linked to the morphometrics of ischial tuberosities of museum samples. These data were respectively collected in the field and in the laboratory, but combining them yields new insights into the ecological function of an anatomical feature, i.e., sitting pads.

Finally, as pointed out by Cheyne (Chapter 11), laboratory studies can provide baseline data for field studies and allow calibration; field studies can feed the laboratory, by indicating what the natural locomotor repertoire of the animals is, in what context particular locomotion patterns occur, what locomotor aspects require further investigation in the laboratory, etc.

The second approach is to "bring the lab to the field" (Williams et al. 2008). In this approach, the same type of questions are asked that are traditionally addressed in laboratory research, but the data are collected in the wild, most often in an effort to increase the relevance of the observed locomotor behavior and guarantee that individuals are performing naturally. Often, this (still) requires invasive laboratory techniques that are brought to the field and it can (and should) be questioned how invasive one can be without impeding the benefits of field-based research. This is a fine balance that will vary for different species and research questions.

The chapters by Blanchard et al. (Chapter 10) and Cheyne (Chapter 11), both based on field research, deal with such questions: what kind of quantitative locomotor data can be reliably collected in the wild, and what kind of data remain bound to the laboratory floor? At the same time, Blanchard et al. point out how rapid technological advances, such as the availability of inexpensive, portable high-speed video recorders, are rapidly blurring the boundaries between fieldwork and laboratory studies. Cheyne (Chapter 11), using field work with gibbons as a case study, brings forward suggestions of how (former) laboratory techniques can be brought effectively into long-term field studies. Importantly, she points out how, with a minimal additional effort, such an approach can foster new insights into a variety of aspects related to an integrated understanding of primate locomotion, including biomechanics and ecology, e.g., knowing the energetic cost of moving on compliant supports.

The third way in which field-based and laboratory-based workers can more closely integrate is to "bring the field to the lab."

This approach aims to include more complexity in experimental setups of laboratory locomotor studies, to accurately reproduce the conditions in the wild (e.g., Stevens et al., Chapter 16, and several other contributors to this volume). It is, of course, impossible to truly bring the complexity of the natural habitat into the laboratory, yet selected aspects of the field *can* be brought to the laboratory and studied, while guaranteeing the full relevance of their origin *in situ*. The contribution by Carlson et al. (Chapter 9) is a good example; in the chapter, the authors performed morphometric analyses on skeletal material collected in the field. This as such is not groundbreaking, but the merit of the study is that the osteological material came with detailed background information of life history of the population (unlike most osteological material available in, e.g., museum collections). In this way, a detailed analysis of long bone structure in different populations of chimpanzees, confined to the laboratory, can be linked to behavioral data collected in the field. Such an approach should be encouraged, for instance, by providing anatomists and biomechanists with well-documented cadaver material from the field (with known life history) and not just from captive populations or museum collections.

Finally, the fourth approach is to truly combine disciplines that were traditionally limited to either the laboratory or the field and use existing, or develop new, techniques for the assessment of the performance and fitness gradient as noninvasively as possible. Several chapters in this volume address how such true integration of primate field and laboratory research can be accomplished.

Watson et al. (Chapter 7) have studied load carrying in humans and apes, with data collected in the field, in zoos, and in the laboratory. Taking the example of human load carrying, they have gathered field observations of carrying behavior in all apes. These observations not only yielded insight in potential carrying modes of our hominin ancestors, but also dictated the protocol for the laboratory-based section of their study. Wunderlich et al. (Chapter 8) explicitly address the link among morphology, performance, and fitness, pointing out that Arnold already argued for an integration of laboratory and field work in his seminal paper of 1983. Wunderlich et al. closely integrate morphological, behavioral, and fitness data collected in the field, and functional analyses collected in the laboratory, of *Propithecus*, exploiting the unique strengths of both approaches. By doing so, they have gained insights that could not have been obtained by either approach in isolation. A good example is how leg shape is biomechanically shown to affect performance and, as a consequence, reproductive success in males, hence their suggestion to expand an understudied topic in primate locomotion: the impact of sexual selection. Guillot (Chapter 13) has further suggestions for tackling the full adaptive process, crucial aspects, e.g., performance measures, heritability studies, that remain understudied in primates when compared to other species.

Youlatos and Gasc (Chapter 14) show that is possible to perform quantitative analysis of kinematics in the field, specifically of red howlers (*Alouatta seniculus*) in primary rain forest, despite technical limitations. The latter are likely to become smaller due to technical advances in sensitivity, autonomy, and cost of video equipment. In Youlatos and Gasc's study, field data support (preliminary) laboratory observations; Stevens et al. (Chapter 16) directly compared field and laboratory data, and also found them to correspond well. Even though the good accordance between laboratory and field data in both studies is reassuring, many contributors state that the obvious decrease of complexity of laboratory setups compared to natural habitat remains a challenge. Shapiro et al. (Chapter 17) demonstrate this point in their contribution, which contains the first quantitative analysis of quadrupedal kinematics of *Saimiri* in the wild. Their study of locomotion on idealized supports, e.g., poles in the laboratory, and natural supports (branches) shows that gait flexibility on less complex supports is reduced, even though basic gait parameters, such as interlimb coordination and duty factors, are similar. Guillot's study of suspensory gait in two cebid species (Chapter 13) is in line with this idea. She has shown that locomotor data collected in the wild may reveal features of gait, such as asymmetries and ways of dealing with pliant and unreliable supports that may be concealed in simplified ("impoverished") laboratory setups. Both of these studies prove the point made by Stevens et al. (Chapter 16), i.e., that we have relatively little information regarding kinematic solutions or locomotor strategies primates employ to navigate their habitats. Laboratory studies should incorporate setups of a higher complexity than is often the case to date, even though an exact replication of natural complexity may remain impossible, or even undesirable in some cases (Stevens et al., Chapter 16). In any case, field workers should try to quantify substrate characteristics such as compliance.

Pontzer et al. (Chapter 15) address the energetic cost of locomotion in chimpanzees. Traditionally, there has been a trade-off between accuracy of energetic cost estimate per distance traveled (best in the laboratory) and an insight in locomotor activities, including time budget and distance traveled (limited to the field). Pontzer et al. review the literature and present concrete ways of combining both, allowing for a more accurate estimate of ranging cost than would be obtained by using biomechanical data or observational data in isolation, while still refraining from invasive experiments in the field, e.g., by using doubly labeled water.

Together, the chapters of this book prove that many primatologists are open to stepping beyond their field (or laboratory) of expertise, by combining work that used to be limited to either field or laboratory settings, or, when theoretical or practical issues prevent doing so, by collaborating intensively. May such interdisciplinary approaches be even strengthened in the future and contribute to an ever increasing understanding of our common interest: an integrated view on primate locomotion.

References

Aerts P, Van Damme R, Van Elsacker L, Duchêne V (2000) Spatio-temporal gait characteristics of the hind-limb cycles during voluntary bipedal and quadrupedal walking in bonobos (*Pan paniscus*). Am J Phys Anthropol 111:503–517.

Arnold SJ (1983) Morphology, performance and fitness. Am Zool 23:347–361.

Chivers DJ (1974) The siamang in Malaya: a field study of a primate in tropical rain forest. Contrib Primatol 4:1–335.

Elftman H, Manter J (1935) Chimpanzee and human feet in bipedal walking. Am J Phys Anthropol 20(1):69–79.

Fleagle JG, Mittermeier RA (1980) Locomotor behavior, body size, and comparative ecology of seven Surinam monkeys.Am J Phys Anthropol 52:301–314.

Fleagle J (1974) Dynamics of a brachiating siamang [*Hylobates* (*Symphalangus*) *syndactylus*]. Nature 248:259–260.

Fleagle JG (1976) Locomotion and posture of the Malayan siamang and implications for hominoid evolution. Folia Primatol 26:245–269.

Fleagle JG (1992) Primate locomotion and posture. In: Jones S, Martin R, Pilbeam D (eds), The Cambridge Encyclopedia of Human Evolution. Cambridge University Press, Cambridge. pp 75–77, 79.

Fleagle JG (1999) Primate Adaptation and Evolution, 2nd ed. Academic Press, San Diego.

Grand TI (1972) A mechanical interpretation of terminal branch feeding. J Mammal 53:198–201.

Mittermeier RA, Fleagle JG (1976) The locomotor and postural repertoires of *Ateles geoffroyi* and *Colobus guereza*, and a re-evaluation of locomotor category semibrachiation. Am J Phys Anthropol 45:235–255.

Mittermeier RA (1978) Locomotion and posture in *Ateles geoffroyi* and *Ateles paniscus*. Folia Primatol 30:161–193.

Muybridge E (1899) Animals in Motion. Chapman & Hall, London.

Napier JR, Napier PH (1967) A Handbook of Living Primates. Academic Press, London.

Richard A (1970) A comparative study of the activity patterns and behavior of *Alouatta villosa* and *Ateles geoffroyi*. Folia Primatol 12:241–263.

Rose MD (1973) Quadrupedalism in primates. Primates 14:337–357.

Rose MD (1976) Bipedal behavior of olive baboons (*Papio anubis*) and its relevance to an understanding of the evolution of human bipedalism. Am J Phys Anthropol 44:247–261.

Williams SH, Vinyard CJ, Glander KE, Deffenbaugh M, Thompson CL (2008) New technology for understanding primate masticatory form, function and evolution: a portable system for assessing jaw-muscle function in the field. Primate Eye 96:296.

Chapter 2
Translating Primate Locomotor Biomechanical Variables from the Laboratory to the Field

Daniel Schmitt

Abstract One of the critical goals of primate evolutionary morphology is to understand the functional anatomy of muscular and osteological features to infer behavior in the fossil record. One of the most productive approaches for testing functional hypotheses is the comparative experimental approach first advocated by Washburn in the early 1950s. Since that time, laboratory-based approaches have provided profound insights into the biomechanics of primate locomotion and helped anthropologists understand important aspects of limb design. However, a lack of connection to naturalistic data collected from the field has limited the full value of these data. This chapter proposes that there are a number of simple variables that can be collected both in the laboratory and the field that reflect important underlying aspects of locomotor biomechanics. These include gait choice, limb phase, and joint yield all of which appear to be associated with joint loading and center of mass movements. Using these measures, this chapter provides a model for the way in which laboratory-based and field-based data may be analyzed to provide a complete perspective on primate locomotion.

Keywords Forces • Gait • Locomotion • Primate

Abbreviations

COM	center of mass
DS	diagonal sequence
KE	kinetic energy
PE	potential energy
TE	total energy

D. Schmitt (✉)
Department of Evolutionary Anthropology, Duke University, Durham, NC, USA
e-mail: Daniel.Schmitt@duke.edu

K. D'Août and E.E. Vereecke (eds.), *Primate Locomotion: Linking Field and Laboratory Research*, Developments in Primatology: Progress and Prospects,
DOI 10.1007/978-1-4419-1420-0_2, © Springer Science+Business Media, LLC 2011

Introduction

For as long as I have been doing research, beginning with my dissertation, whenever I give a talk there is always someone who comes up to me after my presentation to discuss all the limitations of data collected in a captive environment. The person always ends by saying: "Wouldn't it be cool to take your force plate and put it in a tree in the wild?" I always agree that it would be cool, but it would also be prohibitively difficult and expensive. Nonetheless, despite the significant pragmatic limitations, understanding the biomechanics of primates in their normal environment is an important goal and over the past fifteen years many anthropologists have been trying to do the next best thing. We have been gathering the kind of information that would allow us to make reasonable connections between the kind of detailed data you can collect in the laboratory and the kind of data that can be collected in the field. That is why the symposium that Evie Vereecke and Kristiaan D'Août organized in 2008 in Edinburgh, Scotland at the meeting of the International Primatological Society and the resulting volume is so exciting. In the chapter that follows I use the occasion of this project to try and see how close we have gotten to moving the force plate, at least metaphorically, into the field. The goal therefore is to make use of more than 15 years worth of laboratory-based studies of primate locomotor biomechanics to identify simple variables that seem to reflect clearly deeper underlying biomechanical patterns.

The intersection between field-based and laboratory-based studies has a long been a sore point in our field, and has been frequently discussed but rarely acted upon. It was discussed in numerous chapters of an important volume in 1979 (Morbeck et al. 1979), a revisiting of that volume (Plavcan et al. 2002), and a recent symposium (Grossman 2006) and has been touched on repeatedly over the years. By way of example, over the past forty years we can look to several important papers that have tried to make quantitative observations of locomotor behavior in the wild and relate those back to biomechanical aspects of primate locomotion. Grand's two seminal papers in 1968 on howler locomotion are strong examples of this approach (Grand 1968a, b). John Fleagle's (1974) paper on gibbon locomotion is another excellent example of how combining field and lab approaches provide new insights. In 1979, Mike Rose and Maryellen Morbeck published profound chapters on locomotor behavior in the field of vervets and black and white colobus monkeys. More recently, Dunbar and Badam's (2000) work on bonnet macaques and Byron and Covert's (2004) work on langurs also serve as models of what can be done and what should be done. Yet, connecting laboratory-based and field-based data remains something we just have not been doing enough. Moreover, as good as these papers are, they lack explicit connection to laboratory-based data that include movement and force or muscle patterns.

There are some clear reasons for this lack of connection between field and laboratory data. First, there has always been an odd resistance among biological anthropologists to data collected in the laboratory (see Fleagle 1979; Lemelin and Schmitt 2007, Schmitt et al. 2008 for a more detailed discussion). The history of laboratory-based research in physical anthropology begins in earnest with the work of Elftman and Manter (1935) which is followed by Sherwood Washburn's (1951a, b) call for a "modern experimental comparative anatomy". Washburn's argument for an objective mechanism of providing quantitative biomechanical data was compelling and timely. The biological sciences were already developing a vibrant field of laboratory-based research. But this experiment-oriented approach was not immediately embraced by our field. Washburn's approach using laboratory-based data to resolve conflicts in scenarios of human evolution was not always well received by his peers at the time. Many of his peers saw the experimental method as a major threat because "they thought it was destroying the evidence" (DeVore 1992: 417).

In spite of the obvious theoretical strengths and 50 years of observed success of an experimental comparative approach, few physical anthropologists today test their functional models with experimental data. There are several reasons for the lack of rigorous testing using laboratory data. Many anthropologists misunderstand how the experimental approach can be used to test functional hypotheses. Too often criticisms are made about small sample sizes, unnatural laboratory conditions, and the highly technical aspect of methods used in the laboratory. These concerns inhibit the willingness of physical anthropologists to collect experimental data and the acceptance of such data when they are presented. In the absence of experimental data, confirmation of a functional model can be achieved only via traditional comparative anatomy, e.g., the prediction that long legs are mechanically critical for leaping primates is confirmed by the observation that other leaping animals have long legs. This mode of checking functional models may lead to correct conclusions, but as Bock (1977), Homberger (1988), and others have noted, this is not always the case. Lauder (1996:56) noted that such conclusions are based on untested assumptions and that:

> ...in our desire to draw conclusions about biological design and to support theoretical views of how organism are built, we have been too willing to make assumptions about the relationship between structure and mechanical function...[and]... we have not often conducted the mechanical and performance tests needed to assess the average quality of organismal design.

Second, and more relevant to this volume, the practice of "field biomechanics" is difficult, and data collected in the field cannot be as accurate or precise as data collected in controlled conditions. Even if we overcome the practical issues of equipment and animal behavior. there are other serious constraints for collecting any acceptable kinematic data, let alone data on substrate reaction forces, oxygen consumption, bone strain, or muscle activity. Most of these constraints concern how the animals move and how well that movement can be defined. Moreover, the equipment and software programs are very expensive.

This last concern has recently been relieved. Very recent innovations have made high-speed high-quality video analysis inexpensive and easy. There are now several commercially available hand-held cameras that can record at least 250 images per second and some can go as high as 1000 frames per second and do so even in low light. This type of camera was unthinkable until recently.

In addition, durable, inexpensive, lightweight computers make storage and manipulation of video output possible in the field. Video data can be edited, split, and filtered with freeware programs like VirtualDubMod (http://virtualdubmod.sourceforge.net/). With this software alone footfall data and velocity can be collected with ease. Further analysis including joint angles segment velocities and accelerations are made possible with DLTDataivewer (http://www.unc.edu/~thedrick/software1.html), a freeware add-on module for Matlab (The Mathworks, Natick, MA). This simple, yet sophisticated, tool written by Ty Hedrick (Hedrick 2008) will allow any researcher to collect coordinate data from multiple cameras. With just these tools we can now collect and analyze video data right in the field.

Variables That Can Be Collected Under Field and Laboratory Conditions

There are, of course, serious limitations regarding which variables we can collect in the field. Although the new technology allows a quite ambitious approach, this chapter begins with a limited and simple set of variables that can connect field and laboratory work. Using video recordings that one can easily collect under field and laboratory conditions, it is possible to examine the following.

Gait Choice

Using simple video analysis techniques (or even in some cases by eye), researchers can record the type (walk, gallop, canter, bound, amble; see Fig. 2.1a–c), context (substrate used and whether the animal uses this gait during normal travel or rapid escape), and the frequency of each gait in the wild. If reliable data exist as to the mechanical or physiological criteria that govern gait choice, it will be possible to infer the underlying mechanical processes. More importantly, field data will provide the relevant context for gait choice. If it can be determined, for example, that a specific gait is particularly efficient or moderates load, then understanding the context in which the individual chooses that gait can reveal some of the priorities associated with gait choice.

Footfall Sequence and Limb Timing

This includes both the binary distinction between diagonal and lateral sequence walking gaits (Fig. 2.2a) as well as quantification of diagonality (Cartmill et al. 2002).

Fig. 2.1 Three gait types commonly used by primates. (**a**) A walking gait in which the hind foot contact is followed by a contralateral forefoot contact. There is no aerial phase and as a result the duty factor (contact time/stride time) for any foot is greater than 0.5. (**b**) A gallop in which forefoot contact is followed by the second forefoot, followed by an aerial phase, and then the contact of the two hindfeet in sequence. The duty factor of any foot is less than 0.5. Many primates adopt a canter (Howell 1944; O'Neill 2008) that is a slow gallop (based on foot contact sequence) but does not have an aerial phase. The images for (**a**) and (**b**) are derived from Schmitt et al. (1994). (**c**) An ambling gait in which hind limb contact is followed by forelimb contact but there is an aerial phase for the two hind limbs and/or two forelimbs. The definition and image are from Schmitt et al. (2006)

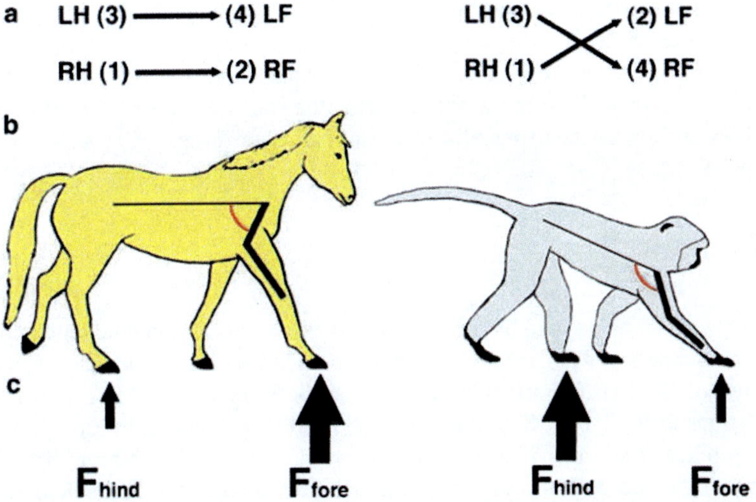

Fig. 2.2 Summary of the commonly accepted differences that are believed to distinguish the walking gaits of most primates from those of most nonprimate mammals. Nonprimates generally use (**a**) lateral sequence walking gaits, (**b**) have a humerus that—at ground contact—is retracted relative to a horizontal axis passing through the shoulder, and (**c**) have greater peak vertical forces on their forelimbs than they do on their hind limbs. Primates show the opposite pattern. (From Schmitt and Lemelin 2002)

Simple video recordings can be easily collected and analyzed for timing. A discrete approach to gait analysis, e.g., walk, trot, amble, canter, gallop, has value but does not necessarily reveal underlying mechanics nor the continuity across categories. But assessing the degree of diagonality does reveal the change across gaits and also allows for inferences concerning the movements of the center of mass (COM) and energy exchange (Griffin et al. 2004; Bishop et al. 2008).

Limb Protraction and Retraction

Limb protraction and retraction is the angle of the humerus or total forelimb and femur or total hind limb relative to the body at the beginning and end of stance, respectively (Larson et al. 1999, 2001) (Fig. 2.2b). Depending on the stiffness of the limb, increased protraction can result in increased vertical oscillations of the COM, which (as discussed later) can influence the energetic costs of movement.

Contact Time

Defined as support phase duration in seconds, this simple variable probably deserves to be part of the section Footfall Sequence and Limb Timing, preceding. It is discussed separately because of its effect on peak loads. For a given force impulse (area under a force curve) applied by the individual at a given speed, increased contact time (Fig. 2.3a) will lower peaks along the curve as the base of the curve is extended while the area remains the same (Schmitt 1998) (Fig. 2.3b). Similarly, a short contact time will result in high peak forces.

Elbow Yield

Elbow yield can be defined by the decrease in elbow angle from touchdown to mid-support (Schmitt 1999; Larney and Larson 2004). This has been seen largely as a measure of limb stiffness that may influence both load (Fig. 2.3c) and possibly oscillations of the COM (Schmitt 1998, 1999, 2003c). Changes in limb yield have been implicated as part of the explanation for the unusual distribution of forces in primates in which peak forces are generally higher on the hind limbs than they are on the forelimbs (Schmitt 1998, 1999, 2003c).

These five variables are chosen because they are easy to quantify with simple video techniques and also appear to reflect important underlying mechanical processes during quadrupedal walking in primates. The point of this chapter is

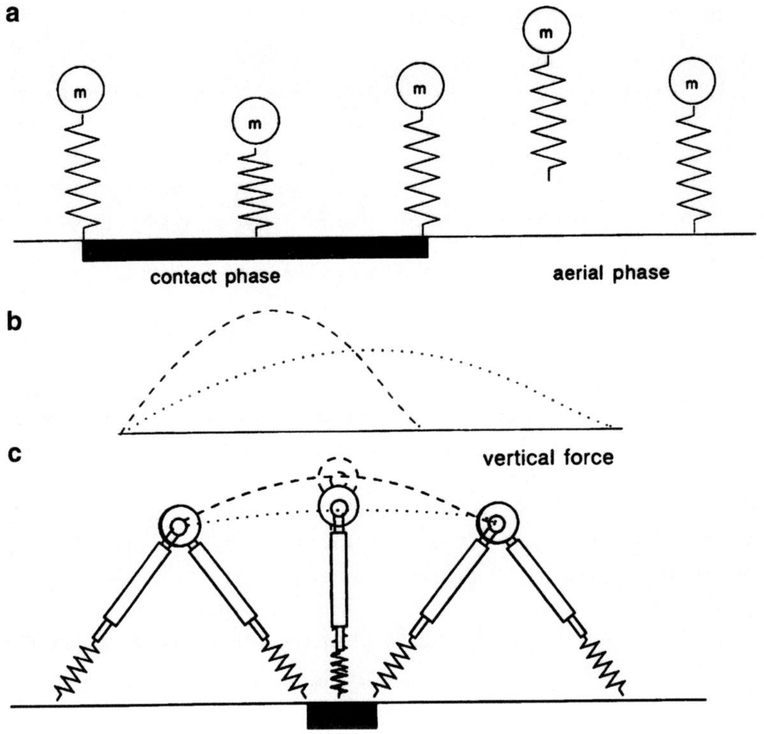

Fig. 2.3 Representation of the mechanical effects of limb compliance. (**a**) Influence of spring stiffness on contact time (shaded in black). If the spring stiffness were reduced, the model bouncing would spend longer in contact with the ground and have a much lower bounce height. (**b**) The model spring applies force to the ground during contact time. If the spring stiffness is reduced and the same force is applied over a longer contact time, the force values at every point in the curve will be reduced. (**c**) If the leg-spring is modeled more realistically, changes in spring stiffness (joint yield) lead to reduced vertical pathways of the center of mass

to illustrate the connection between simple measures of gait mechanics and deeper underlying processes. These data can then be put into the broader service of understanding the potential selective value of locomotor choices in primates. Even small changes in contact or oscillations of the COM on a stiff versus yielding leg can have profound effects on potentially critical aspects of animal fitness.

In the laboratory, it is possible to examine an entire biomechanical system, break it down into its constituent parts, collect complex variables using force plates and high-speed video, and then calculate important performance measures like load and cost. Once a clear relationship between simple variables like gait choice, contact time, joint yield, and more complex variables and performance measure is established, it is then possible to take that information into the field and use the simple measures as a surrogate for the performance measures.

Performance Measures Examined

This section focuses on the extent to which simple measures of primate gait reflects variation in three important performance measures: 1) *stability*, defined in this case as the ability to maintain a secure relationship with the substrate, i.e., not fall down; 2) *load*, defined in this case as peak external forces in the sagittal plane; and 3) *energetic costs* of locomotion.

Stability

Staying safe while moving and avoiding sudden disruptions has to be an important performance variable. Falls and other disruptions in movement can lead to an animal being captured or to serious injury. There is a wide range of ways to define "stability." Schmitt et al. (2006) used stability as another way of saying "security," or, more simply, a way that animals can avoid falling down. This chapter focuses on two ways in which stability can be regulated: 1) by choosing gaits or timing of footfalls that maintain as many limbs as possible in contact with the substrate (Cartmill et al. 2002) and 2) avoiding any whole-body aerial phase (Schmitt et al. 2006).

Peak Ground Forces

Avoiding catastrophic limb fracture as well as smaller microfractures and joint damage is another critical performance variable. It is well known that by combining limb biomechanical properties and limb posture, animals consistently produce safety factors that allow resistance to fracture two to four times higher than needed for normal locomotion (Biewener 1989, 1990). Primates, like other mammals, may make choices in gait (speed, footfall timing, and gait type) that moderates peak loads on their limbs to maintain appropriate safety factors with a minimal amount of bony material (Schmitt 1999).

Metabolic Costs

This last variable deserves some further expansion as it has not been commented on extensively in the anthropological literature. The energetic costs of locomotion are a critical, though not exclusive (Bishop and Schmitt 2008), selective factor in determining many details of primate gait. All animals must perform muscular work to resist joint collapse and to move themselves forward. Thus, work has to be done to accelerate and decelerate the COM. The ways in which the COM oscillates in nonprimate bipeds and quadrupeds has been a critical variable, both in defining gaits as speed increases and in understanding the adaptive trade-offs between stability and efficiency of any gait (see Cavagna et al. 1977; Biewener 2003; Ahn et al. 2004; Griffin et al. 2004). It appears that most animals follow the same basic underlying principles for walking and running. Nonhuman primates, however, may represent an exception to this broadly conserved pattern (Schmitt 2003c).

Direct measures of COM movements in primates are very limited. Wells and Wood (1975) described the movements of the COM during leaping in vervet monkeys. Using video recordings, Vilensky (1989) provided data on the COM in macaques at a wide range of speeds on a treadmill. His data suggested that changes in limb kinematics minimized the movements of the COM at different speeds. Few studies, with the exception of Cavagna et al. (1977) and Kimura (1990, 1991, 1996), have used force platform data to infer the behavior of the COM for either the whole animal or individual limb girdles.

In contrast to primate studies, the analysis of COM movements is a common method of analysis in biomechanical studies of other animals including cockroaches, crabs, frogs, lizards, ostriches, penguins, sheep, horses, dogs, and humans (Manter 1938; Cavagna et al. 1976, 1977; Full and Tu 1990; Full 1991; Farley et al. 1993; Farley and Ko 1997; Griffin and Kram 2000; Alexander 2003; Biewener 2003; Ahn et al. 2004; Griffin et al. 2004; Bishop and Schmitt 2008). The data derived from these studies allow researchers to explore underlying mechanics of various mammalian gaits and allows for comparison across a wide variety of taxa. They have revealed that the walking and running gaits of many animals, regardless of phylogeny and morphological design, can operate with the same basic mechanical principles and that gaits may be defined by those principles.

When modern humans walk, we vault over relatively stiff lower limbs in such a way that our COM is at its lowest point at heel-strike and rises to its highest point at midstance (Fig. 2.4) (Cavagna et al. 1976; Lee and Farley 1998). This type of inverted pendulum gait is common to almost all known quadrupeds and bipeds (Cavagna et al. 1976, 1977; Alexander 1977; Heglund et al. 1982; Blickhan and Full 1987, 1992; Full and Tu 1990; Full 1991; Gatesy and Biewener 1991; Farley et al. 1993; Farley and Ko 1997; Griffin and Kram 2000; Alexander 2003; Biewener 2003; Ahn et al. 2004). In contrast, when humans and other animals run, their limbs operate more like they were part of a spring-mass system in which the limb is compressed during support phase, and the COM is at its lowest point during the middle of stance.

COM movements can be directly calculated from force curves of all three components of the ground reaction forces exerted by an animal. Forces reflect accelerations of the body's COM. Those accelerations can be easily converted to velocities and displacements from which both kinetic energy (KE) and potential energy (PE) can be calculated. The mathematics for calculating these movements are detailed in textbooks (Biewener 2003) and recent publications (Ahn et al. 2004).

The relationship between the PE and KE of the COM provide information about the relative efficiency of different gaits. If the fluctuations of the PE and KE are

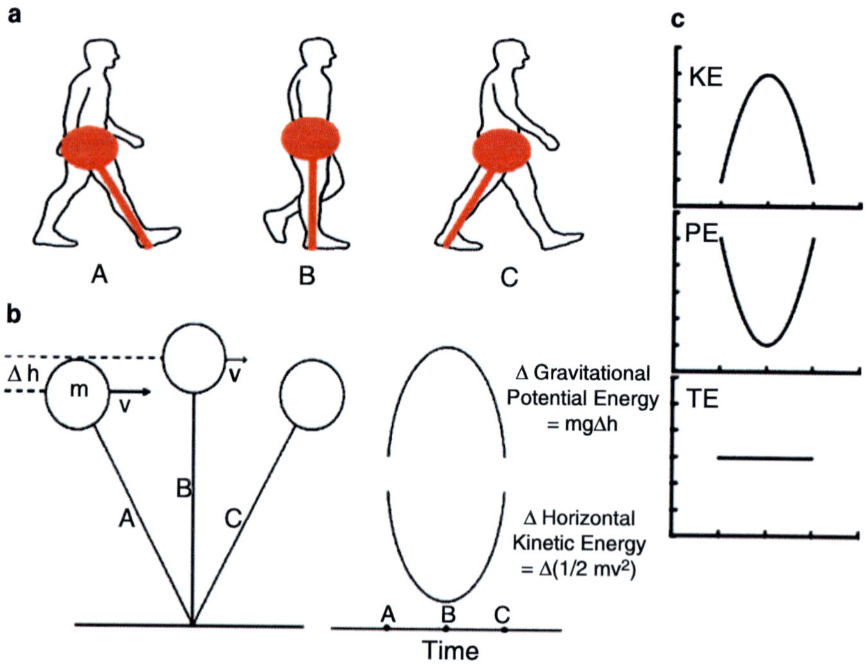

Fig. 2.4 Relationships between leg stiffness and energy exchange. (**a**) As modern humans walk, their legs can be modeled as massless rods with the center of mass (COM) centered on the hip. As people walk, the COM is at its lowest point at heel-strike (point A), its highest point at midsupport (point B), and returns to a low point before toe-off (point C). (**b**) The relationship between potential energy (PE) and kinetic energy (KE) during the stride. At heel-strike (point A) PE is low and KE is high. At midsupport (point B) PE is high while KE is low. PE drops and KE rises as the step shifts toward toe-off (point C). (**c**) When PE and KE are equal in magnitude and out of phase then the total energy (TE) does not fluctuate and energy recovery is 100%

mirror images of each other (have the same amplitude, shape, and opposite direction) (Fig. 2.4), then PE stored at midstance can be converted to KE during the second half of stance. The kinetic energy can be used to drive the COM forward. If the leg is held stiff, as many animals do, then the COM will also be driven upward and regain PE that can be again converted into KE. This mechanism of energy recovery reduces the amount of muscular work required to accelerate and decelerate the COM and thus may reduce the costs of locomotion. This amount of exchange is referred to as the *percentage of recovery*. This is a minimum estimate of the energy used by the system. In walking, PE and KE are similar in amplitude and shape but are largely out of phase. As a result, as much as 70% of the energy needed to move forward can be conserved through an exchange of PE and KE (Cavagna et al. 1977; Griffin et al. 2004). During running, in contrast, KE and PE are in phase and cannot exchange energy. During running much of this energy is converted to elastic energy in spring elements of the lower limb. This exchange mechanism works when fluctuations in PE and KE are equal in magnitude. Excessively large amounts of PE due to high vertical oscillations would be wasted.

Similarly, excessively low amounts of PE due to a smooth, nonoscillating gait would be insufficient for generation of enough KE. These constraints may explain the relatively stiff-legged oscillated gaits that occur in many legged mammals and even explain the waddle of penguins (Griffin and Kram 2000).

The Connection Between Biomechanical Variables and Performance Measures

Gait Choice, Stability, and Energy Efficiency

The choice of gait has a clear effect on stability, peak loads, and energetic costs for primates. As a result, field-based researchers can connect gait choice in the wild to deeper mechanical considerations. As most quadrupedal mammals increase speed, they shift from walking gait with no aerial phase (Fig. 2.1a) to a running gait with a whole-body aerial phase (Fig. 2.1b) (Howell 1944; Muybridge 1887/1957; Gambaryan 1974; Hildebrand 1985; Cartmill et al. 2002). At their fastest speeds, quadrupedal mammals generally use a running gait such as a gallop (Howell 1944; Muybridge 1887/1957; Gambaryan 1974; Hildebrand 1985). But at speeds between that of a walk and a gallop, quadrupedal mammals often use symmetrical running gaits that have an aerial phase and in which the feet strike down in diagonal pairs (trot) or unilateral pairs (pace) (Muybridge 1887/1957; Howell 1944; Hildebrand 1967, 1985; Gambaryan 1974; Cartmill et al. 2002;). These gaits are faster than walking gaits but still provide relatively longer periods of support by both a forelimb and a hind limb than does galloping (Cartmill et al. 2002).

Unlike other mammals, primates almost never adopt a running trot or pace (Hildebrand 1967, 1985; Rollinson and Martin 1981; Vilensky 1989; Demes et al. 1990, 1994; Preuschoft 1990; Schmitt et al. 2006). Instead, Schmitt et al. (2006) showed that primates adopt a highly unusual running gait called an "amble" (Fig. 2.1c). These gaits are referred to as "grounded running gaits" (Rubenson et al. 2004) because they do not involve a whole-body aerial phase. Ambles are exhibited by almost all primates (Schmitt et al. 2006) as well as certain breeds of horses (Muybridge 1887/1957; Barrey 2001; Biknevicus et al. 2003, 2004) and elephants (Gambaryan 1974; Hutchinson et al. 2003). Schmitt et al. (2006) argued that ambling ensures continuous contact of the body with the substrate while dramatically reducing vertical oscillations of the COM. This may explain why ambling appears to be preferable to trotting for extremely large terrestrial mammals such as elephants and for arboreal mammals like primates that move on unstable branches.

But in addition to the increased security of consistent handholds, the use of an amble reduces the whole-body force and vertical oscillations of the COM (Fig. 2.5). This latter phenomenon has two potential outcomes. First, it reduces the movements that may shake branches and/or make the animal more noticeable. Second, it may shift the COM

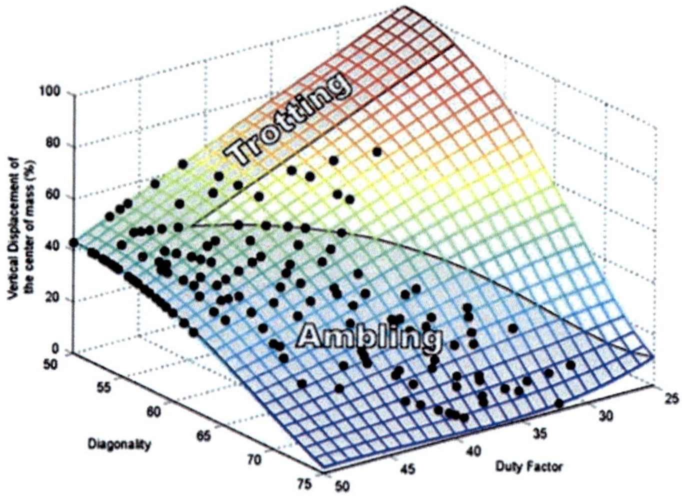

Fig. 2.5 Vertical displacement of the center of mass (COM) across a range of different diagonality and duty factor combinations, computed from a mathematical model of the vertical component of the force exerted by the foot on the ground (Alexander and Jayes 1978). Vertical displacement data are presented for diagonalities of 50%–75% and duty factors of 50%–25%. Empirical data (black dots) are plotted at the actual diagonality: duty factor combinations used by the primates. Ambling results in smaller vertical oscillations of the COM because the individual limb forces are more uniformly distributed throughout the stride, which reduces the peak force acting on the body. (From Schmitt et al. 2006)

movements from a pendular mechanism seen in walking to a bouncing gait like that of most runs without adopting an aerial phase (Biknevicius et al. 2003, 2004).

It has long been known that primates exhibit unusual footfall patterns using a diagonal sequence (DS) gait in which the contralateral forefoot follows contact of the hindfoot, i.e., right hind, left fore, left hind, left fore contact order (Fig. 2.2a). This pattern has been recognized since Muybridge (1887/1957) first filmed locomotion of the baboon and was later described by Magne de la Croix (1936) as the "pithecoid gait." The adaptive value of DS gaits has been the subject of enormous debate, most of which center on the value of this footfall pattern for stability in an arboreal environment (Gray 1968; Rollinson and Martin 1981; Vilensky and Larson 1989; Meldrum 1991; Cartmill et al. 2002, 2007a, b; Shapiro and Raichlen 2005, 2007). However, with the exception of a recent study by Wallace and Demes (2008), nobody has ever demonstrated a clear relationship between DS gaits and the unusual pattern of force distribution in primates (forelimb peak forces are greater than hind limb in most primates (Kimura et al. 1979; Kimura 1985; Demes et al. 1994; Schmitt and Lemelin 2002) (Fig. 2.2c). In fact, Schmitt and Lemelin (2004) found the opposite in the slender loris, which had relatively high forelimb peak forces and a strong preference for DS gaits. As a result, it is currently not clear how much of primate limb biomechanics or energetic costs can be inferred from the footfall pattern.

However, recent research has suggested that if we define diagonality (the delay between ipsilateral hind and fore contacts) quantitatively as did Hildebrand (1967) and Cartmill et al. (2002), we can identify a deeper underlying pattern related to COM oscillations and energetic costs. One of the problems Griffin et al. (2004) identified was that, in a quadruped, it was possible that the movements of the pendulums that define the fore- and hindquarters would cancel each other out. As one half was rising, the other half might be falling. In other words, PE and KE would be out of phase in the front and back half of the animal. In that situation, the movements of the whole body COM would be reduced and the exchange of energy would be impossible. Griffin and colleagues (2004) showed that dogs could have effective energy exchange and reduced muscular work under specific circumstances. The first was that more weight be borne by either the fore- or hindquarters. Dogs meet this criterion because 60% of the body weight is experienced by the forelimbs (Griffin et al. 2004). In addition, exchange can be enhanced by appropriate timing of footfalls. In a trot (50% diagonality) or pace (100% diagonality), where two limbs are moving simultaneously, fore- and hindquarters can work in unison (Cartmill et al. 2002). In gaits with more evenly spaced footfalls and diagonality close to 25% (singlefoot of Hildebrand 1967), the fore- and hindquarters work against each other and reduce effective energy exchange. Griffin et al. (2004) showed that dogs adopt diagonalities close to 15%. Similarly, Bishop and Schmitt (2008) showed that as cats approached diagonalities near 25%, they experienced reduced energy exchange.

This relationship may explain part of a complex pattern reported by Cartmill et al. (2002). Figure 2.6 shows the relationship between diagonality and duty factor in a wide variety of mammalian species. Although primates occupy a completely different section of the locomotor space than other mammals, both groups form clusters that are consistent with predictions about energy exchange. A diagonality of 50% means that the contralateral pairs of limbs are contacting the ground simultaneously. Many nonprimate mammals use trotting gaits but primates rarely do so, often adopting ambles (not included in Fig. 2.6 but illustrated in Fig. 2.1c) instead (Schmitt et al. 2006). Diagonalities of 25% and 75% represent evenly spaced footfalls carried out in a lateral or diagonal sequence respectively. It is this pattern of footfalls that Griffin et al. (2004) noted would inhibit energy exchange. The data from Cartmill et al. (2002) confirm that most nonprimate mammals deviate from this value, moving toward 15%, as did Griffin's dogs, or toward 40%. It is worth noting that the common marmoset, which is unusual in primates in adopting a lateral sequence (LS; in which the ipsilateral forefoot follows contact of the hindfoot) gait (Schmitt 2003a), deviates toward 40% diagonality rather than 15%. Cartmill et al. (2002) also included a large number of primates in their sample, and the graph shows that they deviate from evenly space footfalls at 75% diagonality and move toward 60%–65%. These data, along with the data on actual recovery on dogs and cats from other studies, suggest strongly that by identifying diagonality we can infer relative levels of energy exchange in animals.

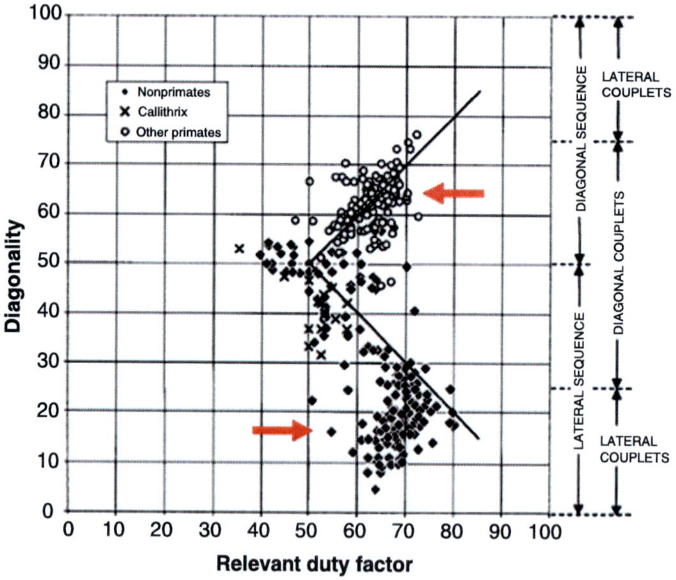

Fig. 2.6 This diagram, derived from Cartmill et al. (2002), shows the relationship between diagonality and duty factor in primates and nonprimate mammals. The arrows show the high concentration of gaits below 25% for nonprimate mammals and below 75% for most primates. As explained in detail in the text, these data suggest that animals choose diagonalities that may allow more effective exchange of potential and kinetic energy

Limb Protraction, Contact Time, and Elbow Yield

The last three variables —limb protraction, contact time, and elbow yield— are best, for the purposes of this chapter, treated together. They have been discussed at length in other publications as separate phenomena to address specific issues. For example, Schmitt and colleagues (Schmitt 1998, 1999, 2003c; Schmitt and Lemelin 2002; Schmitt and Hanna 2004) have described in detail the force implications for decreased limb stiffness. They showed that as elbow yield increased, contact time also increased and peak force decreased. They also noted that increased elbow yield and contact time resulted in decreased oscillations of the body, as inferred from movements of the shoulder. There are certainly subtle exceptions to this pattern of shoulder movement as with *Lemur catta* (Franz et al. 2005), but overall the theory and the model have agreed well in a way that allows field data on contact time and elbow yield to have productive value for inferring deeper biomechanics.

Similarly, Larson and colleagues (Larson 1998; Larson et al. 1999, 2001) and Lemelin and Schmitt (2007) have also described in detail patterns of forelimb protraction and the implications for movement on fine branches. These researchers suggested, among other things, that increased protraction will increase step length and decrease stride frequency. In this case, they argued, the number of limb contacts that would oscillate the branch would be reduced. The same is true for elongated contact time and is consistent with a model first developed by Demes et al. (1990) and expanded by Schmitt (1998, 1999, 2003b, c) and Schmitt et al. (2006). In this model, many of the locomotor choices made by primates appear to be connected to reduced oscillations of the branch on which they are running, thus increasing stability and stealth.

But the converse is also true. Primates certainly may choose gait and limb mechanics that moderate oscillations of the branch, but the same choices may also moderate oscillations of the COM. It has already been discussed in the preceding text that ambling reduces vertical oscillations of the COM relative to trotting. Schmitt (1998, 1999, 2003c) also argued that elbow flexion reduced the vertical movements of the shoulder. He assumed that this was reflected in movements of the COM. In that context, it might be assumed that primates will have lower vertical oscillations for any walking or running gait and therefore have relatively low exchanges of potential and kinetic energy than most animals. It is worth remembering that the PE component of the exchange of potential and kinetic energy is driven largely by vertical oscillations of the COM. The low energy recovery found in crouch-walking cats (Bishop and Schmitt 2008), support this hypothesis.

However, recent data collected by Schmitt and his colleagues Kristin Bishop and Matthew O'Neill have raised questions about such a simple model (Schmitt et al. 2007, 2008). Contrary to expectations, *Lemur catta* does not show substantially reduced energy exchange compared to other animals (Fig. 2.7). With a mean of 49% (±11.3%) energetic recovery when walking on the ground, the value for energy exchange is only somewhat lower than in dogs (Cavagna et al. 1977; Griffin et al. 2004) and much higher than in cats (Bishop and Schmitt 2008). Moreover, the value of 50.2% (±10.2%) for locomotion on a raised horizontal pole does not differ significantly from the ground value. This is in contrast with the clean predictions Schmitt (1998, 1999, 2003b, c) had argued for, but also reveals a far more interesting pattern. Primates have relatively long limbs (Alexander et al. 1979,

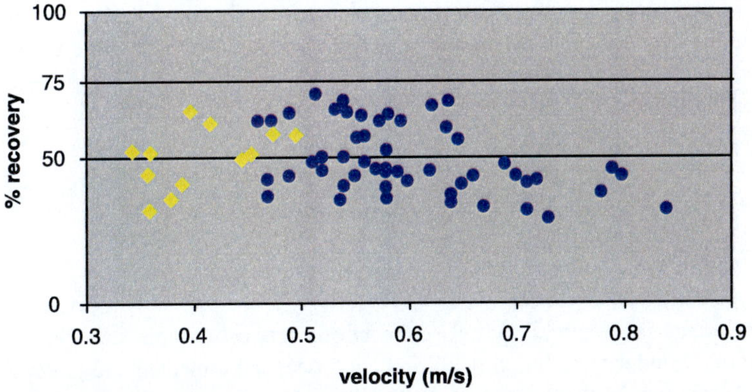

Fig. 2.7 Percentage recovery versus speed in *Lemur catta* walking on the ground (circles) and on a raised horizontal pole (diamonds). There is no significant relationship between recovery and speed and no significant difference between value for locomotion on the pole and the ground

Alexander and Maloiy, 1984), and they protract those limbs at touchdown more than most animals do (Larson 1998; Larson et al. 1999, 2001) and often to a greater degree on arboreal supports (Schmitt 1998, 1999, 2003b; Schmitt and Hanna 2004). What this may mean is that by combining long and highly protracted limbs that would, if held stiff, lead to high oscillations of the forequarters, and deep elbow yield, primates are able to achieve both 1) reasonable exchange of potential and kinetic energy and 2) increased contact time leading to reduced peak forces and reduced stride frequency.

Stride Frequency and Locomotor Cost

Reduced stride frequency can now be seen as an important variable in its own right. Kram and Taylor (1990), and most recently Matt O'Neill (2009), have shown that much of the costs of locomotion can be explained by the inverse of contact time in many mammals in general and primates specifically. Or put more simply, as contact time goes up, stride frequency goes down, and the costs of locomotion are reduced. This model was first elucidated by Taylor (1985), who suggested that frequency of activation of stance phase antigravity muscles influenced the cost of locomotion. The model is of course more complex than that and can be influenced by posture and muscle volume as well (Roberts et al. 1998a, b; Pontzer 2007a, b; Sockol et al. 2007; O'Neill 2008). But the empirical data suggest broadly that walking is "priced by the step" (Alexander and Ker 1990). Thus, whereas pendular exchange can reduce some of the muscular work during locomotion, decreased stride frequency can reduce the frequency of that work for a given speed.

Another source of cost for each step is the work done to move the COM from generally downward, as the animal approaches foot-strike, to generally upward as it pivots on the limb. Ruina et al. (2005) and Usherwood et al. (2007) have both shown that this is an important part of the energetic cost of movement. Although some of that cost is mitigated by the exchange of PE and KE, there remains a substantial cost to such re-directions. This is becoming an important topic of discussion (see Baumgertner et al. 2009a, b and Wunderlich et al.,Chapter 8) and is beyond the scope of this chapter. But it is worth noting that reducing stride frequency reduces, by definition, the number of times the COM must be redirected.

Conclusion

This chapter has attempted to draw connections between simple variables that can be recorded under field-based conditions and deeper biomechanical patterns seen in the laboratory. In that sense the goal has been fairly simple and was summarized at each waypoint in the chapter. In each case, a simple measure was paired with a performance measure. But taken as a whole, rather than as individual variables, a larger picture emerges. In the laboratory we can understand the mechanical principles that govern gait choice. It seems increasingly clear that the variation in gait, footfall pattern, and limb kinematics can be explained by deeper underlying mechanical advantages including increased stability, reduced peak loads, and possible opportunities for regulating energetic costs of locomotion.

When laboratory-based researchers interpret these data they do so in an explicitly selective context. For example, Cartmill et al. (2002) have argued that DS gaits allow animals to move and forage on thin branches. Similarly, Schmitt and Lemelin (2002) argued that the pattern of forelimb versus hind limb force distribution was associated with the same environment. Both studies argued that these features of primate locomotion were associated with the origin of primates. These arguments are based on readings of the literature about the ecology of primates and other animals. But these theories can be fully tested only by direct measures of behavior in the wild. If we predict a gait type will be used in a specific context, then field-based researchers can test that model. This chapter has acted as if the information would flow from the laboratory to the field. That is, of course, my bias. But the reality is that these data travel in both directions. Field-based data serve to test out selective hypotheses and allow us to develop the appropriate tests to follow in the laboratory. The purpose of the symposium in Scotland, this chapter, and this volume is to promote the two-way communication of laboratory-based and field-based researchers to develop and test adaptive hypotheses about primate locomotion.

References

Ahn AN, Furrow E, Biewener AA (2004) Walking and running in the red-legged running frog, *Kasina maculata*. J Exp Biol 207:399–410.

Alexander RMcN (1977) Mechanics and scaling of terrestrial locomotion. In: Pedley T (ed), Scale Effects in Animal Locomotion. Academic Press, London, pp 93–110.

Alexander R. McN., Jayes AS, Maloiy GMO, Wathuta EM (1979) Allometry of the limb bones of mammals from shrews (*Sorex*) to elephant (*Loxodonta*). J Zool (Lond) 194:539–552.

Alexander RM, Jayes AS (1978) Vertical movements in walking and running. J Zool Lond 185:27–40.

Alexander RMcN, Maloiy GM (1984) Stride lengths and stride frequencies of primates. J Zool Lond 202:577–582.

Alexander RMcN, Ker RF (1990) Running is priced by the step. Nature 346:220–221.

Alexander RMcN (1992) A model of bipedal locomotion on compliant legs. Philos Trans R Soc Lond B 338:189–198.

Alexander RMcN (2003) Principles of Animal Locomotion. Princeton University Press, Princeton, NJ.

Barrey E (2001) Inter-limb coordination. In: Back W, Clayton HM (eds), Equine Locomotion. WB Saunders, London. pp 77–94.

Baumgartner RE, Wunderlich RE, Schmitt D (2009a) Collisional costs of bipedalism in primates. Am J Phys Anthropol (Suppl) 48:91.

Baumgartner RE, Wunderlich RE, Schmitt D (2009b) Collisional mechanics during sifaka bipedalism. Integrative and Comparative Biol 49 (Suppl 1):e-193.

Biewener AA (1989) Scaling body support in mammals: limb posture and muscle mechanics. Science 245:45–48.

Biewener AA (1990) Biomechanics of mammalian terrestrial locomotion. Science 250:1097–1103.
Biewener AA (2003) Patterns of mechanical energy change in tetrapod gait: pendula, springs and work. J Exp Zool 305A:899–911.
Biknevicius AR, Mullineaux DR, Clayton HM (2003) Taking the walk for a run: locomotor mechanics of lateral sequence singlefoot gaits. Integr Comp Biol 43:988.
Biknevicius AR, Mullineaux DR, Clayton HM (2004) Ground reaction forces and limb function in tolting horses. Equine Vet J 36:743–747.
Bishop KL, Schmitt D (2008) Evolutionary tradeoffs between stealth and efficiency during walking in cats. PLoS ONE 3(11):e3808.
Blickhan R, Full RJ (1987) Locomotion energetics of the ghost crab. II. Mechanics of the centre of mass during walking and running. J Exp Biol 130:155–174.
Blickhan R, Full RJ (1992) Mechanical work in terrestrial locomotion. In: Biewener AA (ed), Biomechanics: A Practical Approach. Oxford University Press, New York, pp 75–96.
Bock WJ (1977) Adaptation and the comparative method. In: Hecht MK, Goody P, Hecht B (eds), Major Patterns in Vertebrate Evolution. Plenum, New York, pp 57–82.
Byron CD, Covert HH (2004) Unexpected locomotor behaviour: brachiation by an Old World monkey (*Pygathrix nemaeus*) from Vietnam. J Zool 263:101–106.
Cartmill M, Lemelin P, Schmitt D (2002) Support polygons and symmetrical gaits in mammals. Zool J Linn Soc 136:401–420.
Cartmill M, Lemelin P, Schmitt D (2007a) Primate gaits and primate origins. In: Dagosto M, Ravosa M (eds), Primate Origins. Kluwer, New York, pp 403–436.
Cartmill M, Lemelin P, and Schmitt D (2007b). Understanding the adaptive value of diagonal sequence gaits in primates: a comment on Shapiro and Raichlen (2005). Am J Phys Anthropol 133:822–825.
Cavagna GA, Thys H, Zamboni A (1976) Sources of external work in level walking and running. J Physiol Lond 262:639–657.
Cavagna GA, Heglund NC, Taylor CR (1977) Mechanical work in terrestrial locomotion: two basic mechanisms for minimizing energy expenditure. Am J Physiol 233:R243–261.
Demes B, Jungers WL, Nieschalk U (1990) Size- and speed-related aspects of quadrupedal walking in slender and slow lorises. In: Jouffroy FK, Stack MH, Niemitz C (eds), Gravity, Posture and Locomotion in Primates. Il Sedicesimo, Florence, pp 175–198.
Demes B, Larson SG, Stern JT, Jungers WL, Biknevicius AR, Schmitt D (1994) The kinetics of primate quadrupedalism: hindlimb drive reconsidered. J Hum Evol 26:353–374.
DeVore I (1992) An interview with Sherwood Washburn. Curr Anthropol 33:411–423.
Dunbar DC, Badam GL (2000) Locomotion and posture during terminal branch feeding. Int J Primatol 21:649–669.
Elftman H, Manter J (1935) Chimpanzee and human feet in bipedal walking. Am J Phys Anthropol 20:69–79.
Farley CT, Glasheen J, McMahon TA (1993) Running springs: speed and animal size. J Exp Biol 185:71–86.
Farley CT, Ko TC (1997) Mechanics of locomotion in lizards. J Exp Biol 200:2177–2188.
Fleagle JG (1974) Dynamics of a brachiating siamang [*Hylobates (Symphalangus) syndactylus*]. Nature 248:259–260.
Fleagle JG (1979) Primate positional behavior and anatomy: naturalistic and experimental approaches. In: Morbeck ME, Preuschoft H, Gomberg N (eds), Environment, Behavior, and Morphology: Dynamic Interactions in Primates. Wenner-Gren Foundation, New York, pp 313–326.
Franz TM, Demes B, Carlson KJ (2005) Gait mechanics of lemurid primates on terrestrial and arboreal substrates. J Hum Evol 48:199–217.
Full RJ, Tu MS (1990) Mechanics of six-legged runners. J Exp Biol 148:129–146.
Full RJ (1991) Animal motility and gravity. The Physiologist 34:S15–S18.
Gambaryan PP (1974) How Mammals Run. John Wiley & Sons, New York.
Gatesy SM, Biewener AA (1991) Bipedal locomotion: effects of speed, size, and limb posture in birds and humans. J Zool Lond 224:127–147.

Grand TI (1968a) The functional anatomy of the lower limb of the howler monkey (*Alouatta caraya*). Am J Phys Anthropol 28:163–181.
Grand TI (1968b) Functional anatomy of the upper limb. Biology of the howler monkey (*Alouatta caraya*). Bibl Primatol 7:104–125.
Gray J (1968) Animal Locomotion. Wiedenfeld and Nicolson, London.
Griffin TM, Kram R (2000) Penguin waddling is not wasteful. Nature 408:929.
Griffin TM, Main RP, Farley CT (2004) Biomechanics of quadrupedal walking: how do four-legged animals achieve inverted pendulum-like movements? J Exp Biol 207:3545–3558.
Grossman A (2006) Propagating Mike's memes: behavioral reconstruction in paleoanthropology: Evol Anthropol 15:158–159.
Hedrick TL (2008) Software techniques for two- and three-dimensional kinematic measurements of biological and biomimetic systems. Bioinspiration Biomimetics 3:034001.
Heglund NC, Cavagna GA, Taylor CR (1982) Energetics and mechanics of terrestrial locomotion. III. Energy changes of the center of mass as a function of speed and body size in birds and mammals. J Exp Biol 97:41–56.
Hildebrand M (1967) Symmetrical gaits of primates. Am J Phys Anthropol 26:119–130.
Hildebrand M (1985) Walking and running. In: Hildebrand M, Bramble DM, Liem KF, Wake DB (eds), Functional Vertebrate Morphology. Harvard University Press, Cambridge, MA, pp 38–57.
Homberger DG (1988) Models and tests in functional morphology: the significance of description and integration. Am Zool 28:217–229.
Howell AB (1944) Speed in Animals. University of Chicago Press, Chicago.
Hutchinson JR, Famini D, Lair R, Kram R (2003) Are fast-moving elephants really running? Nature 422:493–494.
Kimura T, Okada M, Ishida H (1979) Kinesiological characteristics of primate walking: its significance in human walking. In: Morbeck M, Preuschoft H, Gomberg N (eds), Environment, Behavior, and Morphology: Dynamic Interactions in Primates. Gustav Fischer, New York, pp 297–312.
Kimura T (1985) Bipedal and quadrupedal walking of primates, comparative dynamics. In: Kondo S, Ishida H, Kimura T (eds), Primate Morphophysiology, Locomotor Analyses and Human Bipedalism. University of Tokyo Press, Tokyo, pp 81–104.
Kimura T (1990) Voluntary bipedal walking of infant chimpanzees. In: Jouffroy FK, Stack MH, Niemitz C (eds), Gravity, Posture and Locomotion in Primates. Il Sedicesimo, Florence, pp 237–251.
Kimura T (1991) Body center of gravity and energy expenditure during bipedal locomotion in humans, chimpanzees and macaques. Primate Rep 31:19–20.
Kimura T (1996) Center of gravity of the body during the ontogeny of chimpanzee bipedal walking. Folia Primatol 66:126–136.
Kram R, Taylor CR (1990) Energetics of running: a new perspective. Nature 346:265–267.
Larney E, Larson SG (2004) Limb compliance during walking: comparisons of elbow and knee yield across quadrupedal primates and in other mammals. Am J Phys Anthropol 125:42–50.
Larson SG (1998) Unique aspects of quadrupedal locomotion in nonhuman primates. In: Strasser E, Fleagle JG, Rosenberger AL, McHenry HM (eds), Primate Locomotion: Recent Advances. Plenum, New York, pp 157–173.
Larson SG, Schmitt D, Lemelin P, Hamrick MW (1999) The uniqueness of primate forelimb posture during quadrupedal locomotion. Am J Phys Anthropol 112:87–101.
Larson SG, Schmitt D, Lemelin P, Hamrick MW (2001) Limb excursion during quadrupedal walking, How do primates compare to other mammals. J Zool Lond 255:353–365.
Lauder GV (1996) The argument from design. In: Rose MR, Lauder GV (eds), Adaptation. Academic Press, San Diego, pp 55–91.
Lee CR, Farley CT (1998) Determinants of the center of mass trajectory in human walking and running. J Exp Biol 201:2935–2944.
Lemelin P, Schmitt D (2007) The origins of grasping and locomotor adaptations in primates: Comparative and experimental approaches using an opossum model. In: M. Dagosto & M. Ravosa (eds.), Primate origins, New York: Springer, pp. 329–380.

Magne de la Croix P (1936) The evolution of locomotion in mammals. J Mammal 17:51–54.
Manter JT (1938) The dynamics of quadrupedal walking. J Exp Biol 15:522–540.
Meldrum DJ (1991) Kinematics of the Cercopithecine foot on arboreal and terrestrial substrates with implications for the interpretation of hominid terrestrial adapatations. Am J Phys Anthropol 84:273–290.
Morbeck ME (1979) Forelimb use and positional adpatation in *Colobus guereza*: integration of behavioral, ecological, and anatomical data. In: Morbeck M, Preuschoft H, Gomberg N (eds), Environment, Behavior, and Morphology: Dynamic Interactions in Primates. Gustav Fisher, New York, pp 95–118.
Muybridge E (1887) Animals in Motion [1957 reprint]. Dover, New York.
O'Neill M (2009) The Structure of Locomotor Cost: Gait, Mechanics and Limb Design in Ringtailed Lemurs *(Lemur catta)*. Ph.D. thesis, The Johns Hopkins University.
Plavcan JM, Kay RF, Jungers WL, van Schaick CL (2002) Reconstructing Behavior in the Primate Fossil Record. Springer, New York.
Pontzer H (2007a) Predicting the energy cost of terrestrial locomotion: a test of the LiMb model in humans and quadrupeds. J Exp Biol 210:484–494.
Pontzer H (2007b) Effective limb length and the scaling of locomotor cost in terrestrial animals. J Exp Biol 210:1752–1761.
Preuschoft H (1990) Gravity in primates and its relation to body shape and locomotion. In: Jouffroy FK, Stack MH, Niemitz C (eds), Gravity, Posture and Locomotion in Primates. Il Sedicesimo, Florence, pp 109–127.
Roberts TJ, Kram R, Weyand PG, Taylor CR (1998a) Energetics of bipedal running. I. Metabolic cost of generating force. J Exp Biol 201:2745–2751.
Roberts TJ, Chen MS, Taylor CR (1998b) Energetics of bipedal running. II. Limb design and running mechanics. J Exp Biol 201:2753–2762.
Rollinson J, Martin RD (1981) Comparative aspects of primate locomotion with special reference to arboreal cercopithecines. Symp Zool Soc Lond 48:377–427.
Rose MD (1979) Positional behavior of some natural populations: some quantitative results of a field study of *Colobus guereza* and *Cercopithecus aethiops*. In: Morbeck ME, Preuschoft H, Gomberg N (eds), Environment, Behavior and Morphology: Dynamic Interactions in Primates. Fischer, New York, pp 75–93.
Rubenson J, Heliams DB, Lloyd DG, Fournier PA (2004) Gait selection in the ostrich: mechanical and metabolic characteristics of walking and running with and without an aerial phase. Proc Biol Sci 271:1091–1099.
Ruina A, Bertram JEA, Srinivasan M (2005) A collisional model of the energetic cost of support work qualitatively explains leg sequencing in walking and galloping, pseudo-elastic leg behavior in running and the walk-to-run transition. J Theor Biol 237:170–192.
Schmitt D (1998) Forelimb mechanics during arboreal and terrestrial quadrupedalism in Old World Monkeys. In: Strasser E, Fleagle JG, Rosenberger AL, McHenry HM (eds), Primate Locomotion: Recent Advances. Plenum Press, New York, pp 175–200.
Schmitt D (1999) Compliant walking in primates. J Zool 247:149–160.
Schmitt D (2003a) Evolutionary implications of the unusual walking mechanics of the common marmoset (*Callithrix jacchus*). Am J Phys Anthropol 122:28–37.
Schmitt D (2003b) Mediolateral reaction forces and forelimb anatomy in quadrupedal primates: implications for interpreting locomotor behavior in extinct primates. J Hum Evol 44:47–58.
Schmitt D (2003c) Insights into the evolution of human bipedalism from experimental studies of humans and other primates. J Exp Biol 206:1437–1448.
Schmitt D, Hanna J (2004) Substrate alters forelimb to hindlimb peak force ratios on primates. J Hum Evol 46:237– 252.
Schmitt D, Larson SG, Stern JT (1994) Serratus ventralis function in vervet monkeys: Are primate quadrupeds unique? J Zool 232:215–230.
Schmitt D, Lemelin P (2002) The origins of primate locomotion: gait mechanics of the woolly opossum. Am J Phys Anthropol 118:231–238.

Schmitt D, Lemelin P (2004) Locomotor mechanics of the slender loris *(Loris tardigradus)*. J Hum Evol 47:85–94.

Schmitt D, Cartmill M, Hanna J, Griffin T, Lemelin P (2006) Adaptive value of ambling gait in primates and other mammals. J Exp Biol 209:2042–2049.

Schmitt D, Pai AK, O'Neill MC, Bishop KL (2008) Center of mass movements in arboreal and terrestrial prosimians. Am J Phys Anthropol, Suppl. 46:187.

Schmitt D, Wall CE, Lemelin P (2008) Experimental comparative anatomy in physical anthropology: the functional anatomy of the skull and the contributions of Dr. William Hylander. In: Vinyard C, Ravosa M, Wall C (eds), Primate Craniofacial Biology. Springer, New York, pp 3–15.

Shapiro LJ, Raichlen DA (2005) Lateral sequence walking in infant *Papio cynocephalus*: implications for the evolution of diagonal sequence walking in primates. Am J Phys Anthropol 126:205–213.

Shapiro LJ, Raichlen DA (2007) A response to Cartmill et al.: Primate gaits and arboreal stability. Am J Phys Anthropol 133:825–827.

Sockol MD, Raichlen DA, Pontzer H (2007) Chimpanzee locomotor energetics and the origin of human bipedalism. PNAS 104:12265–12269.

Taylor CR (1985) Force development during sustained locomotion: a determinant of gait, speed and metabolic power. J Exp Biol 115:253–262.

Usherwood JR, Williams SB, Wilson AM (2007) Mechanics of dog walking compared with a passive, stiff-limbed, 4-bar linkage model, and their collisional implications. J Exp Biol 210:533–540.

Vilensky JA (1989) Primate quadrupedalism: how and why does it differ from that of typical quadrupeds? Brain Behav Evol 34:357–364.

Vilensky JA, Larson SG (1989) Primate Locomotion: Utilization and control of symmetrical gaits. Annu Rev Anthropol 18:17–35.

Wallace IJ, Demes B (2008) Symmetrical gaits of *Cebus apella*: implications for the functional significance of diagonal sequence gait in primates. J Hum Evol 54:783–794.

Washburn SL (1951a) The analysis of primate evolution with particular reference to the origin of man. Cold Spring Harbor Symp Quant Biol 15:67–78.

Washburn SL (1951b) The new physical anthropology. Trans NY Acad Sci (Series II) 13:298–304.

Wells JP, Wood GA (1975) The application of biomechanical motion analysis to aspects of green monkey (*Cercopithecus a. sabaeus*) locomotion. Am J Phys Anthropol 43:217–226.

Chapter 3
Studying Captive Ape Locomotion: Past, Present, and Future

Evie E. Vereecke, Kristiaan D'Août, and Peter Aerts

Abstract Most of our knowledge regarding primate locomotion, especially biomechanical aspects, results from studies of captive populations, typically in laboratories or zoos. Obviously, a controlled environment facilitates the acquisition of high-quality locomotion data; it has done so in the past and will continue to do so in the future. We start by outlining how primate locomotion has been studied in captive settings, and the sort of insights such studies have yielded. We draw examples from our own research on hylobatids (*Hylobates lar, Symphalangus syndactylus*) and bonobos (*Pan paniscus*), carried out using integrated setups in a zoo environment. Locomotion is highly variable in these hominoid species; even in a captive setup, it is inevitably less complex than in the natural habitat. Neither species uses a human-like stiff-legged (inverted pendulum) type of terrestrial locomotion. Bonobos use a highly crouched posture both in bipedal and quadrupedal terrestrial locomotion; lar gibbons use a bouncing gait with potential for energy saving mechanisms in the knee and in the foot. Aspects of arboreal locomotion have been, or are being studied in the three species, using stiff substrates and overhead supports. Next, we discuss some shortcomings of working outside of the natural habitat, *ex situ*. They pertain most clearly to the limited number of subjects (a result of availability issues and the high level of detail required) and to the relative lack of complexity in the substrates used. Especially during arboreal locomotion, new research lines should be (and are being) started in which spatial complexity and compliance are incorporated in the experimental setup. We are currently using this approach to study jumping off branches and for brachiation in hylobatids. Finally, we make some suggestions of how field work can help meet some of the limitations intrinsic to *ex situ* studies. Locomotor field studies are complementary to *ex situ* studies in their capacity to study larger sample sizes (albeit in lesser detail) in their natural environment, thereby documenting, preferably quantitatively, the natural

---

K. D'Août (✉)
Laboratory for Functional Morphology, Department of Biology, University of Antwerp, Belgium
e-mail: kristiaan.daout@ua.ac.be

K. D'Août and E.E. Vereecke (eds.), *Primate Locomotion: Linking Field and Laboratory Research*, Developments in Primatology: Progress and Prospects,
DOI 10.1007/978-1-4419-1420-0_3, © Springer Science+Business Media, LLC 2011

locomotion repertoire, unbiased by human-made setups. Specifically, field studies are crucial for describing the locomotor modes that are actually used by the species studied, and for providing an ecological framework for an integrated approach of primate locomotion.

Keywords Biomechanics • Bipedalism • Bonobo • Captive • Gibbon • Hominidae • Locomotion • Quadrupedalism • Techniques • Wild

Introduction

Hominoid primates have been studied for some time (for overview, see Gibbs et al. 2002 on anatomy, Ward 2007 on locomotion, McGrew 1998 on socioecology, and references therein). They belong to the same superfamily as humans (Hominoidea) and have, therefore, been proposed as good models for understanding our own evolution. Research has focused mainly on the two most striking characteristics of humans as opposed to other animals: their elaborate cognitive capacities and their habitual bipedalism. We have been studying several hominoid species over the past 10 years, in order to contribute to a better understanding of their locomotor dynamics and to give us a better insight in the evolution of bipedalism in our hominin ancestors. Such a research line relies heavily on the acquisition of detailed anatomical, kinematic, and kinetic data, and we have, therefore, constructed various experimental setups that enabled us to collect these data. For practical reasons, the setups were constructed in a zoo environment and typically consisted of a relatively simple and uniform substrate. Thanks to the careful design and incorporation of several high-tech devices, the setups enabled us to collect a large amount of high-quality locomotor data on free-ranging, captive hominoids. In this chapter, we present some examples of the insights we have gained from such studies.

Though it is clear that a captive environment provides unique opportunities for the study of primate locomotion, it also has inherent shortcomings. In the second part of this chapter, we present some of the issues we, and other authors, have met while using locomotor data collected solely in captivity. We will also give some suggestions of how *ex situ* research can be taken one step further, for instance, by modifying the experimental setup to reflect better the complexity of the natural substrate. Yet, we are aware that the true complexity of the natural habitat will never be fully mimicked in a captive environment. In addition, captive populations are smaller than natural populations, and may not represent all age classes and all locomotor modes (or not to the same extent) as wild populations. Therefore, locomotor research in the field is crucial for a full understanding of the animal's locomotion in its ecological context (see also Chapter 1). The final part of this chapter highlights some topics that call for (quantitative) observations from the field.

Research *Ex Situ*: Accomplishments

High-Tech Experimental Setups

In captive studies, advanced technical devices and setups can be utilized and experimental conditions controlled, which enables an in-depth investigation of primate gait mechanics. Often the choice for a captive or lab-based approach is dictated by the need for detailed biomechanical data, and, ultimately, by the question posed. In our research, the choice for a zoo-based approach was led mainly by the urge to collect synchronized recordings of kinetics, kinematics, and plantar pressures during locomotion of untrained apes to get a full (and fundamental) insight in their terrestrial gait mechanics. Even with the current technological advances (see also Chapter 1), this would have been impossible to execute in the field. Moreover, the simplified captive environment has huge benefits: it yields high-quality data that are easy to interpret owing to the relative lack of confounding factors. In addition to the possibility of collecting detailed locomotor data, working *ex situ* also has benefits for the anatomical and morphometrical aspects of locomotor research. Animals may be measured and weighed during medical interventions or routine check-ups and cadavers of subjects with a well-documented life history will become available over time, enabling detailed morphological analysis (gross dissection, muscle moment arm measurements, determination of segment inertial properties, etc.).

We conducted our research mainly in the Wild Animal Park of Planckendael (Belgium), a relatively large local zoo with free-ranging conditions. Access to the ape populations and the ability to install high-tech setups in the enclosures was possible because of our affiliation with the Centre for Research and Conservation, the research department of the Royal Zoological Society of Antwerp. The commitment of the zoo management and staff to foster research was crucial for our studies. Fortunately, organizations such as the World Association of Zoos and Aquaria (WAZA) and the European Association of Zoos and Aquaria (EAZA) are actively encouraging zoos to engage in or facilitate scientific research. The WAZA Conservation Strategy states: "Zoos and aquariums are fully and actively integrated into the research community and into public consciousness and understanding of science, as serious, respected scientific institutions that make significant contributions and sound scientific decisions for wildlife worldwide." (WAZA 2005). The EAZA Research Strategy states: "EAZA and its Research Committee will work towards the vision and engage in the mission to encourage and support all European zoos and aquariums to adopt a scientific approach and participate in worthwhile and ethical research. To pursue the vision and achieve the mission, individual zoos should provide the necessary facilities, tools, and staff to conduct effective research and to develop a thriving scientific culture (…)" (Reid et al. 2008).

The setups we established were designed with continuous input from zoo staff. Care was taken to use materials that were harmless, ape-proof, and visitor-friendly. Since our setups were installed inside the ape enclosures, we had to adhere to the zoo regulations, meaning that no direct interaction with the animals was allowed.

This made data collection slightly more cumbersome, but guaranteed the recording of spontaneous, unrestricted locomotion. The setups used to investigate terrestrial locomotion of bonobos (*Pan paniscus*) and gibbons (*Hylobates lar*) were very similar and consisted of an integrated walkway with built-in force plates (AMTI) and pressure plates (footscan, RSScan), covered with a non-skid rubber sheet to make the setup water- and ape-proof. The devices were synchronized with up to four S-VHS cameras (50 Hz), to enable a full biomechanical analysis. The data acquisition was controlled manually from a computer station, located outside the enclosure. After a couple of days of habituation, both setups proved to be very effective, and the apes regularly walked over the walkway, at self-chosen speed and gait. A full description of the setups is given in D'Août et al. 2001 (for the bonobo setup) and Vereecke et al. 2005 (for the gibbon setup).

Locomotion in Bonobos

Since the early 1990s, bonobo research has been a strong focus at the Royal Zoological Society of Antwerp, which is also European studbook keeper for the species. Initially, studies focused on their social behavior, but in the late 1990s, our research group initiated a biomechanical project to investigate their terrestrial locomotion.

Bonobos and chimpanzees are the closest relatives to humans, with which they share more than 98% of DNA (Wildman et al. 2003), and of these two species, it had been proposed in the literature (mostly on morphological bases) that bonobos may be the most suitable models for early hominins (Zihlman et al. 1978). Therefore, we initially focused on bonobos to evaluate their anatomy and bipedal gait in a human evolutionary context (for arboreal locomotion, see further).

Anatomical data have shown that bonobos differ from chimpanzees in several morphological aspects that are probably important for locomotion (Doran 1993). However, recent studies have indicated that the morphological differences between the two species of *Pan* may be more subtle than previously anticipated. Payne et al. (2006) have shown that hind limb muscles of both species are remarkably similar, and Schoonaert et al. (2006) have shown that body (not head) mass distribution and segment lengths are very similar between the two species, even though chimpanzees are overall heavier than bonobos. Detailed anatomical data like this rely on the availability of subjects with documented backgrounds and good dissection facilities, and would be impossible to gather in the wild.

Apart from these morphological observations, recent studies have proposed that orang-utan type locomotion, i.e., hand-assisted "orthograde scrambling" on flexible branches, might have been ancestral for hominins (Thorpe et al. 2007; Crompton et al. 2008). Studies like this, and the morphological findings mentioned earlier, strongly support the case that no single extant model (including the bonobo) is a perfect stand-in for early hominins. As suggested by Aerts et al.

(2000), a comparative approach with as many species as possible should be employed to get insight into the locomotor dynamics of hominins. Together, these studies from the field and the lab urge us to reconsider what we call a "model species," because, in the absence of a perfect stand-in, the choice of a "model" may depend on the nature of the research question asked. In other words, bonobos may be the best "models" for some research questions, e.g., How could Australopithecines have walked?, whereas orang-utans may be particularly suitable for others, e.g., might/could orthograde scrambling have been a precursor of terrestrial bipedality? Other (nonhominoid) primate "models" are proving their merits, as illustrated in Ogihara et al. (Chapter 4) and Berillon et al. (Chapter 5) with their studies on Japanese monkeys and olive baboons, respectively.

Our kinesiological recordings of terrestrial locomotion in a group of captive bonobos have enabled us to compare several aspects of their bipedal and quadrupedal locomotion. In both modes, the hips and knees remain flexed throughout the gait cycle (D'Août et al. 2002). Differences between both modes are reflected in several ways. Spatiotemporally, bipedal locomotion involves smaller strides at a higher frequency to attain a given velocity (Aerts et al. 2000). This corresponds to smaller hip excursion ranges in bipedalism, but the knee and ankle show similar excursions in both gaits (D'Août et al. 2002). Plantar pressure distribution differs between both locomotor modes too, with heel impacts being smaller and a more plantar foot posture in bipedal locomotion (Vereecke et al. 2003).

We have also studied arboreal locomotion in bonobos, as these are primarily arboreal animals (Doran 1993), and locomotion in their most frequently used habitat certainly merits to be functionally understood. In addition, recent studies have proposed an alternative hypothesis for the precursor of upright straight-legged locomotion: orthograde scrambling (see earlier) as observed in wild Sumatran orang-utans (Thorpe et al. 2007; Crompton et al. 2008). These authors suggest, based on observations in the wild and fossil evidence, that bipedalism originated in an arboreal ape, and is probably an ancestral condition for (extant) hominoids (including hominins and humans). In light of this, studying arboreal locomotion in bonobos may be a particularly good way to evaluate hypotheses about the origin of hominin habitual bipedalism.

To study arboreal locomotion in bonobos, we constructed a new experimental setup in the bonobo enclosure (Schoonaert et al. 2006) that allowed for synchronous measurements of 3D-kinematics (by using four cameras operating at 50 Hz) and 3D-ground reaction forces on an instrumented section of a wood-covered steel beam. Substrate complexity was incorporated in the setup by using an adjustable pole which could be inclined from 0° (horizontal) to 90° (vertical).

Preliminary analyses (Schoonaert et al., unpublished) show that substrate reaction forces vary substantially between slopes, especially for the forelimb, which is loaded in compression on low slopes and in tension during vertical climbing. Spatiotemporal gait parameters were relatively uniform on different slopes, and involved smaller strides (but at a higher frequency) than on a horizontal arboreal support.

Locomotion in Gibbons

Unlike the more generalistic bonobos, gibbons are justly considered specialist brachiators (see further), but they are also the most bipedal of the nonhuman primates (Tuttle 1972; Schmid and Piaget 1994). Gibbons alternate brachiation with short bipedal bouts atop branches, and terrestrial bipedalism is occasionally used to cross gaps in the forest canopy (typically in fragmented forests). It is this combination of brachiator morphology, high frequency of bipedalism, and hominoid status that makes the bipedal gait of gibbons of particular interest for anthropologists; comparing the bipedal dynamics of an ape lacking morphological adaptations to bipedalism to that of a highly specialized bipedal ape, i.e., human, may shed light on the evolution of bipedalism in our hominin ancestors.

Although several researchers have looked into gibbon bipedalism, collecting kinetics, kinematics and electromyographic (EMG) data of one or two trained subjects (Prost 1967; Ishida et al. 1976, 1978, 1984, 1985; Kimura et al. 1977; Stern and Susman 1981; Okada and Kondo 1982; Okada et al. 1983; Yamazaki and Ishida 1984; Okada 1985b; Shapiro et al. 1987; Shapiro and Jungers 1988, 1994; Schmid and Piaget 1994), no full biomechanical insight was attained. Therefore, given our interest in hominin bipedalism, and our expertise with studies regarding bonobo locomotion, we initiated a zoo-based project on (terrestrial) bipedalism in gibbons. We expanded the existing data set with an extensive study of bipedalism in a group of free-ranging white-handed gibbons (*Hylobates lar*) (Vereecke et al. 2005, 2006a–c), and recorded kinetics, kinematics, and plantar pressures simultaneously.

One of the findings of this study was that, despite lacking any bipedal specializations, gibbons are adept bipeds. The lack of bipedal adaptations in their musculoskeletal system does result in a bipedal gait that is quite different from that of humans, but does not restrict their capabilities for bipedalism (both terrestrially and arboreally). This observation has important implications for the evolution of bipedalism in hominins, as it suggests that particular morphological adaptations, e.g., medial foot arch, adducted hallux, lumbar lordosis, are not a requirement to achieve a bipedal gait. Yet, morphology does influence the characteristics of the adopted bipedal gait. This should be taken into account when reconstructing the locomotor behavior of hominins from fossil remains.

Another important observation is that gibbons typically adopt a running or crouching gait during bipedalism, with in-phase fluctuation of potential and kinetic energy and a downward movement of the COM during midstance (Vereecke et al. 2006c). These dynamics are compatible with an energy-saving mechanism via elastic energy storage and recoil in the hind limb tendons, as seen in human running, even though bipedal bouts generally lack an aerial phase. Our latest results indeed suggest that, in gibbon bipedalism, elastic energy can be stored and released in tendons and ligaments of the leg and foot, enhancing the efficiency of locomotion (Vereecke and Aerts 2008). It seems that, despite merely being a "byproduct" of brachiation, bipedalism is an efficient means of locomotion for gibbons.

To us, this is a good example of how *ex situ* research can contribute to a better understanding of gait mechanisms and energetics, which, in turn, may be crucial for field researchers in interpreting the observed locomotor behavior.

Research *Ex Situ*: Current Issues and How to Deal with Them

Even though the previous section has shown that studies outside of the natural habitat have practical advantages, and are probably crucial to obtain a detailed insight in primate gait mechanics, they also have several shortcomings. It is becoming more and more acknowledged that such studies may be lacking in sample size, habitat complexity, and overall resemblance to the natural habitat, which is especially important when dealing with arboreal locomotion. Besides, the locomotor behavior observed (and studied) in captivity is not necessarily the type of behavior that matters most in the wild and is not always representative for the locomotion of wild animals. More specifically, primates are renowned for their locomotor generality (notwithstanding exceptions) and the ability to deal with variation in their habitat, whereas laboratory setups have typically been as simple as possible. In the case of primates, even the social context influences locomotion, e.g., spontaneous bipedal walking in wild Japanese macaques (Okada 1985a-b).

In this section we argue that, especially for species of which we have obtained a good fundamental understanding of their (steady) locomotion under relatively simple substrate conditions, we should now take one step further and include substrate complexity and gait variability in our locomotor studies. This should be done by using a setup that mimics the natural habitat (especially the substrate) more closely, giving results that are more representative of naturalistic behavior, while keeping the setup simple enough so that the data can be unequivocally interpreted in kinesiological terms.

Substrate Complexity

Many recent publications have, in fact, pointed out the necessity to simulate the natural environment in studies of captive animal locomotion (e.g., Dickinson et al. 2000). Currently, there is a trend to approach natural conditions by incorporating different substrate characteristics, such as compliance, inclination, orientation, and diameter, in the experimental setup (Demes et al. 1995, 1999; Nakano 2002; Paskins et al. 2007). This is led by the common observation that animal locomotion typically occurs in a very variable and complex 3D environment, which is very much unlike the rigid and uniform situation of most gait labs and other captive settings. This is especially true for arboreal animals, such as primates, who typically use compliant substrates of varying inclination and orientation to travel across the forest canopy.

The lack of substrate variability in our studies of bipedal locomotion in bonobos and gibbons, two highly arboreal apes, is certainly a limitation. However, the used setup was a good approximation of the conditions for terrestrial bipedalism, which is observed in wild gibbons and bonobos, and was justified to get a fundamental insight in their locomotor dynamics. Extrapolation of our results to arboreal bipedalism should be carried out with caution, as the gait dynamics might be influenced by the substrate characteristics. The confounding effect of substrate is probably quite small in gibbon bipedalism, which typically occurs on thick branches or on the ground (Baldwin and Teleki 1976; Fleagle 1976; Gittins 1983; Sati and Alfred 2002). A preliminary study comparing the kinematics of "terrestrial" and "arboreal" hylobatid bipedalism (Vereecke et al. 2006c) indeed found only minor differences between bipedalism on a raised pole (Fig. 3.1) and on a walkway, e.g., stride lengths were longer on the pole compared to the walkway. Still, gibbon bipedalism should ideally be studied using an instrumented, compliant (and inclined) pole to closely mimic the conditions in their natural habitat.

In our bonobo studies, we included some additional substrate complexity by using an instrumented pole with variable inclination to study arboreal locomotion (see earlier). Including more substrate complexity, e.g., compliance, was, however, technically impossible, as the setup had to be able to withstand estimated peak impacts of more than 5000 N. This is a drawback, as substrate compliance probably influences the arboreal gait of this large primate (see also Thorpe et al. 2007).

Fig. 3.1 A simple setup for the study of gibbon locomotion. The setup consists of a horizontal pole of know diameter. Note the presence of fixed landmarks on the pole and on the ground, allowing for 3D reconstruction of subject kinematics when filmed with multiple cameras. Such simple measures greatly increase the potential for a quantitative analysis, and may be feasible in some field studies as well

Given the importance of substrate compliance and variability, we have incorporated these factors in the experimental setup of two recently initiated studies on arboreal locomotion of gibbons.

Field observations indicate that gibbons are proficient jumpers, crossing gaps of 10 m in the forest canopy, and that they make use of branch compliance to propel themselves (Carpenter 1964). Clearly, it is crucial to incorporate substrate compliance in a study on gibbon jumping. Using a springboard setup, consisting of a rigid pole, attached to a force plate via its fulcrum, and provided with a spring, we are not only be able to determine their jumping mechanics, but also if gibbons indeed make use of branch compliance to enhance their jump performance (Channon et al. 2010 and in press Fig. 3.2). Recently, we have also created a new setup to study the bio- and neuromechanics of brachiation in a complex and unpredictable environment. The setup consists of a rigid beam to which handholds with an integrated force transducer (Kistler) and spring-damper-system can be bolted (Fig. 3.3). This enables registration of forces, as well as variation of compliance, orientation, and position of the handholds. This research will provide additional insight into the dynamics of brachiation, which has so far been studied solely in controlled laboratory settings using rigid substrates (Andrew and Groves 1976; Jungers and Stern 1980, 1981, 1984; Preuschoft and Demes 1984; Yamazaki 1990; Bertram and Chang 1996, 2001; Chang et al. 1997, 2000; Bertram et al. 1999; Betram and Chang 2001; Usherwood and Bertram 2003; Usherwood et al. 2003; Bertram 2004).

Fig. 3.2 Setup for studying jumping in gibbons. The setup consists of an aluminium pole with adjustable compliance, mounted onto a force plate. In addition (not visible), two high-speed video cameras are installed (for a full description of the setup see Channon et al. 2010 and in press)

Fig. 3.3 Still video image, showing part of a setup for studying brachiation in siamangs. Handholds can easily be modified, oriented differently, and put at different heights. 3D Force transducers (not visible) are built into three handholds

Gait Variability

Next to the lack of substrate complexity, most captive studies largely fail to address gait variability, though arboreal locomotion is typically irregular and unsteady. Fortunately, this has now been widely acknowledged and we are seeing more and more studies that focus on nonlinear and unsteady locomotion (Walter 2003; Demes et al. 2006; Carlson and Judex 2007; Moreno et al. 2008; Walter and Carrier 2009; Demes and Carlson 2009).

During our experiments, we mostly worked with untrained free-ranging primates, allowing free choice of pace and gait type. Yet, only steady locomotor bouts were selected for analysis, omitting fast irregular gaits, such as the galloping gait of bonobos (infrequently observed in captivity; pers. obs.), despite being very interesting biomechanically. In our study on arboreal locomotion, the bonobos were trained to climb the structure, and tended to move fast and steady, which again contrasts to field observations that typically categorize arboreal locomotion as irregular and opportunistic (Susman et al. 1985).

We strongly suggest that, in the future, an effort should be made to collect biomechanical data on a large variety of—often unsteady—locomotion. To attain this, the technical challenge of incorporating compliance and spatial complexity in the experimental setup should be addressed. Even in simple setups, a wider range of locomotor modes should be studied than has been considered to date. For example, despite clear differences between bipedal and quadrupedal locomotion in bonobos, we have found that these apes display large gait versatility, and that intermediate forms of locomotion exist between "typical" bipedal and "typical" quadrupedal strides (D'Août et al. 2004). Future studies should also investigate tripedal, unsteady, and loaded locomotion, i.e., load or infant carrying, in bonobos and other primates. In addition, higher importance should be given to the collection of quantitative locomotor data in wild primates, providing crucial information on the occurrence and relevance of the full range of adopted gait types. When working in a captive context, the setup dictates, to a large extent, what types of locomotor behavior will be performed by the animals. Therefore, the choice should be deliberate and the potential influence of the setup itself on the gathered data must be borne in mind. For instance, terrestrial bipedal behavior was frequently observed in our bonobo study group: 3.9% of spontaneous locomotor bouts were bipedal and this percentage increased to 18.9% in a food gathering context (D'Août et al. 2004). These percentages seem high when compared to another captive study, where bipedal locomotion was less than 0.01% of total time budget (calculated from Dielentheis et al. 1996). It has yet to be established to what extent the external conditions in captivity influence the occurrence and frequency of specific locomotor modes.

Need for Field Studies

Earlier, we have discussed the advantages and drawbacks of locomotor studies outside of the field. While scientific and technological advances will foster new approaches and more complex settings, the full complexity of the natural habitat will never be achieved in laboratories and zoo settings. Field studies will continue to give a more complete picture of primate locomotion and we want to underline that for a further advancement of our knowledge of primate locomotion it is crucial to integrate both approaches. Further, we discuss some issues we encountered in our studies that, in our view, -can be fully tackled only via quantitative, field-based locomotor research.

Locomotor Behavior in Captivity: Representative of the Wild?

The properties of the environment strongly influence the locomotion of animals and this is probably especially important in an experimental setting, as outlined earlier. Field studies are necessary to document whether the locomotor modes studied elsewhere are representative for the animal's locomotor behavior in the wild. Even though "atypical" locomotor modes can be extremely suitable in order to gain an insight in the biomechanical capacities of animals (e.g., Scholz et al. 2006, for vertical jumping in bonobos), it is necessary to know the frequency and the context in which the studied locomotor mode occurs in the wild. As an example, bipedal walking in bonobos is observed quite frequently in captivity (see earlier), even on arboreal and inclined supports. We know that bipedal locomotion is also observed in wild bonobos (Susman et al. 1980; Kano 1992; Thompson 2002), but the frequency may be rather low during terrestrial locomotion, i.e., 0.3%, even though this may be an underestimation due to the effect of habituation (Susman 1984; Doran 1993). In an arboreal context (see further), bipedal locomotion appears to be more frequent (Susman et al. 1980). Yet, percentages of arboreal bipedal bouts vary considerably, even within one study site, i.e., Lomako (Congo), where percentages range from 1.2% (Doran 1993) to 6% (Susman et al. 1985) to as high as 10% (Susman 1984). These figures are in the same order of magnitude as observed in captive studies, but even in the field studies mentioned, it has been suggested that the increasing degree of habituation of the bonobos may confound the observations. Technological advances, such as unmanned cameras with long autonomy, may facilitate opportunities for gathering locomotor data from unhabituated populations and increasing sample size, expanding current knowledge of locomotor behavior in the wild (e.g., Nishida 1972; Kano 1979, 1983; Kuroda 1979; Doran 1993).

Which Gaits Should Be Studied Ex Situ?

Other observations relate to the diversity of locomotor modes adopted by primates. Indeed, captive bonobos frequently engage in jumping (Scholz et al. 2006), galloping, and various forms of unsteady locomotion. All these modes reflect gait versatility but have been poorly understood. For example, in the case of vertical squat jumping in bonobos, it is not known whether it occurs in the wild at all, as the rare literature accounts report only leaping and jumping to lower supports (Susman et al. 1980). Another example is oblique walking in bonobos, which we observed in our captive study group during both bipedal and quadrupedal bouts. Preliminary analyses (D'Août et al. 2003) suggest that walking obliquely is not an effect of the experimental setup, and that there is a subject preference as to the locomotor asymmetry. This observation merits further study in more detail and in a natural context: is oblique walking also typical in the field? Does locomotor asymmetry correlate with anatomical and functional asymmetry, or even with behavioral laterality (as has been described in primates) and handedness? Another puzzling observation was the use of a knuckle-walking gait on our arboreal setup, at inclines of up to 60°. Knuckle-walking has also been observed in the wild on thick (>30 cm) boughs (Susman et al. 1980), but detailed and quantitative accounts of the occurrence of arboreal knuckle-walking are lacking. Yet, this type of hand use, considered an adaptation for terrestrial locomotion, in an arboreal context certainly merits more attention both in the field and in the laboratory.

Additional quantitative data of the locomotor behavior of wild primates need to be collected to indicate which locomotor modes occur most frequently within the locomotor repertoire (possibly documenting locomotor modes that have been unnoticed *ex situ*), and captive studies should then focus on the analysis of the most "important" locomotor modes.

Substrate Use and Substrate Properties in the Wild

As shown earlier, experimental setups have mostly been rigid, whereas natural substrates (especially in the arboreal niche) are compliant, and therefore captive research should tackle compliant substrates as well. We consider that there are two ways to do this, corresponding to the "bringing the field to the lab" and "bringing the lab to the field" approach.

In the first case, field records should be made of the total range of mechanical substrates available to the animals, and of the ones that are selected (and to what extent) by the animal during its daily activities. Information about substrate size, diameter, orientation, and inclination should be collected, and, where possible, material properties should be assessed. This can be done quite easily by calibrating the used substrates (see also Cheyne, Chapter 11). Evaluating to what extent the ecologically relevant substrate will influence, e.g., an animal's mechanical energy can be made by feeding these quantitative data into lab-based studies. An alternative, and equally interesting, approach could be to control substrate properties (such as compliance or size) in captivity in a naïve way and evaluate how the mechanics of locomotion are altered (or not) by changing these properties. By doing so, it may be possible to establish "mechanically optimal" substrate properties, e.g., a specific compliance may enable an animal of a specific mass to exploit elastic recoil from the substrate. Findings like this would generate hypotheses that can be tested in the field. As an example we refer to gibbon jumping, which has been discussed earlier. Field reports mention that gibbons jump from compliant substrates—in a similar way as human springboard jumpers—enabling them to cross large gaps in the forest canopy. Quantitative information about jumping distance, loss of height, substrate properties, and oscillations are, however, lacking. Even though gibbon jumping is well suited for a study in captivity, as parameters can be changed one by one and make the analysis less complex, several aspects can be studied only in the wild; maximal performance will be difficult, if not impossible, to record in captivity owing to spatial restrictions and lack of natural social behavior (predator avoidance, competitors).

We strongly recommend taking account of substrate properties in both laboratory- and field-based locomotor studies, and to exchange data between the two approaches. Only by integrating field and experimental data will we be able to obtain a good understanding of substrate use in (arboreal) primates.

Energetics: Importance of Efficiency?

Earlier, we discussed several important aspects of primate locomotion, such as substrate use and gait variability, that call for field research. Another important topic for which field investigation is crucial is energetics. Although lab research offers many possibilities for the study of locomotor energetics, this is again conducted mostly in standardized conditions, e.g., treadmill, focusing on steady locomotion (cf. Crompton et al. 1993) and using trained (and often overweight) subjects. Such information can be used to interpret and understand the "cost" of different locomotor modes occurring in the wild, yet the controlled conditions make it difficult to assess the actual efficiency of these locomotor modes in the animal's ecology. Therefore, quantitative data on locomotor behavior (see earlier) and energetics in wild primates are essential.

The way in which locomotor energetics can be studied in the wild is addressed in detail in Chapter 10 (Blanchard et al.), and several other chapters also discuss the opportunities to collect data on energetics in the field. In combination with detailed experimental studies, this will enable us to compare the cost of different locomotor modes and to assess the most efficient way for the animal to get from A to B (see also Pontzer et al. 2009). By contrasting the cost of these gait types against the total energy budget (which may, in some cases, be measured using doubly labeled water; e.g., Nagy 1987), the importance of that particular gait for the animal's fitness can be investigated. Only with that knowledge will we be able to provide a full evolutionary framework for primate locomotion (Leonard and Robertson 1997).

Conclusions

Laboratory studies are particularly suitable for the detailed study of different aspects of primate locomotion, and, as such, have contributed significantly to our understanding of primate locomotor mechanics and energetics. However, incorporation of more naturalistic, and hence more complex, substrate conditions, as well as the inclusion of unsteady and "atypical" locomotor modes, is needed to advance our understanding of primate locomotion further. In addition, qualitative and quantitative locomotor data from the field are essential to provide the ecological framework, allowing us to get a full insight into adaptive processes of primate evolution.

References

Aerts P, Van Damme R, Van Elsacker L, Duchêne V (2000) Spatio-temporal gait characteristics of the hind-limb cycles during voluntary bipedal and quadrupedal walking in bonobos (*Pan paniscus*). Am J Phys Anthropol 111:503–517.

Andrew P, Groves CE (1976) Gibbon and brachiation. In: Rumbaugh DM (ed), Gibbon and Siamang: A Series of Volumes on the Lesser Apes, Vol. 4: Suspensory Behavior, Locomotion, and Other Behaviors of Captive Gibbons: Cognition. Karger, Basel, pp 167–218.

Baldwin LA, Teleki G (1976) Patterns of gibbon behavior on Hall's Island, Bermuda: A preliminary ethogram for *Hylobates lar*. Gibbon Siamang 4:21–105.

Bertram JEA, Chang YH (1996) Gait change in gibbon brachiation. Am Zool 36:20.

Bertram JEA, Ruina A, Cannon CE, Hui Chang Y, Coleman MJ (1999) A point-mass model of gibbon locomotion. J Exp Biol 202:2609–2617.

Bertram JEA, Chang YH (2001) Mechanical energy oscillations of two brachiation gaits: measurement and simulation. Am J Phys Anthropol 115:319–326.

Bertram JEA (2004) New perspectives on brachiation mechanics. Yrbk Phys Anthropol 47:100–117.

Carlson KJ, Judex S (2007) Increased non-linear locomotion alters diaphyseal bone shape. J Exp Biol 210(17):3117–3125.

Carpenter CR (1964) A field study in Siam of the behavior and social relations of the gibbon (*Hylobates lar*). In: Carpenter CR (ed), Naturalistic Behavior of Nonhuman Primates. Pennsylvania State University Press, University Park, pp 145–271.

Chang YH, Bertram JEA, Ruina A (1997) A dynamic force and moment analysis system for brachiation. J Exp Biol 200:3013–3020.

Chang YH, Bertram JEA, Lee DV (2000) External forces and torques generated by the brachiating white-handed gibbon (*Hylobates lar*). Am J Phys Anthropol 113:201–216.

Channon AJ, Crompton RH, Gunther MM, D'Août K, Vereecke EE (2010) The biomechanics of leaping in gibbons. Am J Phys Anthropol 143:403–416.

Channon AJ, Gunther MM, Crompton RH, D'Août K, Preuschoft H, Vereecke EE (in press) The Effect of Substrate Compliance on the Biomechanics of Gibbon Leaps. J Exp Biol

Crompton RH, Sellers WI, Guenther MM (1993) Energetic efficiency and ecology as selective factors in the saltatory adaptation of prosimian primates. Proc R Soc Lond B254(1339): 41–45.

Crompton RH, Vereecke EE, Thorpe SK (2008) Locomotion and posture from the common hominoid ancestor to fully modern hominins, with special reference to the last common panin/hominin ancestor. J Anat 212(4):501–543.

D'Août K, Aerts P, De Clercq D, Schoonaert K, Vereecke E, Van Elsacker L (2001) Studying bonobo (*Pan paniscus*) locomotion using an integrated setup in a zoo environment: preliminary results. Primatologie 4:191–206.

D'Août K, Aerts P, De Clercq D, De Meester K, Van Elsacker L (2002) Segment and joint angles of the hind limb during bipedal and quadrupedal walking of the bonobo (*Pan paniscus*). Am J Phys Anthropol 119(1):37–51.

D'Août K, Vereecke E, Schoonaert K, Aerts P (2003) Asymmetrical aspects of bipedal and quadrupedal walking in bonobos (*Pan paniscus*). Am J Phys Anthropol S36:83.

D'Août K, Vereecke E, Schoonaert K, De Clercq D, Van Elsacker L, Aerts P (2004) Locomotion in bonobos (*Pan paniscus*): differences and similarities between bipedal and quadrupedal terrestrial walking, and a comparison with other locomotor modes. J Anat 204(5):353–361.

Demes B, Jungers WL, Gross TS, Fleagle JG (1995) Kinetics of leaping primates: influence of substrate orientation and compliance. Am J Phys Anthropol 96(4):419–429.

Demes B, Fleagle JG, Jungers WL (1999) Takeoff and landing forces of leaping strepsirhine primates. J Hum Evol 37(2):279–292.

Demes B, Carlson KJ, Franz TM (2006) Cutting corners: the dynamics of turning behaviors in two primate species. J Exp Biol 209(5):927–937.

Demes B, Carlson KJ (2009) Locomotor variation and bending regimes of capuchin limb bones. Am J Phys Anthropol DOI: 10.1002/ajpa.21020.

Dickinson MH, Farley CT, Full RJ, Koehl MAR, Kram R, Lehman S (2000) How animals move: an integrative view. Science 288:100–106.

Dielentheis TF, Hofstetter AM, Niemitz C (1996) Bipedality in captive bonobos (*Pan paniscus*). Primate Rep 44:7.

Doran DM (1993) Comparative locomotor behavior of chimpanzees and bonobos: the influence of morphology on locomotion. Am J Phys Anthropol 91(1):83–98.

Fleagle JG (1976) Locomotion and posture of the Malayan siamang and implications for hominid evolution. Folia Primatol 26:245–269.

Gibbs S, Collard M, Wood B (2002) Soft-tissue anatomy of the extant hominoids: a review and phylogenetic analysis. J Anat 200(1):3–49.

Gittins PS (1983) Use of forest canopy by the agile gibbon. Folia Primatol 40:134–144.

Ishida H, Kimura T, Okada M, Yamazaki N (1976) Kinesiological aspects of bipedal walking in gibbons. In: Preuschoft H, Chivers DJ, Brockelman WY, Creel N (eds), The Lesser Apes: Evolutionary and Behavioral Biology. Edinburgh University Press, Edinburgh, pp 135–145.

Ishida H, Okada M, Tuttle RH, Kimura T (1978) Activities of hindlimb muscles in bipedal gibbons. In: Chivers DJ, Joysey KA (eds), Recent Advances in Primatology, Vol. 3: Evolution. Academic Press, London, pp 459–462.

Ishida H, Kimura T, Okada M, Yamazaki N (1984) Kinesiological aspects of bipedal walking in gibbons. In: Preuschoft H, Chivers DJ, Brockelman WY, Creel N (eds), The Lesser Apes: Evolutionary and Behavioral Biology. Edinburgh University Press, Edinburgh, pp 135–145.

Ishida H, Kumakura H, Kondo S (1985) Primate bipedalism and quadrupedalism: comparative electromyography. In: Kondo S, Ishida H, Kimura T, Okada M, Yamazaki N, Prost JH (eds), Primate Morphophysiology, Locomotor Analyses and Human Bipedalism. University of Tokyo Press, Tokyo, pp 59–79.

Jungers WL, Stern JT Jr (1980) Telemetered electromyography of forelimb muscle chains in gibbons (*Hylobates lar*). Science 208:617–619.

Jungers WL, Stern JT (1981) Preliminary electromyographical analysis of brachiation in gibbon and spider monkey. Int J Primatol 2(1):19–33.

Jungers WL, Stern JT (1984) Kinesiological aspects of brachiation in lar gibbons. In: Preuschoft H, Chivers DJ, Brockelman WY, Creel N (eds), The Lesser Apes: Evolutionary and Behavioral Biology. Edinburgh University Press, Edinburgh, pp 119–134.

Kano T (1979) A pilot study on the ecology of pygmy chimpanzees, *Pan paniscus*. In: Hamburg DA, McCown ER (eds), Perspectives on Human Evolution, Vol. 5: The Great Apes. Benjamin/Cummings, Menlo Park, CA, pp 123–135.

Kano T (1983) An ecological study of the pygmy chimpanzees (*Pan paniscus*) of Yalosidi, Republic of Zaire. Int J Primatol 4(1):1–31.

Kano T (1992) The last ape: pygmy chimpanzee behavior and ecology. Stanford University Press, Stanford, CA.

Kimura T, Okada M, Ishida H (1977) Dynamics of primate bipedalism as viewed from the force of the foot. Primates 18:137–147.

Kuroda S (1979) Grouping of the pygmy chimpanzees. Primates 20:161–183.

Leonard WR, Robertson ML (1997) Comparative primate energetics and hominid evolution.Am J Phys Anthropol 102(2):265–281.

McGrew WC (1998) Culture in nonhuman primates? Annu Rev Anthropol 27:301–328.

Moreno CA, Main RP, Biewener AA (2008) Variability in forelimb bone strains during non-steady locomotor activities in goats. J Exp Biol 211(7):1148–1162.

Nagy KA (1987) Field metabolic rate and food requirement scaling in mammals and birds. Ecol Monogr 57(2):111–128.

Nakano Y (2002) The effects of substratum inclination on locomotor patterns in primates. Z Morphol Anthrop 83(2–3):189–199.

Nishida T (1972) Preliminary information of the pygmy chimpanzees (*Pan paniscus*) of the Congo Basin. Primates 13:415–425.

Okada M, Kondo S (1982) Gait and EMGs during bipedal walk of a gibbon (*Hylobates agilis*) on flat surface. J Anthropol Soc Nippon 90:325–330.

Okada M, Yamazaki N, Ishida H, Kimura T, Kondo S (1983) Biomechanical characteristics of hylobatid walking on flat surfaces. Ann Sci Nat Zool 5:137–144.

Okada M (1985a) Primate bipedal walking: comparative kinematics. In: Kondo S, Ishida H, Kimura T, Okada M, Yamazaki N, Prost JH (eds), Primate Morphophysiology, Locomotor Analysis and Human Bipedalism. University of Tokyo Press, Tokyo, p 47.

Okada M (1985b) Experimental studies on primate positional behavior. Rec Prog Nat Sci Jpn 8:71–78.

Paskins KE, Bowyer A, Megill WM, Scheibe JS (2007) Take-off and landing forces and the evolution of controlled gliding in northern flying squirrels *Glaucomys sabrinus* Exp Biol 210(8):1413–1423.

Payne RC, Crompton RH, Isler K, Savage R, Vereecke EE, Günther MM, Thorpe SKS, D'Août K (2006) Morphological analysis of the hindlimb in apes and humans. Part I: Muscle architecture. J Anat 208:709–724.

Pontzer H, Ocobock C, Shumaker RW, Raichlen DA (2009) Daily energy expenditure in orangutans measured using doubly labeled water. Am J Phys Anthropol 138(S48):328–329.

Preuschoft H, Demes B (1984) Biomechanics of brachiation. In: Preuschoft H, Chivers DJ, Brockelman WY, Creel N (eds), The Lesser Apes: Evolutionary and Behavioral Biology. Edinburgh University Press, Edinburgh, pp 96–118.

Prost JH (1967) Bipedalism of man and gibbon compared using estimates of joint motion. Am J Phys Anthropol 26:135–148.

Reid GMcG, Macdonald AA, Fidgett AL, Hiddinga B, Leus K (2008) Developing the research potential of zoos and aquaria. The EAZA Research Strategy. EAZA Executive Office, Amsterdam.

Sati, JP, Alfred JRB (2002) Locomotion and posture in Hoolock gibbon. Ann Forest 10:298–306.

Schmid P, Piaget A (1994) Three-dimensional kinematics of bipedal locomotion. Z Morphol Anthropol 80(1):79–87.

Scholz M, D'Août K, Bobbert MF, Aerts P (2006) Vertical jumping performance of bonobo (*Pan paniscus*) suggests superior muscle properties. Proc Roy Soc Lond B 273:2177–2184.

Schoonaert K, D'Août K, Aerts P (2006) A dynamic force analysis system for climbing of large primates. Folia Primatol 77:246–254.

Shapiro LJ, Jungers WL, Stern JT Jr (1987) Function of back and trunk muscles during bipedal walking in chimpanzee and gibbon. Am J Phys Anthropol 72:253.

Shapiro LJ, Jungers WL (1988) Back muscle function during bipedal walking in chimpanzee and gibbon: implications for the evolution of human locomotion. Am J Phys Anthropol 77:201–212.

Shapiro LJ, Jungers WL (1994) Electromyography of back muscles during quadrupedal and bipedal walking in primates. Am J Phys Anthropol 93:491–504.

Stern JT, Susman RL (1981) Electromyography of the gluteal muscles in *Hylobates*, *Pongo*, and *Pan*: implications for the evolution of hominid bipedality. Am J Phys Anthropol 55:153– 166.

Susman RL, Badrian NL, Badrian AJ (1980) Locomotor behavior of *Pan paniscus* in Zaire. Am J Phys Anthropol 53:69–80.

Susman RL (1984) The locomotor behavior of *Pan paniscus* in the Lomako Forest. In: Susman RL (ed), The Pygmy Chimpanzee. Evolutionary Biology and Behavior. Plenum Press, New York, pp 369–393.

Susman RL, Badrian N, Badrian A, Handler NT (1985) Positional behavior and feeding ecology of the pygmy chimpanzee (*Pan paniscus*): first year results of the Lomako Forest pygmy chimpanzee project. Natl Geogr Soc Res Rep 20:725–739.

Thompson JAM (2002) A model of the biogeographical journey from Proto-pan to *Pan paniscus*. Primates 44(2):191–197.

Thorpe SKS, Holder RL, Crompton RH (2007) Origin of human bipedalism as an adaptation for locomotion on flexible branches. Science 316(5829):1328–1331.

Tuttle RH (1972) Functional and evolutionary biology of hylobatid hands and feet. In: Rumbaugh DM (ed), Gibbon and Siamang: A Series of Volumes on the Lesser Apes, Vol. 1. Karger, Basel, pp 136–206.

Usherwood JR, Bertram JEA (2003) Understanding brachiation: insight from a collisional perspective. J Exp Biol 206(10):1631–1642.

Usherwood JR, Larson SG, Bertram JEA (2003) Mechanisms of force and power production in unsteady ricochetal brachiation. Am J Phys Anthropol 120(4):364–372.

Vereecke EE, D'Août K, De Clercq D, Van Elsacker L, Aerts P (2003) Dynamic plantar pressures during terrestrial locomotion of bonobos (*Pan paniscus*). Am J Phys Anthropol 120:373–383.

Vereecke EE, K D'Août, P Aerts (2005) Functional analysis of the gibbon foot during terrestrial bipedal walking: plantar pressure distributions and three dimensional ground reaction forces. Am J Phys Anthropol 128:659–669.

Vereecke EE, D'Août K, Aerts P (2006a) Speed modulation in hylobatid bipedalism: a kinematical analysis. J Hum Evol 51(5):513–526.

Vereecke EE, D'Août K, Aerts P (2006b) The dynamics of hylobatid bipedalism: evidence for an energy-saving mechanism? J Exp Biol 209:2829–2838.

Vereecke EE, D'Août K, Aerts P (2006c) Locomotor versatility in the white-handed gibbon (*Hylobates lar*): a spatiotemporal analysis of the bipedal, tripedal, and quadrupedal gaits. J Hum Evol 50(5):552–567.

Vereecke EE, Aerts P (2008) The mechanics of the gibbon foot and its potential for elastic energy storage during bipedalism. J Exp Biol 211(23):3661–3670.

Walter RM (2003) Kinematics of 90 degrees running turns in wild mice. J Exp Biol 206(10):1739–1749.

Walter RM, Carrier DR (2009) Rapid acceleration in dogs: ground forces and body posture dynamics: J Exp Biol 212(12):1930–1939.

Ward CV (2007) Postcranial and locomotor adaptations of hominoids. In: Henke W, Tattersall I (eds), Handbook of Paleoanthropology, Vol. 2. Springer, New York, pp 1011–1030.

WAZA (2005). Building a Future for Wildlife – The World Zoo and Aquarium Research Strategy. WAZA Executive Office, Bern.

Wildman DE, Uddin M, Liu G, Grossman LI, Goodman M (2003) Implications of natural selection in shaping 99.4% nonsynonymous DNA identity between humans and chimpanzees: enlarging genus *Homo*. Proc Natl Acad Sci USA 100(12):7181–7188.

Yamazaki N (1990) The effects of gravity on the interrelationship between body proportions and brachiation in the gibbon. In: Jouffroy FK, Stack MH, Niemitz C (eds), Gravity, Posture and Locomotion in Primates. Editrice Il Sedicesimo, Florence, pp 157–172.

Yamazaki N, Ishida H (1984) A biomechanical study of vertical climbing and bipedal walking in gibbons. J Hum Evol 13:563–571.

Zihlman AL, Cronin JE, Cramer DL, Sarich VM (1978) Pygmy chimpanzee as a possible prototype for the common ancestor of humans, chimpanzees and gorillas. Nature 275:744–746.

Chapter 4
Experimental and Computational Studies of Bipedal Locomotion in the Bipedally Trained Japanese Macaque

Naomichi Ogihara, Eishi Hirasaki, and Masato Nakatsukasa

Abstract Japanese macaques trained as performing monkeys have a remarkable ability to walk bipedally. This shift toward bipedalism in an inherently quadrupedal primate could be regarded as a modern analogue for the evolution of bipedal locomotion and offers an interesting model for understanding the emergence of human bipedalism. Here, we review the results of our recent experimental studies of bipedal locomotion in trained Japanese macaques to explore the unique characteristics of their bipedal locomotion. We then describe the development of an anatomically based, whole-body musculoskeletal model of the Japanese macaque for biomechanical analyses and predictive simulations of locomotion. Such constructive studies of locomotion based on a mathematical model of the musculoskeletal system allow computational investigation of the mechanisms of bipedal walking. To understand the diversification of locomotor patterns in primates and the emergence of erect bipedal walking in the human lineage, field studies of locomotion and performance, in terms of selection, are indispensable. Our approach allows quantification of functional consequences of morphological variations in the musculoskeletal system, hopefully helping to identify the causal relationship between morphology and performance and as such complement data collected in the wild.

Keywords Gait • Modeling • Musculoskeletal system • Primate primate

Abbreviations

BW	body weight
COM	center of mass
CPG	central pattern generation
KE	kinetic energy
PCSA	physiological cross-sectional area
PE	potential energy

N. Ogihara (✉)
Keio University, Japan
e-mail: ogihara@mech.keio.ac.jp

K. D'Août and E.E. Vereecke (eds.), *Primate Locomotion: Linking Field and Laboratory Research*, Developments in Primatology: Progress and Prospects,
DOI 10.1007/978-1-4419-1420-0_4, © Springer Science+Business Media, LLC 2011

Introduction

Performing monkeys have a long history as a popular attraction in Japan. Trained Japanese macaques (*Macaca fuscata*) entertain audiences by exhibiting their remarkable physical capabilities through feats such as jumping over hurdles, walking on stilts and balancing on a ball, all of which are performed bipedally (Fig. 3.1a). Performing monkeys must first demonstrate the ability to stand upright on their hind limbs and walk bipedally for a sufficiently long period of time before engaging in training for performances. This initial training in bipedal posture and locomotion, starting at approximately the age of 1 year, forms a critical foundation for future performances. Acquisition of bipedal posture and locomotion is the key adaptation for the performing macaques.

This shift in positional behavior towards bipedalism for an inherently quadrupedal primate could be regarded to some extent as a modern analogue for the evolution of bipedal locomotion (Hayama et al. 1992; Nakatsukasa 2004; Hirasaki et al. 2006). Comparative studies of primate and human bipedal locomotion allow us to analyze how structural and morphological differences in locomotor musculoskeletal apparatuses can affect the biomechanical characteristics of bipedal locomotion. In particular, bipedal locomotion in Japanese macaques trained as bipedal performing monkeys offers an interesting model for understanding emergence of human bipedalism and has been thoroughly investigated in the field of physical anthropology.

Previous studies on bipedally trained macaques focused mainly on their morphological adaptations for habitual bipedalism. These studies demonstrated that human-like lordosis of the lumbar spine occurs, and joint morphologies (Preuschoft et al. 1988; Hayama et al. 1992) and long-bone strength are also modified to adapt to the mechanical requirements of habitual bipedalism (Nakatsukasa and Hayama 1991; Nakatsukasa et al. 1995). However, although kinematic and kinetic analyses of bipedal locomotion have been conducted for ordinary captive Japanese macaques that have received only a minimum amount of training to induce bipedal walking for experiments (e.g. Ishida et al. 1974; Kimura 1977, 1985; Yamazaki et al. 1979; Kimura et al. 1983; Okada 1985; Yamazaki 1985; Ishida 1991), performing monkeys have never been systematically and comparatively evaluated for these effects.

Recently, we examined the kinematics and kinetics of bipedal walking in the bipedally trained performing monkeys and compared them to those in ordinary (less trained) monkeys to clarify the locomotor characteristics of elaborated bipedal walking in trained monkeys (Hirasaki et al. 2004; Ogihara et al. 2005, 2007). We here first review our current findings on the kinematic and kinetic characteristics of the locomotion of trained monkeys, then discuss the unique mechanisms of bipedal locomotion for inherently quadrupedal primates. Further, we have developed an anatomically based, whole-body musculoskeletal model of the Japanese macaque to explore fully the mechanisms of their bipedal walking (Ogihara et al. 2009). Here, we briefly describe the constructed musculoskeletal model, and introduce our ongoing computational studies of locomotion, i.e., analytical and synthetic studies of locomotion, based on this musculoskeletal model.

To understand the diversification of locomotor patterns in primates and the emergence of erect bipedal walking in the human lineage, field studies of locomotion and performance, in terms of selection, are indispensable. For instance, locomotor efficiency is an important performance factor because it affects ranging and foraging strategies. However, analyses of causal relationships between musculoskeletal morphology, locomotor kinematics, and energetics are very difficult to execute in the field. Our *ex situ* approach allows interpretation of functional consequences of morphological variations in the musculoskeletal system on locomotor performance observed in wild environments, providing profound insights on the evolution of primate locomotion and human bipedalism.

Kinematics of Bipedal Locomotion in Trained and Ordinary Japanese Macaques

To capture kinematic characteristics, we videotaped bipedal locomotion in trained Japanese macaques at Suo Monkey Performance Association (Kumamoto, Japan) using two cameras in lateral view while the monkeys walked on a treadmill (Hirasaki et al. 2004; Ogihara et al. 2005) (Fig. 4.1b). We attached a piece of reflective tape to each of the following five points: acromion, hip joint, knee joint, ankle joint, and head of the fifth metatarsal. We selected only bipedal sequences in which the monkey walked steadily for analysis. We digitized the landmarks via

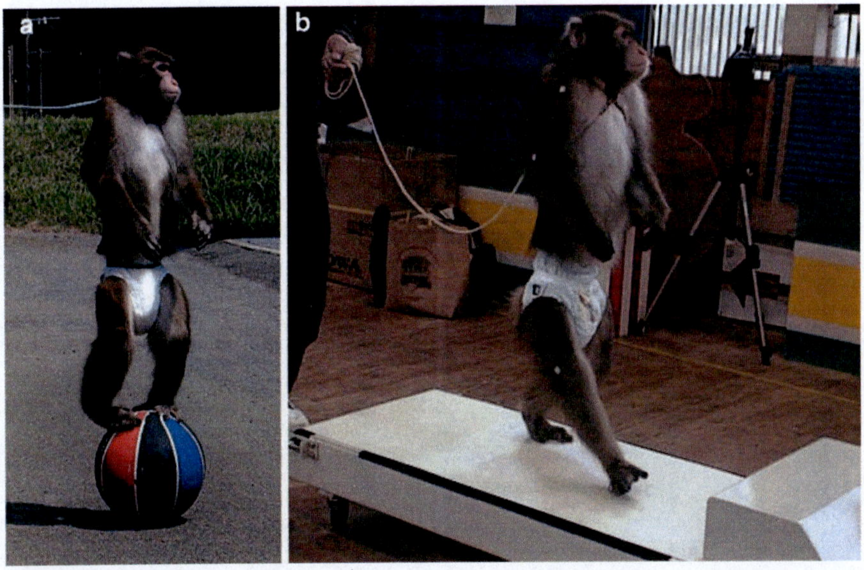

Fig. 4.1 A performing monkey balancing on a ball (**a**) and walking on a treadmill (**b**)

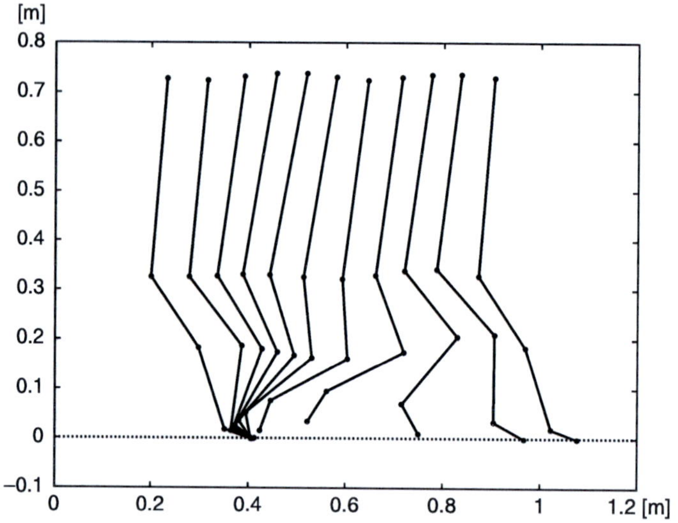

Fig. 4.2 Stick diagram of bipedal locomotion in a performing monkey walking on a treadmill at 3 km/h. Traced every 10% of the gait cycle

three-dimensional motion analysis software to obtain joint angle profiles. For comparison, we measured the bipedal locomotion of ordinary (less trained) Japanese macaques at Osaka University in the same manner, but using a stationary walkway instead of a treadmill.

Comparisons of the kinematic profiles of the trained and ordinary monkeys revealed that the trained monkeys walk with a longer stride length and lower stride frequency. In addition, their hip, knee, ankle joints are in relatively more extended positions throughout stance phase; thus, their locomotion can be characterized as a relatively less "bent-hip, bent-knee" compared to that of the ordinary monkeys (Hirasaki et al. 2004), although joint angle flexion is still higher than that in normal human walking. Further, the hip joint moves sinusoidally in the trained monkeys, and this trajectory is highest in the middle of a stance phase (Fig. 4.2), but this was not observed in the less trained monkeys. In human walking, the body vaults over the supporting leg, which is maintained in extended posture as an inverted pendulum (Fig. 4.3a); thus, the center of mass (COM) is elevated at the midstance phase and lowered at the double-support phase (Fig. 4.3b). Conversely, horizontal velocity is lowest at the midstance phase and highest at the double-support phase. Therefore, as the COM moves forward and then down in the late stance phase, stored potential energy (PE) is released and exchanged for kinetic energy, which is then converted back into PE as the COM is elevated. This mutual exchange of the two energy types is called an inverted-pendulum mechanism and conserves up to 70% of the mechanical energy of each stride in human walking, indicating that this is the fundamental mechanism of energy conservation in human bipedal walking (Cavagna et al. 1977; Ortega and Farley 2005; Massaad et al. 2007). Our kinematic analyses suggested that bipedal walking with a human-like energy conservation

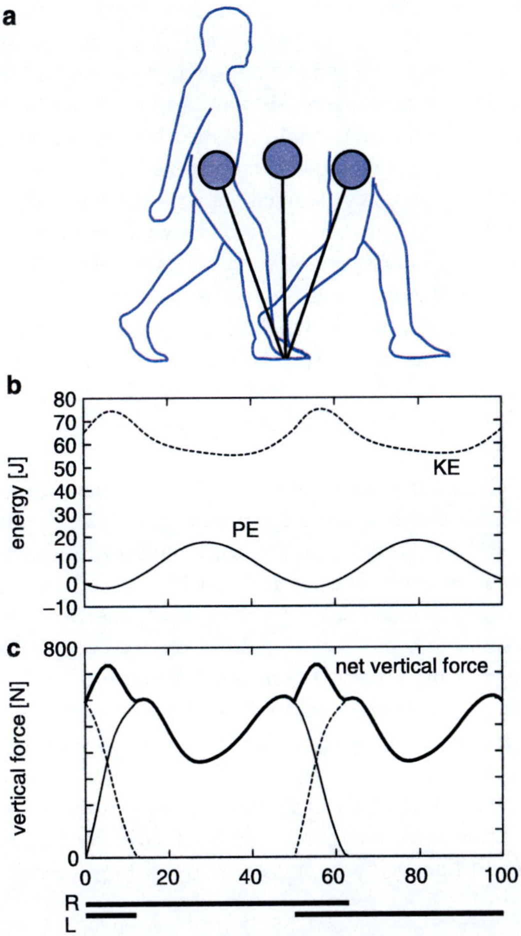

Fig. 4.3 Biomechanical characteristics of human bipedal walking. (**a**) Inverted-pendulum mechanism. (**b**) Changes in the potential and kinetic energies (PE and KE, respectively) of the body COM. (**c**) Vertical ground reaction force profile in human walking (Winter 1990)

mechanism can be acquired by inherently quadrupedal monkeys through training (Hirasaki et al. 2004; Ogihara et al. 2005).

Ground Reaction Force Profiles of Bipedal Locomotion in Trained and Ordinary Japanese Macaques

It is not clear how trained monkeys acquire the inverted-pendulum mechanism. In humans, the oscillation of the body COM during normal walking is realized by generating a vertical ground reaction force profile with two peaks (Fig. 4.3c).

Because of this two-peaked force profile, the net vertical force applied to the COM from the ground is highest at the double-support phase and lowest at the midstance phase, creating this distinctive feature of normal human walking. To clarify whether trained monkeys generate a two-peaked force profile similar to that in humans, as well as to characterize better the dynamics of their bipedal locomotion, we collected their three-dimensional ground reaction force profiles (Ogihara et al. 2007).

Trained and ordinary monkeys walked across a force platform set in a wooden walkway at a self-selected speed. We recorded signals from the force platform via a computer through A/D interfaces. We simultaneously videotaped bipedal locomotion to measure cycle duration and duty factor, i.e., stance phase duration/cycle duration to exclude sequences in which the animals walked too slow or too fast; we excluded steps wherein stance phase duration was one standard deviation outside the range of all trials of that monkey from the analysis.

The averaged vertical ground reaction force profiles of the trained and ordinary monkeys against stance time (0% = foot contact, 100% = toe-off) are shown in Figure 4.4. Typically, nonhuman primates, including Japanese macaques, generate a single-peaked force curve when they walk bipedally (Schmitt 2003), and our result was in accordance with these previous findings. Surprisingly, the trained monkeys, who were expected to generate a double-peaked force curve because of the above-mentioned kinematic characteristics, also generated a single-peaked vertical ground reaction force profile; we did not observe human-like double peaks. However, the shape of the force curve differed between the two groups; the force curve was relatively trapezoidal or parabolic in the ordinary monkeys and relatively triangular in the trained monkeys and the timing of the vertical peak was shifted toward early stance phase.

Oscillatory motions of the COM calculated based on the measured force curves suggest that the trained monkeys are able to induce greater fluctuations in the vertical ground force than ordinary monkeys by shifting the timing of generating

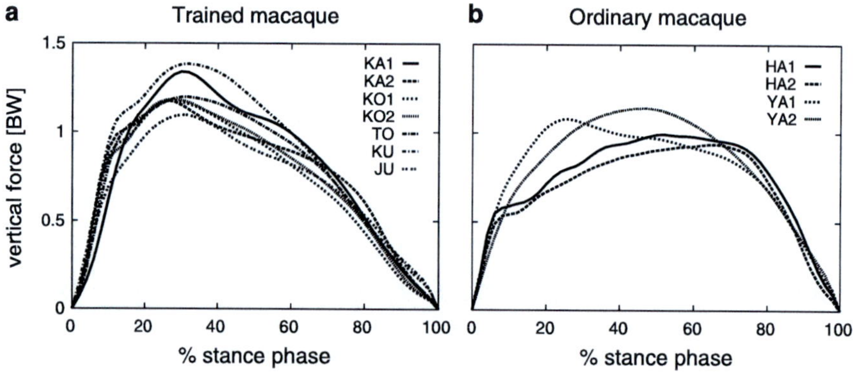

Fig. 4.4 Vertical ground reaction forces for a single bipedal step in Japanese monkeys (Ogihara et al. 2007). The curves were averaged per subject in (**a**) trained and (**b**) ordinary monkeys. Forces were normalized by body weight (BW)

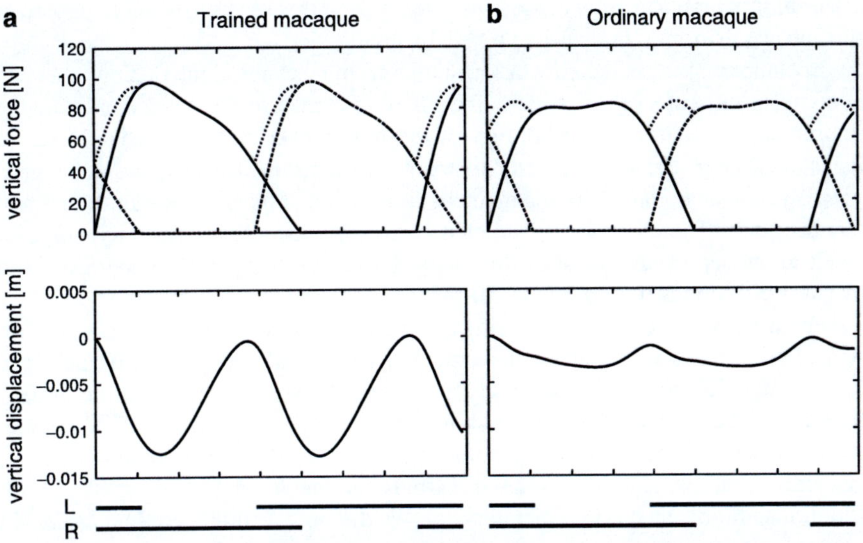

Fig. 4.5 Schematic comparisons of net vertical force acting on the COM and resultant vertical fluctuation of the COM in (**a**) trained and (**b**) ordinary monkeys. The vertical fluctuation of the COM is obtained by numerically integrating the equation of motion. Thick solid line = net vertical force, thin solid and dotted lines = forces exerted by the right and left feet, respectively

peak vertical force toward early stance phase (Fig. 4.5). However, the vertical COM position in the trained monkeys was highest in late stance phase just before foot contact, unlike normal human walking, suggesting that the energetic advantage based on the inverted-pendulum mechanism is probably limited in Japanese macaques.

Musculoskeletal Model of Japanese Macaques

Our kinematic and kinetic studies of bipedal walking in highly trained Japanese macaques demonstrated that although human-like kinematic features of bipedal locomotion are acquired through their extensive training, these monkeys do not exhibit human-like two-peaked vertical ground reaction force when walking bipedally. The bipedal walking mechanics of the trained monkeys were fundamentally different from that of both the ordinary (less trained) monkeys and humans. It is possible that structural and/or functional differences in the locomotor neuro-control system prevents Japanese macaques from generating human-like bipedal walking. However, the basic neuronal circuits for generating locomotion are considered to be comparable in humans and nonhuman primates (Dietz 2002). Further, the fact that bipedal locomotion in Japanese macaques can be adaptively altered *a posteriori* through training also implies that the nervous system is not the constraining factor.

Instead, structural alteration of the locomotor apparatuses seems to be more important for the acquisition of human-like bipedal walking.

Locomotion is a challenging task resulting from mechanical interactions between a complex musculoskeletal system and the physical world. To understand how the morphological and structural design of the musculoskeletal system facilitates or constrains the generation of bipedal locomotion, biomechanical analyses of locomotion using an anatomically based mathematical model of the musculoskeletal system are necessary. Therefore, we constructed a three-dimensional, whole-body musculoskeletal model of the Japanese macaque based on computed tomography and cadaver dissections (Ogihara et al. 2009).

For a realistic representation of body motion, the fresh cadaver of an adult male Japanese macaque underwent whole-body computed tomography. We obtained a total of 1935 cross-sectional images, from which we constructed three-dimensional surface models of the entire skin surface and skeletal system. The skeleton was divided into the following bone segments: head, thorax, lumbus, pelvis, scapula, humerus, ulna, radius, hand (carpals and metacarpals), femur, tibia (with fibula), and foot (tarsals and metatarsals). We did not consider pedal phalanges in the present version of the model, although they seem to be kinematically important for generation of bipedal walking. A bone coordinate system embedded in each of the bones was defined by principal axes. We approximated each joint as a combination of hinge joints, joint centers and rotational axes estimated by joint morphology based on joint surface approximation using a quadric function. Joint centers were determined from the apex of the fitted quadric surface and the two principal radii of curvature at the apex (Ogihara et al. 2009). We determined rotational axes from the principal directions in which the principal curvatures occurred. A kinematic skeleton of the Japanese macaque was thus mathematically described as a chain of 20 links connected by revolute joints, as illustrated in Figure 4.6 (Ogihara et al. 2009).

We represented joints connecting the trunk segments as 3-degree-of-freedom (DOF) joints. The scapula is usually modeled to be immobile with respect to the thorax, although the relationship between the scapula and forelimb is functionally equivalent to that of the femur and hind limb, thus representing an important element for propulsion (Fischer 2001; Schmidt and Krause, Chapter 6). A new approach was thus used to model mathematically translational motion of the scapula along the rib cage using three revolute joints. Shoulder (glenohumeral), elbow, radioulnar, and wrist joints were modeled as 3-, 1-, 1-, and 2-DOF joints, respectively. We represented hip, knee, and ankle joints as 3-, 1-, and 2-DOF joints, respectively. The total number of DOFs for the skeletal system was 47. As a result of morphologically accurate description of the joint kinematics based on quadric function approximation, rotational axes of the joints did not coincide with bone coordinate axes, unlike robots or conventional biomechanical models.

We dissected the fresh cadaver of a female Japanese macaque to describe mathematically the path of each muscle and its associated capacity to generate force. We carefully exposed each muscle in the fore- and hind limb and observed the points of origin and insertion. We then removed the muscle and mass and systematically

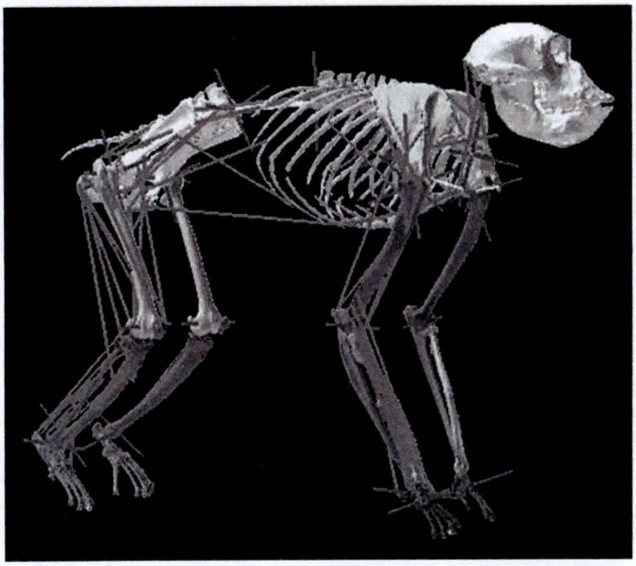

Fig. 4.6 Anatomically based whole-body musculoskeletal model of the Japanese monkey

recorded fascicle length to calculate physiological cross-sectional area (PCSA). The path of each muscle was defined using a series of points connected by line segments. We assumed the capacity of each muscle to generate force to be proportional to PCSA.

Musculoskeletal Motion During Bipedal Locomotion in Trained Japanese Macaques

If the musculoskeletal model described in the preceding text could be matched to the temporal history of the three-dimensional digitized marker coordinates, all body skeletal motion could be reconstructed. Thus, we first scaled the musculoskeletal model to the size of the trained monkey in the video based on segment lengths, and adjusted the joint angles frame-by-frame to minimize the sum of the distances between corresponding (model vs. measured) markers while minimizing deviations of joint angles from the midpoints of the ranges of joint rotations (Ogihara et al. 2009).

Figure 4.7 illustrates the result of model matching to the temporal history of 16 digitized marker coordinates (8 per side) during bipedal walking of a trained Japanese macaque (9 years old). The markers were placed at the acromion, elbow joint, wrist joint, head of the fifth metacarpal, hip joint, knee joint, ankle joint, and head of the fifth metatarsal. The whole-body kinematics of bipedal walking by a trained Japanese macaque was successfully reconstructed using the musculoskeletal

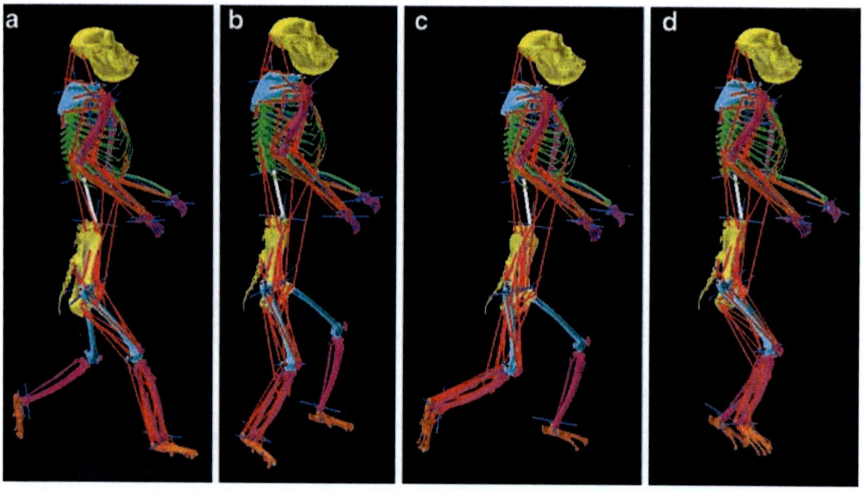

Fig. 4.7 Reconstruction of the whole-body kinematics of a trained Japanese monkey during bipedal walking. (**a**) at the time of foot-contact, (**b**) mid-stance phase, (**c**) toe-off, and (**d**) mid-swing phase

model and the model-based matching technique. By introducing kinematic constraints defined by joint morphology, the present study yielded natural and anatomically reasonable skeletal motion from a limited number of external markers.

From this reconstructed musculoskeletal motion, we can clearly observe that the knee joint is fully extended at the time of foot contact (Fig. 4.7a) in the bipedal walking of the trained monkey, but is largely flexed at the time of toe-off (Fig. 4.7c). In human bipedal walking, the knee joint is extended so that the thigh and shank lie on the same line at the time of both heel contact and toe-off to structurally stiffen the leg. However, in the bipedal walking of the trained Japanese macaque, the leg is structurally compliant at the time of toe-off because the knee is relatively flexed. As a result, effective toe-off with an extended, stiffened leg, i.e., generation of the second peak of the vertical ground reaction force profile, is not achieved.

Based on the estimated skeletal kinematics of Japanese macaques, we calculated changes in muscle length and contractile velocity during bipedal and quadrupedal walking in trained monkeys. Our preliminary analyses of the changes in muscle length suggested that trained Japanese macaques flex their knee joint at the time of toe-off due to restricted hip joint mobility. In bipedal walking, hip muscles such as the psoas major and tensor fasciae latae are stretched, particularly at the time of toe-off, when the hip is most extended. This restriction in the range of hip joint motion seems to prevent Japanese macaques from generating effective push-off, and consequently, generation of human-like walking is quite impossible for an inherently quadrupedal primate. Morphological preadaptation of hip joint mobility is therefore a fundamental determinant of the acquisition of human-like bipedal walking. Although the accuracy of estimated musculoskeletal motion may be limited by a number of potential error factors, such as possible skin motion artifacts and differences in body proportions between the videotaped and modeled macaques, locomotor analyses based on an anatomical musculoskeletal model greatly facilitate interpretation of dynamical behaviors of the musculoskeletal system during locomotion, offering insights into the functional relationship between positional/ locomotor behaviors and musculoskeletal morphology. Furthermore, if a realistic three-dimensional skeletal model is available for registration, it should be possible to reconstruct its skeletal motion reliably from uncalibrated video sequences based on a similar model-based matching technique (e.g., Krosshaug and Bahr 2005). This method therefore seems to be expandable for application in field studies of primate locomotion in which fixations and calibrations of cameras are generally difficult.

Computer Simulation of Locomotion

Incorporation of biomimetic models of locomotor nervous systems with the aforementioned musculoskeletal model could autonomously generate anatomically natural locomotor patterns for a given musculoskeletal system by forward dynamics, allowing predictive simulation studies of locomotion. This would enable evaluation and prediction of changes in kinematics, kinetics, and energetic of locomotion resulting from virtual alterations of the musculoskeletal system, such as deformation of pelvic morphology or modification of muscle length, size, or disposition. Therefore, a forward dynamics simulation of a musculoskeletal model would be an effective way to examine hypotheses and scenarios of the origin and evolution of human bipedalism (Sellers et al. 2003, 2005). Bipedal locomotion of early hominins may also be estimated in a biomechanically reliable manner using the same technique. Further, computer simulation studies based on biologically relevant neuromusculoskeletal modeling have gained particular attention in investigating mechanisms underlying the emergence of adaptive locomotion in animals (Frigon and Rossignol 2006; Pearson et al. 2006).

To develop a biologically plausible predictive simulation, we are currently trying to construct a model of a biological sensorimotor nervous system. Animal locomotion, including that of primates, is generally accepted as being generated by a rhythm-generating neuronal network in the spinal cord known as the central pattern generator (CPG), with locomotion evoked by stimulus input from the brain stem. The rhythmic signals produced by the CPG are then mutually coordinated with afferent proprioceptive information for generation of locomotion. We attempt to model such a neuronal process mathematically for synthetic study of locomotion. Although still at a preliminary stage, we aim to conduct simulation studies using the present model to investigate predictively how changes in morphology and structure of the musculoskeletal system affect locomotor mechanics in Japanese macaques, which may provide profound insights into the origin and evolution of human bipedalism.

Conclusion

Fundamental difficulties exist in attempts to clarify the origin and evolution of human habitual bipedal locomotion based solely on morphological analyses of hominin fossils as such fossils are spatiotemporally very scarce. Herein, we introduce our experimental and computational studies of bipedal locomotion in trained Japanese macaques as an attempt to complement our understanding of the evolution of human bipedal locomotion. Investigating Japanese monkeys trained for bipedal locomotion based on an anatomically relevant musculoskeletal model and predictive simulations will help elucidate the acquisition process of human bipedalism.

Acknowledgments The authors thank Evie Vereecke and Kristiaan D'Août for inviting us to join the symposium Primate Locomotion: Linking In Situ and Ex Situ Research at the 2008 Meeting of the International Primatological Society in Edinburgh, UK, and to contribute to the present volume. We also thank all the staff at the Suo Monkey Performance Association for their generous collaboration in the experiment. This study was supported by a Grant-in-Aid for Scientific Research on Priority Areas "Emergence of Adaptive Motor Function through Interaction between Body, Brain and Environment" from the Japanese Ministry of Education, Culture, Sports, Science and Technology.

References

Cavagna GA, Heglund NC, Taylor CR (1977) Mechanical work in terrestrial locomotion: two basic mechanisms for minimizing energy expenditure. Am J Physiol 233:243–261.

Dietz V (2002) Do human bipeds use quadrupedal coordination? Trends Neurosci 25:462–467.

Fischer MS (2001) Locomotory organs of mammals: New mechanics and feed-back pathways but conservative central control. Zoology 103:230–239.

Frigon A, Rossignol S (2006) Experiments and models of sensorimotor interactions during locomotion. Biol Cybern 95:607–627.

Hayama S, Nakatsukasa M, Kunimatsu Y (1992) Monkey performance: The development of bipedalism in trained Japanese monkeys. Acta Anat Nippon 67:169–185.

Hirasaki E, Ogihara N, Hamada Y, Kumakura H, Nakatsukasa M (2004) Do highly trained monkeys walk like humans? A kinematic study of bipedal locomotion in bipedally trained Japanese macaques. J Hum Evol 46:739–750.

Hirasaki E, Ogihara N, Nakatsukasa M (2006) Primates trained for bipedal locomotion as a model for studying the evolution of bipedal locomotion. In: Ishida H, Tuttle R, Pickford M, Ogihara N, Nakatsukasa M (eds), Human Origins and Environmental Backgrounds. Springer, New York, pp. 149–156.

Ishida H (1991) A strategy for long distance walking in the earliest hominids: Effect of posture on energy expenditure during bipedal walking. In: Coppens Y, Senut B (eds), Origine(s) de la Bipedie chez les Hominids. CNRS, Paris, pp. 9–15.

Ishida H, Kimura T, Okada M (1974) Patterns of bipedal walking in anthropoid primates. In: Kondo S (ed), Proceedings from the Symposia of the Fifth Congress of the International Primatological Society. Japan Science Press, Tokyo, pp. 287–301.

Kimura T (1977) Dynamics of primate bipedal walking as viewed from the force of foot. Primates 18:137–147.

Kimura T (1985) Bipedal and quadrupedal walking of primates: comparative dynamics. In: Kondo S (ed), Primate Morphophysiology, Locomotor Analyses and Human Bipedalism. University of Tokyo Press, Tokyo, pp. 81–104.

Kimura T, Okada M, Yamazaki N, Ishida H (1983) Speed of the bipedal gaits of man and nonhuman primates. Ann Sci Natur Zool Paris 5:145–158.

Krosshaug T, Bahr R (2005) A model-based image-matching technique for three-dimensional reconstruction of human motion from uncalibrated video sequences. J Biomech 38: 919–929.

Massaad F, Lejeune TM, Detrembleur C (2007) The up and down bobbing of human walking: a compromise between muscle work and efficiency. J Physiol (Lond) 582:789–799.

Nakatsukasa M (2004) Acquisition of bipedalism: the Miocene hominoid record and modern analogues for bipedal protohominids. J Anat 204:385–402.

Nakatsukasa M, Hayama S (1991) Structural strength of the femur of bipedally trained monkey. J Anthropol Soc Nippon 99:289–296.

Nakatsukasa M, Hayama S, Preuschoft H (1995) Postcranial skeleton of a macaque trained for bipedal standing and walking and implications for functional adaptation. Folia Primatol 64:1–29.

Ogihara N, Usui H, Hirasaki E, Hamada Y, Nakatsukasa M (2005) Kinematic analysis of bipedal locomotion of a Japanese macaque that lost its forearms due to congenital malformation. Primates 46:11–19.

Ogihara N, Hirasaki E, Kumakura H, Nakatsukasa M (2007) Ground-reaction-force profiles of bipedal walking in bipedally trained Japanese monkeys. J Hum Evol 53:302–308.

Ogihara N, Makishima H, Aoi S, Sugimoto Y, Tsuchiya K, Nakatsukasa M (2009) Development of an anatomically based whole-body musculoskeletal model of the Japanese macaque (*Macaca fuscata*). Am J Phys Anthropol 139:323–338.

Okada M (1985) Primate bipedal walking: comparative kinematics. In: Kondo S (ed), Primate Morphophysiology, Locomotor Analyses and Human Bipedalism. University of Tokyo Press, Tokyo, pp. 47–58.

Ortega JD, Farley CT (2005) Minimizing center of mass vertical movement increases metabolic cost in walking. J Appl Physiol 99:2099–2107.

Pearson K, Ekeberg O, Buschges A (2006) Assessing sensory function in locomotor systems using neuro-mechanical simulations. Trends Neurosci 29: 625–631.

Preuschoft H, Hayama S, Gunther MM (1988) Curvature of the lumbar spine as a consequence of mechanical necessities in Japanese macaques trained for bipedalism. Folia Primatol 50:42–58.

Schmitt D (2003) Insights into the evolution of human bipedalism from experimental studies of humans and other primates. J Exp Biol 206:1437–1448.

Sellers WI, Dennis LA, Crompton RH (2003) Predicting the metabolic energy costs of bipedalism using evolutionary robotics. J Exp Biol 206:1127–1136.

Sellers WI, Cain GM, Wang WJ, Crompton RH (2005) Stride lengths, speed and energy costs in walking of *Australopithecus afarensis*: using evolutionary robotics to predict locomotion of early human ancestors. J R Soc Interface 2:431–441.

Winter DA (1990) Biomechanics and Motor Control of Human Movement. John Wiley & Sons, New York.

Yamazaki N (1985) Primate bipedal walking: computer simulation. In: Kondo S (ed), Primate Morphophysiology, Locomotor Analyses and Human Bipedalism. University of Tokyo Press, Tokyo, pp. 105–130.

Yamazaki N, Ishida H, Kimura T, Okada M (1979) Biomechanical analysis of primate bipedal walking by computer simulation. J Hum Evol 8:337–349.

Chapter 5
In What Manner Do Quadrupedal Primates Walk on Two Legs? Preliminary Results on Olive Baboons (*Papio anubis*)

Gilles Berillon, Kristiaan D'Août, G. Daver, G. Dubreuil, F. Multon, G. Nicolas, and B. de la Villetanet

Abstract Olive baboons (*Papio anubis*) are typically described as highly specialized for quadrupedal locomotion. Yet, they regularly and spontaneously walk bipedally as well. In what manner do they do this, when compared to other primates and humans? This question is of interest with regard to the origin of bipedalism in hominids. As a first step in understanding bipedal locomotion in baboons, we here present novel kinematic data, achieved using a custom-built setup that allows to measure individual locomotor parameters in a population of 55–60 captive olive baboons housed at the Primatology Station of the National Centre for Scientific Research (CNRS, France) using a high-speed digital recording system (100 fps) and a walkway (Podium). Within our population, we observed bipedality mainly in infant and subadult individuals: we present the sagittal motion parameters that we collected on a sample of 10 males and females of ages between 6 months and 5.5 years. As far as angular trajectories of the trunk and the lower limb joints are concerned, olive baboons walk bipedally in a rather stereotyped, compliant manner with a semiplantigrade stance phase, a trunk that is slightly tilted forward and immobile forelimbs kept forward in a parasagittal plane. Some small variations can be depicted, especially in the younger individuals of the sample. Among other "quadrupedal" primates of which the bipedal locomotion has been kinematically analyzed, the kinematics of bipedal walking of *Papio anubis* more closely resembles those described in *Macaca fuscata*. In the broader framework of our study, numerous transversal and longitudinal analyses are in progress on data as varied as noninvasive anatomical investigations, kinematics, kinetics, and paedobarography.

Keywords Bipedalism • Captivity • Kinematics setup • Locomotion

---

G. Berillon (✉)
UPR 2147 CNRS, Dynamique de l'Evolution Humaine, Paris, France
e-mail: gilles.berillon@evolhum.cnrs.fr

K. D'Août and E.E. Vereecke (eds.), *Primate Locomotion: Linking Field and Laboratory Research*, Developments in Primatology: Progress and Prospects,
DOI 10.1007/978-1-4419-1420-0_5,

Introduction

Baboons are committed quadrupedal primates (e.g., Rose 1973; Fleagle 1988). Yet, positional behavioral studies demonstrate that they occasionally walk bipedally in the wild (e.g., Rose 1976, 1977; Wrangham 1980; Hunt 1989). In what manner do they walk bipedally? Considering that many other phylogenetically distant nonhuman primates can walk bipedally despite being anatomically specialized to other modes of locomotion, this question is relevant with regard to the conditions under which bipedalism developed in hominins (reviews: Schmitt 2003; Crompton et al. 2008). Characterizing the anatomical conditions that allow complex motions, such as bipedal locomotion, relies on a detailed description of the bipedal gait, as is the case in humans (Inman et al. 1981; Aiello and Dean 1990; Winter 1991 for a synthesis of anatomical background). Yet, as far as kinematics and kinetics are concerned, bipedal locomotion in baboons is very little documented to date (Okada 1985). Following the very early descriptive analysis of chimpanzee bipedal walking by Elftman and Manter (Elftman and Manter 1935a, b; Elftman 1944), fundamental comparative biomechanical data became available on bipedal locomotion in few nonhuman primates, especially in chimpanzees, gibbons, bonobos, and macaques (Jenkins 1972; Ishida et al. 1974; Kimura et al. 1979; Yamazaki et al. 1979; Okada 1985; Kimura 1985, 1990; Tardieu et al. 1993; Li et al. 1996). This field of research has recently expanded thanks to kinematical and kinetic cross-analyses dedicated to precise taxa. Japanese macaques (*Macaca fuscata*), which are phylogenetically relatively close to baboons (Groves 2001), have been studied intensively by Japanese research teams (Nakatsukasa et al. 1995, 2004, 2006; Hirasaki et al. 2004; Ogihara et al. 2005, 2007). More distant from baboons but closer to humans, bonobos (*Pan paniscus*) (Aerts et al. 2000; D'Août et al. 2001, 2002; Vereecke et al. 2003, 2004) and white-handed gibbons (*Hylobates lar*) (Vereecke et al. 2005, 2006a, b) have, like macaques, been very intensively analyzed via laboratory-based experimental protocols developed specifically by these teams. Moreover, morphological studies (Crompton et al. 1996; Isler et al. 2006; Payne et al. 2006; Schoonaert et al. 2007) provide a technical and conceptual basis that allows testing of locomotor hypotheses in early hominins (Alexander 1984; Crompton et al. 1998; Kramer and Eck 2000; Wang et al. 2003; Sellers et al. 2003, 2004, 2005; Nagano et al. 2005; Raichlen et al. 2008). To take these studies further, fundamental data on bipedalism from any other living nonhuman primates, such as baboons, acquired through appropriate protocols and repeated observations of numerous individuals, could make valuable contributions to this field.

We observed spontaneous bipedal locomotion in a captive population of olive baboons at the Primatology Station of the National Centre for Scientific Research (CNRS, France). We therefore initiated a research program designed to analyze bipedal and quadrupedal locomotion in the genus *Papio* at this station (with approval from the ethics committee of the Midi-Pyrénées Regional Authority). In 2007, we designed a protocol for repeated measurements of individual parameters of both motion and anatomy within a captive population of baboons. In this chapter, we report the first preliminary results: two-dimensional kinematical data of bipedal locomotion in olive baboons (*Papio anubis*).

Materials and Methods

Subjects

We organized the experiment in an open-air park with a stable population of 55–60 baboons (5–6 births annually). Within this population, we selected 10 *Papio anubis* for which we recorded both bipedal and quadrupedal locomotion at the same stages in their individual development; this condition allowed us to limit the effect of individual variations when bipedal and quadrupedal data were compared. The baboons were between 0.67 and 5.39 years old, with an individual body mass of 2.9–15.2 kg. Individual information, including general anatomical data, is given in Table 5.1. Two facts explain the age distribution of our sample: 1) in this park, both adult males and females stand upright, but among them only the females walk bipedally, yet only in rare cases and essentially when carrying babies or food on their belly; and 2) infants, up to the age of 4 months, are carried mainly by their mothers.

Protocols and Study Site

Full details of the experimental protocol is presented in Berillon et al. (2010). The protocol for motion capture and analysis is based on high-speed multicamera video recordings of the displacement of anatomical points while the baboons are walking on a flat horizontal surface. It builds on protocols developed by teams from Japanese universities (Hirasaki et al. 2004) and from the University of Antwerp (D'Août et al. 2001, 2002), and on the experience of some of our team in motion analysis in humans (F. Multon and G. Nicolas) and other animals

Table 5.1 Composition of the sample used in the study

Name	Medal	Code	Sex	Age (yr)	Mass (kg)
Chris	854	V792BA	M	0.67	2.9
Chantal	139	V908I	F	1.09	4.1
Babar	632	V916F	M	1.58	5.4
Alf	643	V894G	M	2.38	7.1
Vinci	568	V896F	M	3.14	8.3
Voltarelle	604	V915F	F	3.28	7.3
Vernie	638	V903D	F	3.28	6.3
Victoire	406	V896E	F	3.82	10.3
Volga	411	V916D	F	3.95	12.5
Tassadite	606	V893E	F	5.39	15.2

(K. D'Août). Because the CNRS Primatology Station is both a breeding and a research center, we were able to adapt the protocols to baboons and to the local environment as well as to develop a number of new research lines. Among other opportunities, we were able to measure the anatomy of the individuals whose locomotion was recorded at different developmental stages, thanks to regular capture under veterinary supervision.

The park has an area of ca. 300 m^2 and communicates with several indoor rooms through small corridors with trapdoors, so that the number of subjects present in the open-air park can be kept under control. This has several advantages: interactions between subjects can be kept within limits during recording times, and it allows both targeted and opportunistic recording. In addition, each subject is monitored from a veterinary point of view and identified by a unique collar for individual follow-up.

The Motion Capture Protocol

A necessary methodological condition was that the protocol should allow the baboons to move in a straight line within a selected area to ensure a well-calibrated video recording with a multiple-camera system. We therefore built a raised walkway in a cleared zone of the park, 80 cm wide, 30 cm high, and 5 m long, to optimize the chances of regular, unconstrained motion. Two spaces were reserved for a force plate and a pressure mat. The walkway was built of concrete and painted in red to ensure a clear contrast with the baboon's fur.

The motion capture system is based on a high-speed video recording system. We used four-colour digital cameras (Basler 602fc), each allowing 100 fps recordings in 656 × 490 pixel resolution for standard use (lower resolutions allow recordings at higher speeds). One sagittal camera and two oblique cameras were dedicated to 3D motion capture within a 2 m-wide field. Another sagittal camera with a 4 m-wide field was used to check *a posteriori* that the speed of the sequence was constant. The four cameras were linked to a PC, a Streamstation specifically built for this research protocol, and used to verify the recording parameters with StreamPix© (Norpix), including synchronization of the four cameras.

To attract the baboons to the walkway, we fixed a 0.5-m^2 mirror at its end. Cameras were located outside the park at an average distance of 3.5 m from the recording field. We selected only sequences for which the speed was constant for the analysis. Finally, no constraints were applied on speed and the baboons walked at their own pace with free hands.

Motion Analysis

To obtain a general characterization of the baboons' movements, we recorded the position of the forelimbs, trunk orientation, and the type of foot contact to the ground during bipedal locomotion.

To describe the motion quantitatively, we unilaterally digitized 19 anatomical reference points by field: the extremity of the muzzle, the center of the eyebrows, the occiput, the dorsal and sagittal point at the base of the neck, the shoulder (acromion), the elbow (center of the epicondyle-epitrochlea segment), the wrist (midpoint), the metacarpophalangeal joint (3rd ray), the extremity of the hand (3rd ray), the dorsal and sagittal points at the base of the tail, the extremity of the tail, the great trochanter, the knee (anterior patellar point), the ankle (center point of the ankle area), the heel (posterior extremity), the midtarsal joint (lateral plantar point), the metatarsophalangeal joint (3rd ray), the extremity of the foot (3rd ray), and the extremity of the opposite foot (3rd ray). We used a calibrating procedure to calculate the 3D coordinates of these reference points in the single reference frame of the walkway. This was done by means of a calibration object 1 m long, 1 m high, and 0.7 m wide.

From these coordinates, we calculated the vertical excursions of the hip, the knee, the ankle and the extremity of the foot, the trunk angle and the joint angles of the hind limb, and, finally, spatiotemporal parameters. Figure 5.1 provides a description of the calculated angles. The evolution of these parameters is expressed in terms of time, as a fraction of the cycle (duration of one stride). The entire process of digitizing the reference points and calculating the motion parameters was conducted with Kwon3D© (Visol). In addition to the individual angular trajectories, we made a chart of the average pattern of the joint angles and their variation at key events in the cycle (right and

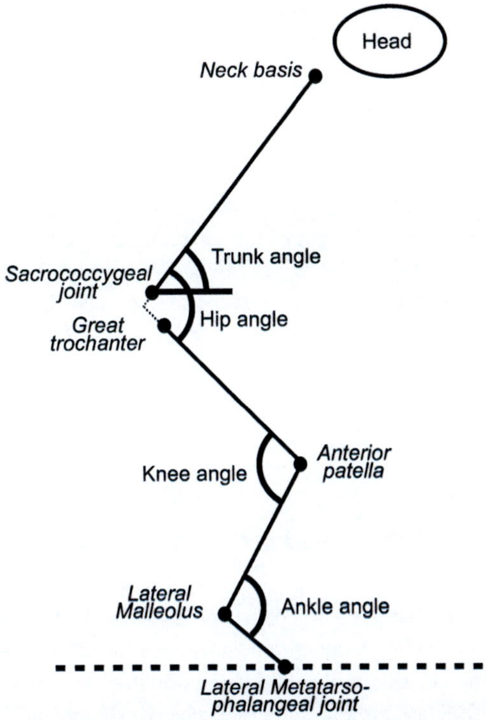

Fig. 5.1 Sagittal trunk and joint angles as calculated from anatomical landmarks

left foot contacts and right and left toe-off) and at their respective local maxima and minima. These events are common to any cycle and to any individual. Graphically, a cross-projection of the averages and variations of the angles (*y*-axis) on the one hand, and of the averages and variations of the events (*x*-axis) on the other hand, gives a representation of the general pattern of bipedal locomotion within the sample studied. In addition, the average, minimum, maximum, and range values are given in Table 5.2.

We also calculated the following spatiotemporal parameters:

Stride length (m), *stride duration* (s), and *stride frequency* (s^{-1}); the stride starts with ground contact of a foot (initial contact) and ends with the subsequent ground contact of the same foot.

The *step length* (m) is defined as the distance between the two feet while both are in contact with the ground.

The *absolute speed* (m s^{-1}) is equal to the ratio between the stride length and the stride duration.

The *duty factor* is defined as "the fraction of the duration of a stride for which each foot remains on the ground" (Alexander 2002).

Because our sample represents a considerable range of ages, and therefore body sizes, spatiotemporal parameters were normalized following the principle of similarity described in Aerts et al. (2000). We calculated dimensionless parameters as follows:

$$\text{Dimensionless speed} = V/\sqrt{(g \times l)} \quad (5.1)$$

$$\text{Dimensionless stride length} = SL/l \quad (5.2)$$

$$\text{Dimensionless stride} = SF \times \sqrt{(l/g)} \quad (5.3)$$

wherein V = absolute speed; SL = absolute stride length; SF = absolute stride frequency; g = gravity; l = the length of the lower leg that we measured on the specimens (distance from the top of the lateral femoral epicondyle to the top of the lateral malleolus). For three specimens in the sample that we could not capture at the time they were video-recorded, we calculated lower leg length from video recordings.

Results

General Features of the Bipedal Locomotor Mode

The stick figure (Fig. 5.2) shows a bipedal stride for individual no. 568. This individual has a bent hip, bent knee gait. A quantified analysis of the whole sample is given in the text that follows. A peculiarity of bipedal locomotion in olive baboons is the position of the two forelimbs: they are almost immobile, placed forward in a parasagittal plane; each individual walks with semiflexed elbows and wrists.

Table 5.2 Average joint angles at different events of the stride (mean ± SD)

	Initial foot contact	Opposite toe-off	Minimum ankle angle	Opposite foot contact	Maximum hip angle	Toe-off	Maximum ankle angle	Minimum knee angle	Final minimum ankle angle	Final foot contact	Average range of motion
%	0	19.4 ± 6.5	34.6 ± 10.0	51.2 ± 2.6	57.9 ± 3.2	69.9 ± 3.4	67.5 ± 4.2	74.7 ± 3.1	84.3 ± 3.4	100	—
Hip angle	104.8 ± 9.8	110.2 ± 8.3	—	131.8 ± 8.1	134.6 ± 9.4	124.5 ± 11.5	—	—	—	105.9 ± 10.8	34 ± 7.1
Knee Angle	124.8 ± 6.6	102.9 ± 5.8	—	85.5 ± 6.1	—	66.8 ± 9.0	—	63.7 ± 8.8	—	127.1 ± 8.4	65.1 ± 8.3
Ankle angle	119.6 ± 8.0	98.8 ± 9.5	89.5 ± 11.7	100.0 ± 13.6	—	122.1 ± 12.1	124.1 ± 13.0	—	100.8 ± 7.6	122.0 ± 8.7	30.5 ± 6.8

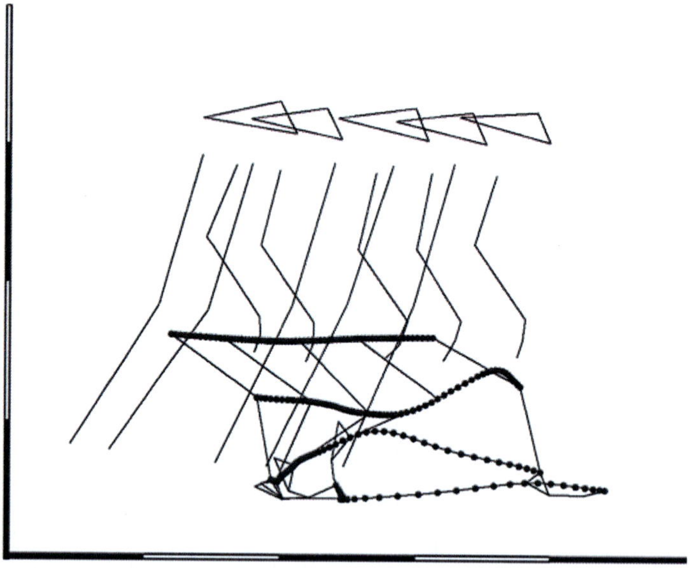

Fig. 5.2 Stick diagram representing the bipedal locomotion of a young male olive baboon (no. 568) at different gait events. From left to right: initial right foot contact, left foot contact, left toe-off, right toe-off, final right foot contact. Scale: each black and white bar corresponds to 20 cm

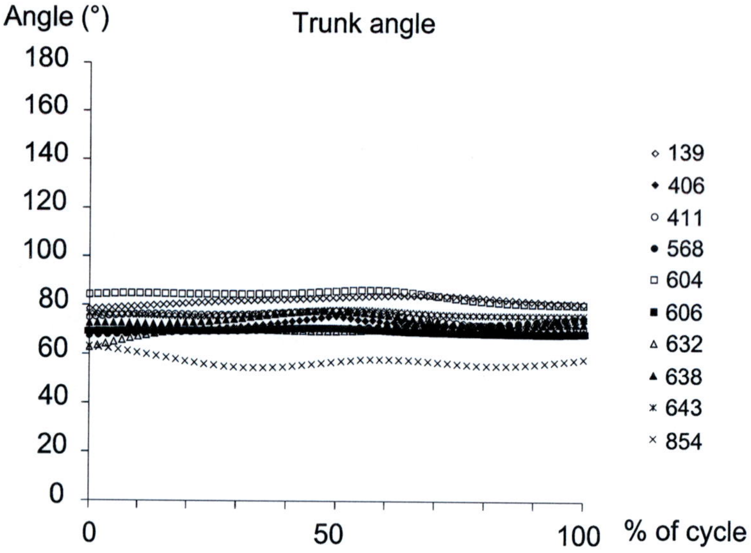

Fig. 5.3 Individual evolution of the trunk angle through a bipedal stride

As illustrated by the stick figure, the trunk is bent slightly forward. Figure 5.3 shows the individual changes in the trunk angle as it moves through a cycle. The erect position of the trunk observed for each individual produced a mean value, for the sample, of the trunk angle of 73.2 ± 7.4°. The mean value for the range of the trunk angle is equal to 6.8 ± 2.97°, which indicates that the trunk angle varies very little in the bipedal gait of baboons. We observed two extreme configurations in no. 854, the youngest individual in the sample, the trunk was bent forward the most (57.3 ± 2.14°; range: 8.53°), and in individual no. 604 the trunk was almost vertical (83.98 ± 1.8°; range: 5.88°).

Type of Foot Contact to the Ground

Figure 5.4 shows the type of foot contact during bipedal locomotion for individual no. 568. Thanks to the high-speed video recordings, we observed that the foot first comes into contact with the ground at its midsection, i.e., from the transverse midtarsal joint to the metatarsophalangeal joint. Then, the toes come into contact with the ground, but the heel never does so. From toe-off of the opposite foot, the midfoot area starts flexing dorsally, then the metatarsophalangeal joint, while the foot lifts off gradually. At the end of the stance phase, the dorsal flexion of the lateral metatarsophalangeal joints is particularly evident. We observed an alternative to this general outline for the two youngest individuals in the sample: foot contact was initiated by the plantar side of the midfoot and toes simultaneously.

Vertical Excursion of Joint Centers

The stick diagram (Fig. 5.2) shows very limited vertical excursion of the hip (greater trochanter) through the cycle. There is a marked vertical excursion of the knee; the knee moves upward during the swing phase and reaches its highest position just before the foot contact. Distally, the extremity of the foot is held very low during the swing phase, while the ankle moves upward; the long axis of the foot becomes almost vertical at the mid-swing phase.

Spatiotemporal Parameters

Individual values as well as average values, standard deviation, and the range of calculated spatiotemporal parameters are given in Table 5.3. The average duty factor (0.7 ± 0.03) shows that the stance phase accounts for ca. 70% of the total cycle duration of a bipedal stride. On average, absolute stride length is equal to 0.52 m, absolute stride duration to 0.69 s, and absolute speed to 0.79 m s^{-1}.

We investigated relationships between speed and other spatiotemporal parameters. Figure 5.5 shows absolute speed as a function of absolute stride length, on the one hand, and absolute stride frequency, on the other hand. It shows that during bipedalism, olive baboons increase their absolute speed by increasing their stride frequency. Because the size variation within the sample is high, we normalized the

Fig. 5.4 Sagittal views of the right foot contact at bipedal gait events for individual no. 568. (**a**) At initial right foot contact. (**b**) At left toe-off. (**c**) At maximal midtarsal dorsal flexion. (**d**) At left foot contact. (**e**) At maximal metatarsophalangeal dorsal flexion. (**f**) At right toe-off

Table 5.3 Individual and average (mean ± SD) lower leg length and spatiotemporal parameters

Individual/code	Lower leg length (m)	Step length (m)	Stride length (m)	Stride duration (s)	Stride frequency (1/s)	Absolute speed (m/s)	Dimension-less stride length	Dimension-less stride frequency	Dimension-less speed	Duty factor
Chris/V792BA	0.12	0.26	0.52	0.51	1.96	1.02	4.32	0.22	0.94	0.67
Chantal/V908I	0.13	0.24	0.45	0.76	1.32	0.60	3.42	0.15	0.52	0.72
Babar/V916F	0.14	0.26	0.56	0.60	1.67	0.94	4.18	0.20	0.82	0.73
Alf/V894G	0.15	0.28	0.50	0.83	1.20	0.60	3.29	0.15	0.49	0.73
Vinci/V896F	0.17	0.28	0.54	0.53	1.89	1.02	3.22	0.25	0.79	0.68
Vernie/V903D	0.16	0.30	0.50	0.61	1.64	0.83	3.15	0.21	0.66	0.67
Voltarelle/V915F	0.15	0.28	0.50	0.58	1.72	0.87	3.37	0.21	0.72	0.64
Victoire/V896E	0.20[a]	0.28	0.56	0.68	1.47	0.83	2.87	0.21	0.60	0.68
Volga/V916D	0.19[a]	0.37	0.55	0.74	1.35	0.75	2.93	0.19	0.55	0.73
Tassadite/V893E	0.20[a]	0.27	0.50	1.05	0.95	0.47	2.51	0.14	0.34	0.72
Mean ± SD	0.16 ± 0.03	0.28 ± 0.03	0.52 ± 0.03	0.69 ± 0.16	1.52 ± 0.32	0.79 ± 0.19	3.33 ± 0.56	0.19 ± 0.04	0.64 ± 0.18	0.70 ± 0.03

[a]Calculated from video recordings

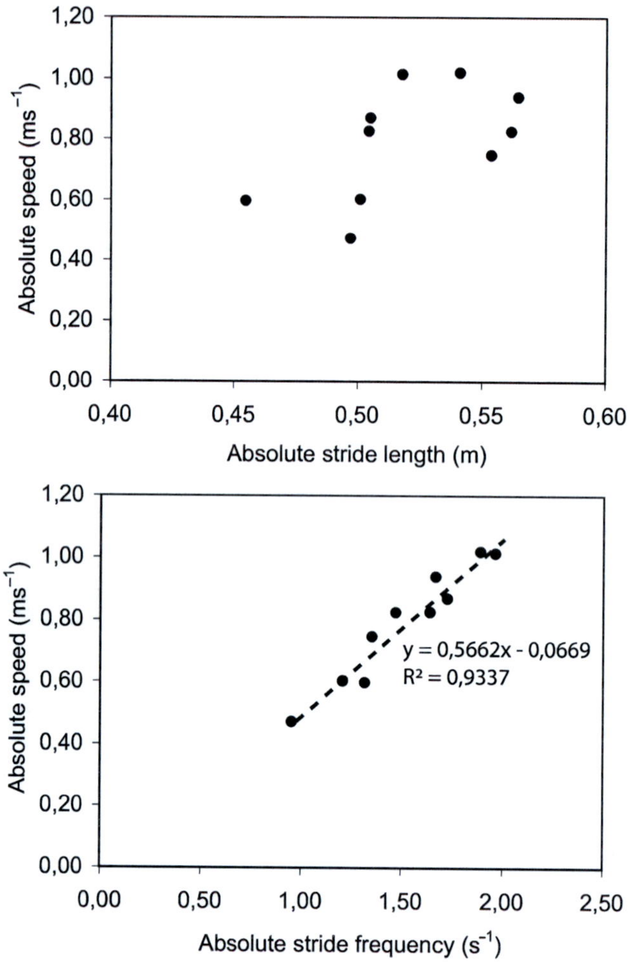

Fig. 5.5 Absolute speed expressed as a function of absolute stride length (**top**) and absolute stride frequency (**bottom**)

data with individual sizes. When plotting these dimensionless data (Fig. 5.6), we see that olive baboons increase their speed by increasing both stride length and stride frequency. There is no clear relationship between duty factor and speed.

Joint Angles

Table 5.3 and Figures 5.7 and 5.8 summarize how individual and average joint angles evolve through time as a fraction of the cycle. Generally speaking, each joint angle evolves through time in a rather similar manner from one individual to another.

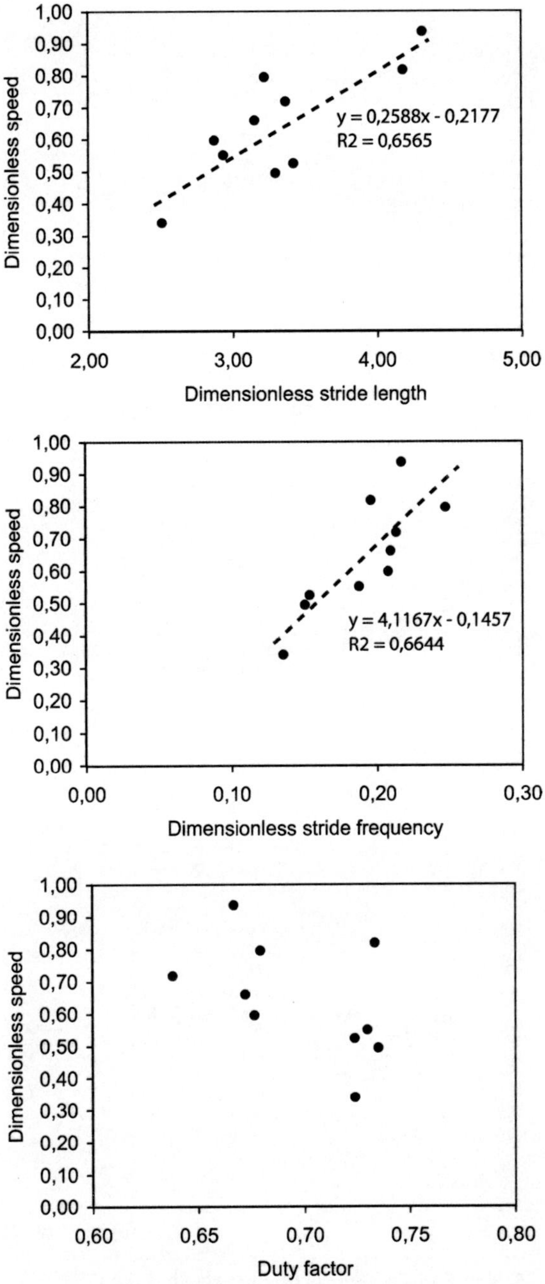

Fig. 5.6 Dimensionless speed expressed as a function of dimensionless stride length (**top**), dimensionless stride frequency (**middle**), and duty factor (**bottom**)

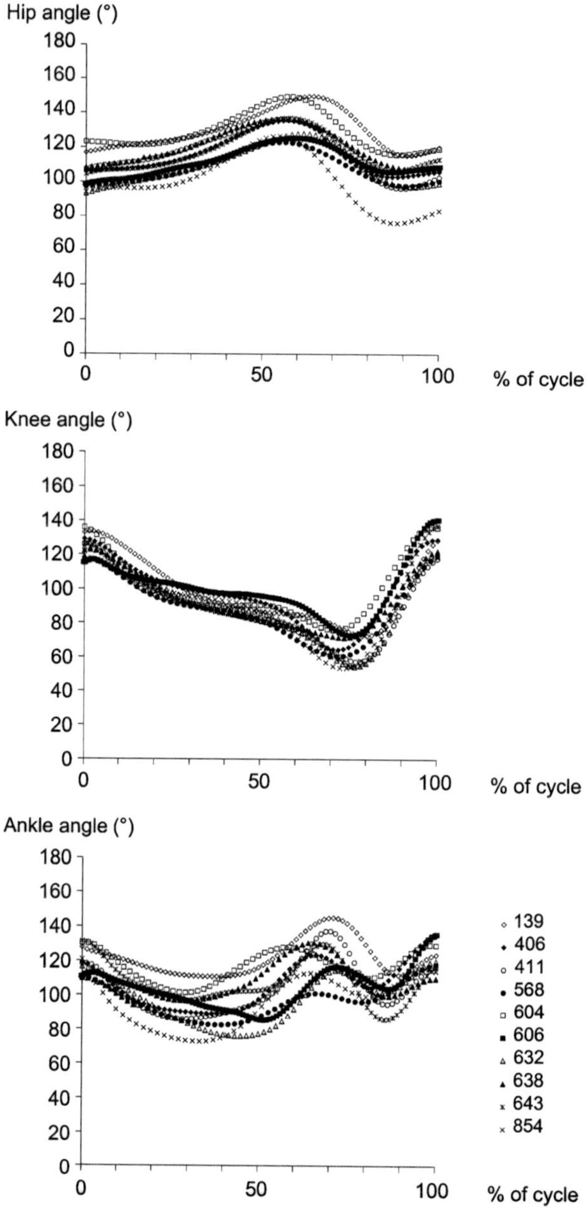

Fig. 5.7 Individual evolution of the joint angles through a bipedal stride

The hip is bent, with a low total range of motion on average (34 ± 7.1°). The curves peak at a point that corresponds to the minimal hip flexion occurring at ca. 60% of the cycle, shortly after ground contact of the opposite foot. Maximal hip flexion at the end of cycle is observed in some individuals. This is obvious in individual no. 854's

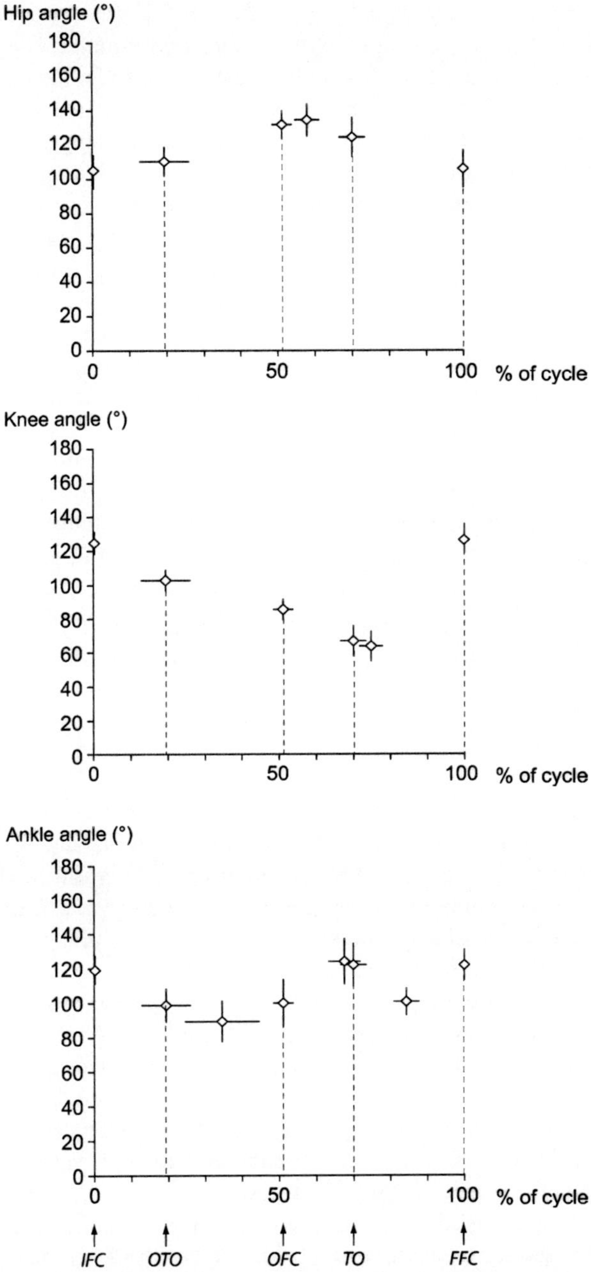

Fig. 5.8 Mean evolution of the joint angles and the events through a bipedal stride. IFC = initial foot contact; OFC = opposite toe-off; OTO = opposite foot contact; TO = toe-off; FFC = final foot contact

trajectory, but not common to all individuals; for some of them, maximal hip flexion occurs just after foot contact with the ground. Finally, we noted a "vertical" dispersion of the hip trajectories that corresponds to that described for the trunk angle, with the diagrams for no. 854 at one extremity and for no. 604 and no. 139 at the other.

The knee is also bent throughout the gait cycle. The knee angle trajectory is very stereotyped with lower variation at each event than that observed for the hip and ankle joints. Minimal knee flexion is reached at the point of foot contact and maximal knee flexion occurs immediately after toe-off. There is a wide range of knee motion (65.1 ± 8.3°). Immediately after foot contact, knee flexion quickly increases for a short time (absorption), and then keeps increasing steadily until the swing phase starts; no decrease or flattening of the trajectory is observed between the two events.

Concerning the ankle angle trajectories, two extension-flexion phases were identified, separated by a peak corresponding to minimal ankle flexion (124.1 ± 13°); the latter is variably expressed in the different individuals. The first phase that corresponds to the stance phase is the most variable, where the joint angle itself is concerned (at minimal ankle angle, 89.5 ± 11.7°; toe-off, 122.1 ± 12.1°) and at the point when maximal ankle flexion is reached (34.6 ± 10.0%). The second phase is shorter and corresponds to the swing phase; this is less variable during its second half. Lastly, the ankle generally extends to values higher than 90°.

Discussion and Perspectives

This chapter describes an experimental protocol for a motion analysis of captive olive baboons, focusing on the sagittal kinematics of their bipedal locomotion. This research line has recently been started and additional data is still needed to characterize fully bipedal locomotion in the genus *Papio*. Nevertheless, the stereotyped behavior of certain calculated motion parameters has enabled us to draw some general characteristics of the bipedal gait in *Papio anubis*.

Generally speaking, the bipedal gait is performed in a fairly similar manner by the individuals in the sample, despite large differences in age and body mass. The associated movements appear to be stereotyped, coordinated, and natural. Olive baboons walk in a compliant manner, with bent hips and knees and relatively little forward tilting of the trunk; this is quite similar to the pattern described for other quadrupedal primates such as sacred baboons, chimpanzees, bonobos, and macaques (e.g., Okada 1985; D'Août et al. 2002; Hirasaki et al. 2004). The position of the foot is semiplantigrade, as observed in macaques walking bipedally (Hirasaki et al. 2004), and the stance phase lasts for 70% of the duration of the cycle on average; this is higher than in humans (Winter 1991) and macaques (Yamazaki et al. 1979; Hirasaki et al. 2004) and similar to the value published for bonobos (Aerts et al. 2000). The ankle is generally flexed at an angle of more than 90°, as in macaques (Hirasaki et al. 2004), while in bonobos the angle is less than 90° (the ankle is more dorsally flexed) (D'Août et al. 2002); this is coherent with the semiplantigrade type of foot contact in cercopithecinae, which is characterized by an elevated position of the proximal part of the foot. Lastly, knee flexion is highly stereotyped, with an angle that increases steadily during the stance phase; this was observed by Okada (1985) on a single sacred baboon (*Papio hamadryas*) and corresponds to the pattern of knee flexion in macaques that are untrained in bipedal locomotion (Hirasaki et al. 2004). In humans, the damping phase that follows foot contact is followed by a phase of reduced knee flexion (Winter 1991), while in chimpanzees and bonobos, knee flexion remains the same (Okada 1985; D'Août et al. 2002).

Apart from joint mobility, the position of the forelimbs, which are almost immobile and kept bent forward, points to a bipedal equilibrium that is clearly distinct from that described in the other taxa. We are not yet able to estimate the position of the segmental and total centers of mass of the subjects studied. We are currently acquiring anatomical data to supplement this aspect of the analysis. Nevertheless, at this stage in our research, taking into account the low mobility of the trunk and the arms and the very low vertical oscillation of the hips, it is reasonable to assume that the vertical oscillations of the center of mass during bipedal walking are very limited.

From a methodological aspect, our sample of 10 subjects was too small to allow us to describe the pattern of a complex motion such as the bipedal locomotion and its variations. For example, the effect of speed cannot really be investigated even though it is a key issue in motion analysis, as illustrated by previous research (Kimura et al. 1983; Alexander 1984, 2004; Schmitt and Larson 1995; Aerts et al. 2000; Danion et al. 2003; Vereecke et al. 2004, 2006b). This could be done with a larger sample, but also by increasing the number of recorded sequences per individual. We are currently working toward this.

Other parameters need to be investigated to obtain a more detailed description of movements. For example, in this study, the dispersion of the models at level of the hip seems to result from variations in the position of the trunk (trunk angle) but not from any significant variation in the kinematics of the joint itself; a breakdown of the joint angles into segment angles, as proposed by others (D'Août et al. 2002), would be useful. But, also the ankle angle, as calculated here and in the other studies, corresponds to the angle between the tarsometatarsal complex —which is considered as a rigid element—and the lower leg. Although this allows comparison with the ankle angle measured in humans, where the tarsometatarsal complex is indeed rigid, it introduces a bias into the analysis. Following Elftman and Manter's research on chimpanzees (Elftman and Manter 1935a, b), many comparative analysis of primates demonstrated that the tarsometatarsal complex in nonhuman primates is not rigid during the ground contact (Lewis 1989; Meldrum 1991; Gebo 1992; Schmitt and Larson 1995; Vereecke and Aerts 2008). Our observations, based on high-speed video recordings, confirm this for olive baboons. In fact, the ankle angle, usually calculated in the sagittal plane (here and previous analysis), is actually a combination of many joints, from the talocrural joint angle to the joint angles of the tarsometatarsal complex. Some of these angles can be calculated from high-speed video recordings. We have collected these data and we present some results in Berillon et al. (2010). In addition, future paedobarographic analysis will positively complete these investigations. Looking further ahead, a more detailed analysis of the movements that takes anatomical reality into account more accurately, even if this limits possibilities for comparison with human movements, will ensure better cross-analysis with plantar pressure and kinetic data for each taxon.

To obtain a more detailed analysis of movements of *Papio*, the data acquisition protocols will need to be optimized. This can be achieved with high-speed, high-resolution video recordings, but also with more precise identification of anatomical reference points. Automatic motion capture procedures based on reflective markers cannot be used with baboons. However, we do have specific possibilities for regular capture of individuals in the population studied, which is not possible in a public environment, e.g., zoos, and even less so in the wild. This enables us to ensure anatomical follow-up and therefore greater accuracy, through palpation to identify the anatomical points. We are also developing a process based on shaving and colouring the baboon's skin at the anatomical points in question. Lastly, radiographs are obtained for each individual during capture to ensure regular calibration of the external data on bone morphology.

Setting up an experimental protocol that combines individual high-quality motion and anatomical data is not realistic in the wild. There are at least some technical reasons for this. For example, the chance that observed individuals walk bipedally in exactly the same area as required for quantified and calibrated three-dimensional measurements is very small. There are some ethical and security reasons as well; regular captures for anatomical measurements of selected individual at different stage of their development may not be conceivable. An experiment such as the one we set up at the CNRS Primatology Station with a captive population of baboons seems to be a good alternative. This chapter has presented a number of preliminary results based on a small sample of young olive baboons that allow us to draw out some general characteristics of the bipedal gait in *Papio anubis*. Numerous transversal and longitudinal cross-analyses are in progress, including noninvasive anatomical investigations, kinematics and kinetics. Although setting up a complete experimental protocol of motion analysis is not realistic in the wild, the need of motion data concerning baboons and any other nonhuman primates in the wild is obvious. To our knowledge, such field protocols have not yet been developed. In that perspective, captive protocols such as the one we developed at the CNRS Primatology Station could be used as an experimental basis; e.g., among our set of variables, some might be more meaningful than others; these variables could be selected for specific protocols that would be adapted to the very selective field conditions.

From an ontogenetic point of view, it should be noted that the ability to walk bipedally is present very early in baboons, as early as 6 months in our sample. This is described in African great apes, both in captivity and in the wild (e.g., Kimura 1990; Doran 1997) but, to our knowledge, no similar observations have been reported for baboons, neither in their natural environment nor in captivity. Generally speaking, the bipedal gait in very young individuals in our sample resembles that of older subjects. Nevertheless, we noted certain peculiarities in our youngest subjects, e.g., concerning the position of the foot or the tilt of the trunk, which could be due to their age. The changing of bipedal locomotion kinematics with age is well documented in humans (Grimshaw et al. 1998; Yaguramaki and Kimura 2002; Hallemans et al. 2006). Concerning nonhuman primates, Kimura has provided the only available kinematic and kinetic data, to our knowledge, on the ontogeny of bipedal walking in nonhuman primates, from a study on infant chimpanzees (Kimura 1990, 1996). As far as development is concerned, based on our observations, the bipedal gait is more frequent in young and juvenile baboons than in adults. This has been commonly observed in African great apes in which the influence of body size on the ontogeny of locomotion has been discussed (Doran 1993). In view of the ontogenetic peculiarities of the bipedal gait in humans and in chimpanzees, and given the close phylogenetic relationship of these two taxa, it would be of palaeontological interest to provide, thanks to the population of baboons we are analyzing and to repeated observations, a cross-referenced kinematic, kinetic and anatomical description of what might be the ontogenetic peculiarities of bipedalism in baboons, which, although phylogenetically distant, may be seen as a primate whose locomotor repertoire could, in some respects, represent that of a possible hominoid ancestor.

As far as the ontogeny of the locomotor behavior is concerned, although bipedal locomotion in baboons has been observed in the wild, nothing is known about its ontogeny and very little about the characteristics of the associated movements. Thanks to analyses of captive populations, we are investigating anatomical conditions of bipedal locomotion through age at an individual level. Though the captive protocol we set up appears valuable to this type of investigations, positional behavioral data collected outside of the natural habitat must be considered with caution because of the inevitable artificial constraints. Setting up a field research that would allow investigating the frequency and the conditions under which baboons walk bipedally in the wild through age such as Doran's research on African great apes would bring very valuable behavioral data.

Acknowledgments We thank the editors of the volume for inviting us to submit this contribution. We are very grateful to the team of the Station of Primatology, especially Georges Digrandi and Valérie Moulin, for their permanent help. The entire motion capture system was set up with the collaboration of P. Trannois (Opto France, France); the *Streamstation* was configured with the help of T. Lemaire. This research is supported by the Fyssen Foundation (Research Grant), and the Groupement de Recherche GDR 2655 of the CNRS (Dir. L. Rosetta).

References

Aerts P, Van Damme R, Van Elsacker L, Duchene V (2000) Spatio–temporal gait characteristics of the hind-limb cycles during voluntary bipedal and quadrupedal walking in Bonobos (*Pan paniscus*). Am J Phys Anthropol 111:503–517.

Aiello LC, Dean C (1990) An Introduction to Human Evolutionary Anatomy. Academic Press, London.

Alexander RM (1984) Stride length and speed for adults, children, and fossil hominids. Am J Phys Anthropol 63:23–27.

Alexander RM (2002) Energetics and optimization of human walking and running: the 2000 Raymond Pearl Memorial lecture. Am J Hum Biol 14:641–648.

Alexander RM (2004) Bipedal animals, and their differences from humans. J Anat 204(5):321–330.

Berillon G, Daver G, D'Août K, Nicolas G, de la Villetanet B, Multon F, Digrandi G, Dubreuil G (2010) Bipedal vs. quadrupedal hindlimb and foot kinematics in a captive sample of olive baboons (*Papio anubis*): setup and preliminary results. Int J Primatol 31(2):159–180.

Crompton RH, Li Y, Alexander RM, Wang WJ, Günther MM (1996) Segment inertial properties of primates: new techniques for laboratory and field studies of locomotion. Am J Phys Anthropol 99:547–570.

Crompton RH, Li Y, Wang W, Günther MM, Savage R (1998) The mechanical effectiveness of erect and "bent-hip, bent-knee" bipedal walking in *Australopithecus afarensis*. J Hum Evol 35(1):55–74.

Crompton RH, Vereecke EE, Thorpe SKS (2008) Locomotion and posture from the common hominoid ancestor to fully modern hominins, with special reference to the last common panin/hominin ancestor. J Anat 212:501–54.

Danion F, Varraine E, Bonnard M, Pailhous J (2003) Stride variability in human gait: the effect of stride frequency and stride length. Gait Posture 18(1):69–77.

D'Août K, Aerts P, De Clercq D, De Meester K, Van Elsacker L (2002) Segment and joint angles of hindlimb during bipedal and quadrupedal walking of the bonobo (*Pan paniscus*). Am J Phys Anthropol 119(1):37–51.

D'Août K, Aerts P, De Clercq D, Schoonaert K, Vereecke E, Van Elsacker L (2001) Studying bonobo (*Pan paniscus*) locomotion using an integrated setup in a zoo environment: preliminary results. Primatologie 4:191–206.

Doran DM (1993) Comparative locomotor behavior of chimpanzees and bonobos: the influence of morphology on locomotion. Am J Phys Anthropol 91:83–98.

Doran DM (1997) Ontogeny of locomotion in mountain gorillas and chimpanzees. J Hum Evol 32(4):323–344.

Elftman H (1944) The bipedal walking of the chimpanzee. J Mammal 25:67–71.

Elftman H, Manter J (1935a) Chimpanzee and human feet in bipedal walking. Am J Phys Anthropol 20:69–79.

Elftman H, Manter J (1935b) The evolution of the human foot, with especial reference to the joints. J Anat 70:56–70.

Fleagle JG (1988) Primate Adaptation and Evolution. Academic Press, New York.

Gebo DL (1992) Plantigrady and foot adaptation in African Apes: implications for hominids origins. Am J Phys Anthropol 89:29–58.

Grimshaw PN, Marques–Bruna P, Salo A, Messenger N (1998) The 3-dimensional kinematics of the walking gait cycle of children aged between 10 and 24 months: cross sectional and repeated measures. Gait Posture 7(1):7–15.

Groves CP (2001) Primate Taxonomy. Smithsonian Institute Press, Washington.

Hallemans A, Clercq DD, Aerts P (2006) Changes in 3D joint dynamics during the first 5 months after the onset of independent walking: a longitudinal follow-up study. Gait Posture 24(3):270–279.

Hirasaki E, Ogihara N, Hamada Y, Kumakura H, Nakatsukasa M (2004) Do highly trained monkeys walk like humans? A kinematic study of bipedal locomotion in bipedally trained Japanese macaques. J Hum Evol 46(6):739–750.

Hunt KD (1989) Positional behaviour in *Pan troglodytes* at the Mahale Mountains and Gombe Stream National Parks. PhD dissertation, University of Michigan.

Inman VT, Ralston HJ, Todd F (1981) Human Walking. Williams & Wilkins, Baltimore.

Ishida H, Kimura T, Okada M (1974) Patterns of bipedal walking in anthropoid primates. In: Kondo S, Kawai M, Ehara A, Kawamura S (eds.), Proceedings from the symposia of the 5th congress of the International Primatological Society, pp. 287–301. Japan Science Press, Tokyo.

Isler K, Payne RC, Günther MM, Thorpe SKS, Li Y, Savage R, Crompton RH (2006) Inertial properties of hominoid limb segments. J Anat 209(2):201–218.

Jenkins FA (1972) Chimpanzee bipedalism: Cineradiographic analysis and implications for the evolution of gait. Science 178(4063):877–879.

Kimura T (1985) Bipedal and quadrupedal walking of primates: comparative dynamics. In: Kondo S (ed), Primate Morphophysiology, Locomotor Analyses and Human Bipedalism, pp. 81–104. University of Tokyo Press, Tokyo.

Kimura T (1990) Voluntary bipedal walking in infant chimpanzees. In: Jouffroy FK, Stack MH Niemitz C (eds), Gravity, Posture and Locomotion in Primates, pp. 237–251. Il Sedicessimo, Firenze.

Kimura T (1996) Centre of gravity of the body during the ontogeny of chimpanzee bipedal walking. Folia Primatol 66(1-4):126–136.

Kimura T, Okada M, Ishida H (1979) Kinesiological characteristics of primate walking: its significance in human walking. In: Morbeck ME, Preuschoft H, Gomberg N (eds), Environment, Behavior, and Morphology: Dynamic Interactions in Primates, pp. 297–311. Gustav Fischer, New York.

Kimura T, Okada M, Yamazaki N, Ishida H (1983) Speed of the bipedal gaits of man and nonhuman primates. Ann Sci Nat Zool Paris 5:145–158.

Kramer PA, Eck GG (2000) Locomotor energetics and leg length in hominid bipedality. J Hum Evol 38(5):651–666.

Lewis OJ (1989) Functional Morphology of the Evolving Hand and Foot. Clarendon Press, Oxford.

Li Y, Crompton RH, Alexander RM, Gunther MM, Wang WJ (1996) Characteristics of ground reaction forces in normal and chimpanzee-like bipedal walking by humans. Folia Primatol 66(1-4):137–159.

Meldrum DJ (1991) Kinematics of the Cercopithecine foot on arboreal and terrestrial substrates with implications for the interpretation of Hominid terrestrial adaptations. Am J Phys Anthropol 84:273–289.

Nagano A, Umberger BR, Marzke MW, Gerritsen KGM (2005) Neuromusculoskeletal computer modeling and simulation of upright, straight-legged, bipedal locomotion of *Australopithecus afarensis* (A.L. 288-1). Am J Phys Anthropol 126(1):2–13.

Nakatsukasa M, Hayama S, Preuschoft H (1995) Postcranial skeleton of a Macaque trained for bipedal standing and walking and implications for functional adaptation. Folia Primatol 64:1–29.

Nakatsukasa M, Ogihara N, Hamada Y, Goto Y, Yamada M, Hirakawa T, Hirasaki E (2004) Energetic costs of bipedal and quadrupedal walking in Japanese macaques. Am J Phys Anthropol 124(3):248–256.

Nakatsukasa M, Hirasaki E, Ogihara N (2006) Energy expenditure of bipedal walking is higher than that of quadrupedal walking in Japanese macaques. Am J Phys Anthropol 131(1):33–37.

Ogihara N, Usui H, Hirasaki E, Hamada Y, Nakatsukasa M (2005) Kinematic analysis of bipedal locomotion of a Japanese macaque that lost its forearms due to congenital malformation. Primates 46(1):11–19.

Ogihara N, Hirasaki E, Kumakura H, Nakatsukasa M (2007) Ground-reaction-force profiles of bipedal walking in bipedally trained Japanese monkeys. J Hum Evol 53(3):302–308.

Okada M (1985) Primate bipedal walking: comparative kinematics. In: Kondo S (ed), Primate Morphophysiology, Locomotor Analyses and Human Bipedalism, pp. 47–58. Tokyo University Press, Tokyo.

Payne RC, Crompton RH, Isler K, Savage R, Vereecke EE, Günther MM, Thorpe DE, D'Août K (2006) Morphological analysis of the hindlimb in apes and humans. I. Muscle architecture. J Anat 208(6):709–724.

Raichlen DA, Pontzer H, Sockol MD (2008) The Laetoli footprints and early hominin locomotor kinematics. J Hum Evol 54:112–117.

Rose MD (1973) Quadrupedalism in primates. Primates 14:337–357.

Rose MD (1976) Bipedal behavior of olive baboons (*Papio anubis*) and its relevance to an understanding of the evolution of human bipedalism. Am J Phys Anthropol 44:247–262.

Rose MD (1977) Positional behaviour of olive baboons (*Papio anubis*) and its relationship to maintenance and social activities. Primates 18:59–116.

Schmitt D, Larson SG (1995) Heel contact as a function of substrate type and speed in Primates. Am J Phys Anthropol 96:39–50.

Schmitt DO (2003) Insights into the evolution of human bipedalism from experimental studies of humans and other primates. J Exp Biol 206:1437–1448.

Schoonaert K, D'Août K, Aerts P (2007) Morphometrics and inertial properties in the body segments of chimpanzees (*Pan troglodytes*). J Anat 210(5):518–531.

Sellers WI, Dennis LA, Crompton RH (2003) Predicting the metabolic energy costs of bipedalism using evolutionary robotics. J Exp Biol 206:1127–1136.

Sellers WI, Dennis LA, Wang WJ, Crompton RH (2004) Evaluating alternative gait strategies using evolutionary robotics. J Anat 204(5):343–351.

Sellers WI, Cain GM, Wang W, Crompton RH (2005) Stride lengths, speed and energy costs in walking of *Australopithecus afarensis*: using evolutionary robotics to predict locomotion of early human ancestors. J R Soc Interface: 1–11.

Tardieu C, Aurengo A, Tardieu B (1993) New method of the three-dimensional analysis of bipedal locomotion for the study of displacements of the body and body-parts centers of mass in man and non-human primates: evolutionary framework. Am J Phys Anthropol 90:455–476.

Vereecke EE, D'Août K, De Clercq D, Van Elsacker L, Aerts P (2003) Dynamic plantar pressure distribution during terrestrial locomotion of bonobos (*Pan paniscus*). Am J Phys Anthropol 120(4):373–383.

Vereecke EE, D'Août K, De Clercq D, Van Elsacker L, Aerts P (2004) The relationship between speed, contact time and peak plantar pressure in terrestrial walking of bonobos. Folia Primatol 75(4):266–278.

Vereecke EE, D'Août K, Van Elsacker L, De Clercq D, Aerts P (2005) Functional analysis of the gibbon foot during terrestrial bipedal walking: plantar pressure distributions and three-dimensional ground reaction forces. Am J Phys Anthropol 128(3):659–669.

Vereecke EE, D'Août K, Aerts P (2006a) Locomotor versatility in the white-handed gibbon (*Hylobates lar*): a spatiotemporal analysis of the bipedal, tripedal, and quadrupedal gaits. J Hum Evol 50(5):552–567.

Vereecke EE, D'Août K, Aerts P (2006b) Speed modulation in hylobatid bipedalism: A kinematic analysis. J Hum Evol 51(5):513–526.

Vereecke EE, Aerts P (2008) The mechanics of the gibbon foot and its potential for elastic energy storage during bipedalism. J Exp Biol 211 (23):3661-3670.

Wang WJ, Crompton R, Li Y, Gunther MM (2003) Energy transformation during erect and "bent-hip, bent-knee" walking by humans with implications for the evolution of bipedalism. J Hum Evol 44(2):563–579.

Winter DA (1991) The Biomechanics and Motor Control of Human Gait: Normal, Elderly and Pathological, 2nd ed. University of Waterloo Press, Waterloo Biomechanics.

Wrangham RW (1980) Bipedal locomotion as a feeding adaptation in Gelada Baboons, and its complicartions for hominid evolution. J Hum Evol 9:329–331.

Yaguramaki N, Kimura T (2002) Acquirement of stability and mobility in infant gait. Gait Posture 16(1):69–77.

Yamazaki N, Ishida H, Kimura T, Okada M (1979) Biomechanical analysis of the primate bipedal walking by computer simulation. J Hum Evol 8:337–349.

Chapter 6
Scapula Movements and Their Contribution to Three-Dimensional Forelimb Excursions in Quadrupedal Primates

Manuela Schmidt and Cornelia Krause

Abstract Primates are endowed with a great repertoire of locomotor and nonlocomotor abilities, for which they rely on their powerful prehensile hind limbs and their skilled and mobile forelimbs. The overall mobility of the forelimb depends to a considerable degree on structure and function of the shoulder region but only scant information about the detailed motion of the primate shoulder is available so far. Therefore, we investigated shoulder movements during locomotion in four arboreal quadruped primates using high-speed cineradiography. Together with previous radiographic studies, four different patterns of combined scapular and humeral excursions can be identified that differ in the degree of dorsolateral scapular motion and mediolateral glenohumeral mobility. It becomes obvious that the overall shoulder mobility in primates is affected by scapular mobility on one hand and glenohumeral mobility on the other hand, but the two components of shoulder mobility likely evolved, to some extent, independently. Finding morphological correlates to the observed motion patterns, however, is difficult because morphometric features likely determine the maximum range of shoulder motion, which is not necessarily the motion range that occurs during locomotor activities of the forelimb. For a better understanding of the great variability in shoulder morphology, future studies both in the field and the laboratory should focus on the overall repertoire of shoulder function in primates. Only the combined effort of both research strategies will enhance our understanding of the evolution of this morphofunctional complex in primates.

Keywords Clavicle • *Eulemur fulvus* • *Microcebus murinus* • *Saguinus oedipus* • *Saimiri sciureus* • Scapula shape shoulder kinematics • Thorax

---

M. Schmidt (✉)
Institut für Spezielle Zoologie and Evolutionsbiologie, Friedrich Schiller Universität Jena, Germany
e-mail: schmidt.manuela@uni-jena.de

K. D'Août and E.E. Vereecke (eds.), *Primate Locomotion: Linking Field and Laboratory Research*, Developments in Primatology: Progress and Prospects,
DOI 10.1007/978-1-4419-1420-0_6, © Springer Science+Business Media, LLC 2011

Introduction

Living primates are endowed with a great repertoire of locomotor and nonlocomotor abilities for which they rely on their powerful prehensile hind limbs and their skilled and mobile forelimbs. The overall mobility of the forelimb depends to a considerable degree on structure and function of the shoulder region (Larson 1993; Chan 2007). Understanding the evolution of shoulder mobility is therefore among the great challenges in reconstructing the evolutionary processes that produced the locomotor diversity of primates. Studying the functional morphology of the primate shoulder has a long tradition in anthropology, with a main focus on the principal differences and similarities between apes and humans. The anatomy of the primate forelimb has been studied extensively with reference to the adaptations for brachiation (Keith 1923; Miller 1932; Ashton and Oxnard 1963, 1964a; Erikson 1963; Corruccini and Ciochon 1976). Most of these classical studies highly profit from the pioneering work of Inman and co-workers on the function of the human shoulder (Inman et al. 1944). But they also document that, for a long time, the functional morphology of the primate shoulder has been based solely on comparative anatomy. Knowledge of shoulder function, e.g., the actual movement of the shoulder region, was limited until the publication of the first *in vivo* study of scapular motion in a nonhuman primate—the spider monkey—using radiographic analysis (Jenkins et al. 1978). Many studies (Inman et al. 1944; Davis 1949; Ashton and Oxnard 1963, 1964a; Müller 1967; Roberts 1974; Fleagle 1976) attempted to relate the shape of the scapula to its mechanical function, either as a means of providing leverage for the scapulohumeral musculature or as a means of absorbing the stress produced by its associated musculature. Roberts (1974) has attempted to relate the shape of the scapula to its function as a dynamic structure when he considered the motion and its spatial limitations along the thoracic wall. He was the first who reported some observations of scapular excursions in a quadrupedal primate, the vervet monkey (*Chlorocebus aethiops*). A detailed kinematic analysis of the shoulder movements of this species, combined with electromyographic recordings, was published in 1994 (Whitehead and Larson 1994). However, the consideration of shoulder movements during locomotor activities must not be necessarily limited to X-ray analyses in the laboratory. Schön Ybarra and Schön (1987) demonstrated that even field observations of positional behavior can yield valuable information about shoulder movements when combined with a thorough morphological investigation. The authors documented that a comprehensive understanding of the relationship between form and function is an important prerequisite for deducing shoulder motion from observations of primate locomotor and nonlocomotor behavior in the field.

The present study investigates shoulder movements during quadrupedal locomotion in four arboreal primates—the mouse lemur, the brown lemur, the cotton-top tamarin, and the squirrel monkey—via cineradiography. Our goal is to find out whether we are able to describe some common principles of scapular motion that can be helpful in estimating shoulder movements in other primate species which are not accessible for laboratory investigation. However, before we can go on to make such generalizations from our observations in the laboratory, we have to consider two important questions: First, how is the behavioral activity that we induced by our experimental setup related to the natural behavioral activities of the species in the field? Or in other words, do we observe a "typical" behavior that reflects a major part of the natural behavioral repertoire? Second, does our experimental setup provide our animals with an adequate surrounding that imitates a naturalistic situation as close as possible? This refers mainly to the chosen substrates and their mechanical properties. But it is also important to consider daily activity patterns of the investigated species, the influence of day light or noise on their behavioral activity, and so on. Today, researchers make conscious effort to simulate natural situations in their laboratory setups. The sorts of information needed can be exemplified on this study. Because the ultimate objective of our studies is the reconstruction of evolutionary transformation processes, we have to select our species according to a phylogenetic hypothesis and to hypotheses about the origin of primates and the presumed morphology of their last common ancestor. The selection of a species for an experimental investigation, therefore, depends on other criteria than the amount of information available for locomotor and substrate preferences and activity behavior. Fortunately, the four species investigated here (or the genera to which they belong) are well represented in field studies of primate behavior (Ashton and Oxnard 1964b; Martin 1972; Walker 1979; Ward and Sussman 1979; Fleagle and Mittermeier 1980; Garber 1980; Fontaine 1990; Arms et al. 2002). So, we could obtain valuable information about the preferred locomotor mode (arboreal quadrupedal walking and running) as well as about preferred substrates, their diameters and inclination. Accordingly, we used raised horizontal poles with different diameters, each adjusted to the known preference of the species. An arboreal analogue of a treadmill was also used. However, it differed from usual treadmills by being made of a tight rope that is driven by the animals walking activity itself. So, it never restrained the animals' performance.

In this study, we attempt to find common principles of shoulder motion that can be applied to other quadrupedal primates that have not been investigated so far. Linking the known and the unknown, however, requires some support from other sources, i.e., comparison of locomotor behavior and the comparative anatomy of the shoulder girdle. For both aspects, comprehensive information is available from the literature. As mentioned earlier, the anatomy of the primate shoulder has been extensively investigated in the past (for references see earlier). Therefore, we take another look on morphological features that influence the mobility of the shoulder and we try to relate them to the current knowledge about shoulder function. Thus, morphometric features of the scapula, clavicle, and thorax, gathered from skeletal material of 78 primate species, are inspected and discussed to find morphological correlates to the observed movements. With this study we hope to provide both the field and the laboratory observer with some useful information about the role of scapular motion for the mobility of the forelimb in primates. On a few examples, we demonstrate how shoulder motion can easily deduced from observable key structures also without highly sophisticated technical support.

Materials and Methods

Subjects

We compared shoulder kinematics in four species of arboreal quadrupedal primates: the gray mouse lemur (Cheirogaleidae: *Microcebus murinus* J.F. Miller 1777), the brown lemur (Lemuridae: *Eulemur fulvus* E. Geoffroy St. Hilaire 1796), the cotton-top tamarin (Callitrichidae: *Saguinus oedipus* Linnaeus 1758), and the squirrel monkey (Cebidae: *Saimiri sciureus* Linnaeus 1758). Note that taxonomists group Callitrichidae and Cebidae into the same family, Cebidae, because of molecular estimates of divergence time (Goodman et al. 1998). We conducted motion analysis on two adult individuals of each species. All subjects were kept in accordance with German animal welfare regulations, and experiments were registered with the Committee for Animal Research of the Freistaat Thüringen, Germany.

We derived criteria for species selection from the hypotheses placing the adaptive origin of primates in a small branch milieu (Napier 1967; Cartmill 1972, 1974; Rose 1973; Sussman 1991; Gebo 2004). Accordingly, the subjects needed to be small in terms of body size and they had to use arboreal quadrupedalism as their preferred locomotor mode.

Motion Analysis

We trained subjects to walk on a raised pole or on a horizontal motor-driven ropemill, an arboreal analogue of a treadmill. We adapted the diameter of the support to the preferred natural substrate of the species (mouse lemur: 10 mm, cotton-top tamarin: 25 mm, squirrel monkey: 30 mm, brown lemur: 50 mm). We obtained data on substrate preferences from several sources (Walker 1979; Garber 1980; Fontaine 1990; Arms et al. 2002).

We collected uniplanar cineradiographs in lateral view at 150 frames/s to visualize shoulder motion. The methods of collecting and processing kinematic variables from cineradiographs have been described in detail elsewhere (Schmidt and Fischer 2000; Schmidt 2005) and are only briefly summarized here. The X-ray equipment consists of an automatic Philips® unit with one X-ray source which applies pulsed X-ray shots (Institut für den Wissenschaftlichen Film, Göttingen). We corrected distortions of the X-ray maps by reference to an orthogonal grid of steel balls (diameter 1.0 mm, with a mesh width of 10.0 mm), filmed before and after each experiment session. We recorded the X-ray images from the image amplifier either onto 35-mm film (Arritechno R35-150 camera) or via a high-speed CCD camera (Mikromak® Camsys; Mikromak Service K. Brinkmann, Berlin, Germany). We then copied X-ray films onto video tape and A/D-converted using a video processing board. These films were then analyzed frame-by-frame to identify previously defined skeletal landmarks (Unimark by R. Voss, Tübingen, Germany). Unimark calculates angles and distances based on the *x* and *y* coordinates of the landmarks, correcting the distortions of the X-ray maps automatically with reference to the *x* and *y* coordinates of the grid.

We calculated scapular angles relative to the horizontal plane. The term "cranial rotation" is used for the cranial displacement of the glenoid fossa relative to the intersection between the spine and the vertebral border of the scapula. We reconstructed movements of the elements outside the parasagittal plane by the foreshortening of the bones and by qualitatively comparing the depicted bony elements with the original bones at hand.

Morphometry

We examined morphometric features that influence the range of motion of the scapula on 171 specimens of 78 primate species (Table 6.1) belonging to the collections of the Phylogenetisches Museum Jena, Museum für Naturkunde Berlin, and Naturhistorisches Museum Bern. We judged the adult status of the specimens on the basis of the fusion of the epiphyses of the long bones. More than 50% of the total sample is composed of wild-collected specimens ($n = 91$). For most of these specimens, body mass values are not associated with the collected material and had to be taken from the literature (Grzimek 1987; Rowe 1996). We used the available head-trunk length to estimate whether the given mean or the maximum body mass values were appropriate to estimate the unknown mass.

We determined the following morphometric features: 1) length of the scapula relative to the cube root of body mass; 2) ratio between length of the scapular spine and maximum width of the shoulder-blade (= scapular index); 3) length of the clavicle relative to the cube root of body mass; 4) length of the clavicle relative to the width of the rib-cage at the first rib; and 5) ratio between height and width of the rib-cage at the fourth rib.

Because more than one specimen was available for most of the species, we had to ask whether we can average the individual values of a certain parameter to calculate a species value. Although contrary meanings exist in the literature, we reject calculating an average parameter for a species whenever individual parameters do not scale isometrically with body size among different-sized conspecifics (Schmidt 2008). In the present study we tested the relative length of the scapula and clavicle as well as the scapular index for intraspecific isometric scaling at the example of four species. For these species, five or more specimens were available with their associated body masses: *Eulemur fulvus* ($n = 6$), *Saguinus oedipus* ($n = 5$), *Papio hamadryas* ($n = 5$), and *Nasalis larvatus* ($n = 5$). Calculation followed standard methods (Schmidt 2008). In all four species, the relative length of the scapula and the relative length of the clavicle scale with strong negative allometry to body mass. Thus, larger specimens have relatively shorter scapulae and clavicles than smaller specimens within the same species. The scapular index behaves differently; it scales allometrically in the brown lemur (positive) and the cotton-top tamarin

Table 6.1 Species examined in morphometry (body mass in grams)

Lemuridae	Callitrichidae	Cercopithecidae
Eulemur coronatus (1250)	*Callimico goeldii* (500)	*Cercopithecus cephus* (2900)
Eulemur fulvus (2500)	*Callithrix argentata* (320)	*Cercopithecus diana* (5000)
Eulemur macaco (2400)	*Callithrix geoffroyi* (250)	*Cercopithecus hamlyni* (3680)
Eulemur mongoz (1685)	*Callithrix jacchus* (481)	*Cercopithecus mona* (2750)
Hapalemur griseus (895)	*Cebuella pygmaea* (130)	*Chlorocebus aethiops* (5500)
Lemur catta (2680)	*Leontopithecus rosalia* (550)	*Erythrocebus patas* (4900)
Varecia variegata (3500)	*Saguinus fuscicollis* (200)	*Lophocebus albigena* (7000)
Lepilemuridae	*Saguinus imperator* (500)	*Macaca fascicularis* (2500)
Lepilemur leucopus (580)	*Saguinus labiatus* (667)	*Macaca mulatta* (9000)
Lepilemur mustelinus (620)	*Saguinus midas* (586)	*Macaca nemestrina* (14500)
Lepilemur ruficaudatus (800)	*Saguinus oedipus* (339)	*Macaca nigra* (4500)
Cheirogaleidae	Aotidae	*Macaca sylvanus* (7513)
Cheirogaleus major (283)	*Aotus nigripes* (825)	*Miopithecus talapoin* (820)
Microcebus murinus (110)	*Aotus trivirgatus* (800)	*Papio hamadryas* (23500)
Microcebus myoxinus (31)	Cebidae	*Theropithecus gelada* (20400)
Microcebus rufus (70)	*Cebus albifrons* (1615)	Colobinae
Loridae	*Cebus apella* (3250)	*Colobus guereza* (9800)
Arctocebus aureus (210)	*Cebus capucinus* (1300)	*Colobus polykomos* (9000)
Loris tardigradus (223)	*Saimiri sciureus* (800)	*Nasalis larvatus* (7000)
Nycticebus coucang (610)	Pitheciidae	*Presbytis melalophus* (6300)
Perodicticus potto (1200)	*Cacajao calvus* (3450)	*Procolobus badius* (6250)
Galagonidae	*Cacajao melanocephalus* (3000)	*Procolobus pennantii* (7000)
Galago alleni (314)	*Callicebus moloch* (800)	*Pygathrix nemaeus* (8000)
Galago senegalensis (193)	*Chiropotes satanas* (2000)	*Trachypithecus obsc.* (6000)
Galagoides demidoff (100)	*Pithecia irrorata* (2500)	Hylobatidae
Otolemur crassicaudatus (1122)	*Pithecia monachus* (1500)	*Hylobates lar* (6000)
Otolemur garnetti (725)	*Pithecia pithecia* (1000)	Hominidae
Daubentoniidae	Atelidae	*Gorilla gorilla* (170 kg)
Daubentonia madagasc. (2500)	*Alouatta seniculus* (6000)	*Pan troglodytes* (35 kg)
Tarsiidae	*Lagothrix lagotricha* (6000)	*Pongo pygmaeus* (40 kg)
Tarsius syrichta (120)		

(negative) but isometrically in the baboon and the proboscis monkey. Therefore, we decided to choose always the largest specimen when more than one specimen was available instead of calculating an average value.

The morphometric features (1–5) are compared across the primate sample by means of a one-way fixed-factor analysis of variance (ANOVA). We used ANOVA to determine the degree of variance of these morphometric parameters among the primate families. We divided the cercopithecid Old World monkeys into the Cercopithecinae and Colobinae, because of the significant differences in locomotor behavior between the two subfamilies (Ashton and Oxnard 1963, 1964a, b; Rowe 1996; Fleagle 1999). Because sample sizes are unequal across the taxa, we employed the GT2 method to compare group means and to calculate lower and upper comparison limits for each sample mean (Sokal and Rohlf 1995).

Though the osteological measurements bear only minor sources of error, given they were always carried out by the same investigator, it should be noted that the confidence of measurements on skeletal rib cages depends largely on the quality of preparation of the specimen. The shrinkage of the cartilaginous ventral parts of the ribs can produce unnatural deformations. Therefore, we were very careful in choosing the material and we excluded all rib cages that seem asymmetrical or depressed or which have very disordered distances between the ribs. However, corresponding measurements on cadavers would always provide a better description of the form of the rib-cage, but unfortunately the number of available cadaveric specimens was too limited to perform "wet" measurements (see Chan 2007).

Results and Discussion

Scapular motions in quadrupedal primates are complex even during walking because they are highly three-dimensional. Each of the three components has a particular function with respect to whole limb excursion. Therefore, it seems reasonable to break down the complexity into its subsets (Fig. 6.1). The *cranio-caudal rotation* of the scapula is a decisive part of the fore-aft motion of the forelimb. This rotation alone produces more than 50% of the step length of the forelimb (Schmidt 2005, 2008). The two other components of scapular motion are *tilting and inwards rotation* (Fig. 6.1). These components are important for sideward directed limb excursions and rotations of the forelimb about its longitudinal axis. In quadrupedal primates, this kind of mobility cannot provided by the glenohumeral joint at least during locomotion when this joint requires muscular stabilization against weight bearing (Roberts 1974; Schmidt 2008).

Scapular Motion in Cranio-caudal Direction and Its Importance for Limb Pro- and Retraction

The cranio-caudal rotation of the scapula does not take place in a "true" joint. Instead, the scapula moves at an instantaneous center of rotation (Fig. 6.1a). It lies close to the intersection of the scapular spine and the vertebral border of the shoulder blade (English 1978a; Jenkins and Weijs 1979; Kuznetsov 1985; Fischer 1994; Whitehead and Larson 1994; Schmidt 2008). Cranio-caudal rotation is guided by a

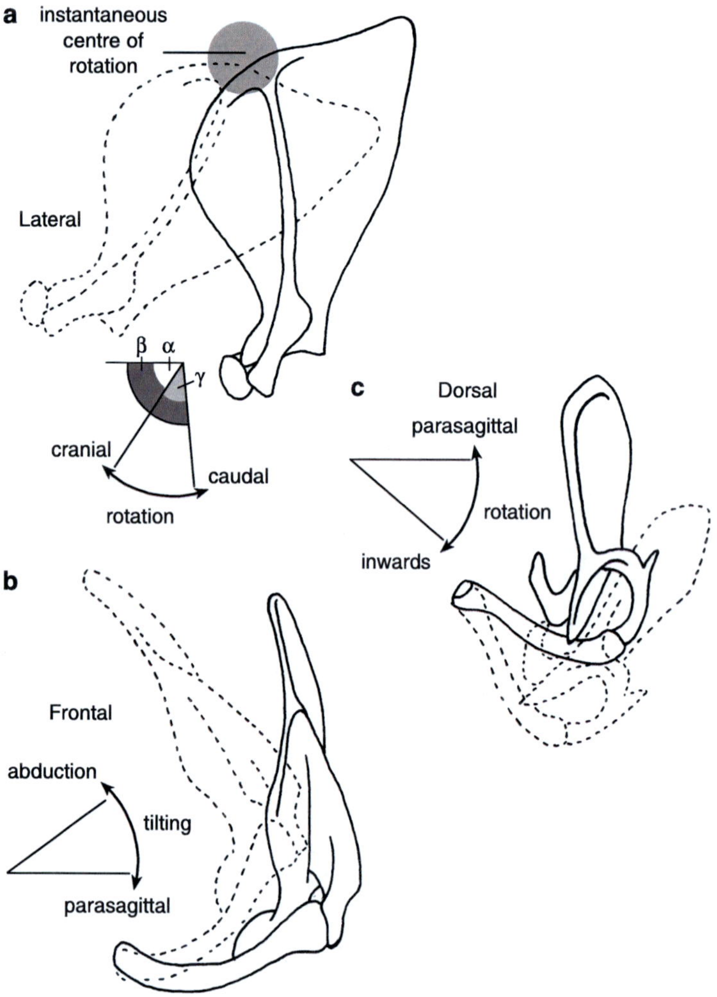

Fig. 6.1 Components of scapular motion in arboreal quadrupedal primates. (**a**) Cranio-caudal rotation in lateral perspective (α = touchdown angle, β = lift-off angle, γ = amplitude of rotation). (**b**) Scapular tilting in frontal perspective. (**c**) Scapular inwards rotation in dorsal perspective

large sample of muscles. Some of them insert onto the scapula, but others insert onto the humerus and move the shoulder blade indirectly, when the glenohumeral joint is fixed. It has been assumed that the scapulothoracic pivot is not fixed but may change during the stride related to the activity of scapulothoracic and humerothoracic muscles (English 1978b). The best approximation of the location of this pivot is obtained by drawing a horizontal line from the hip joint of the hind limb to the forelimb. The most proximal pivots of fore- and hind limb should be on the same horizontal level, at least during symmetrical gaits, assigning the limbs the same functional length (= distance between proximal pivot and support contact) (Schmidt 2008).

Table 6.2 Scapula excursions: angles at touchdown, lift-off, and the amplitude of excursion

	Touchdown angle (°)	Lift-off angle (°)	Amplitude (°)
	Mean ± SD (*n*)	Mean ± SD (*n*)	Mean ± SD (*n*)
Microcebus murinus	41 ± 7 (76)	87 ± 6 (92)	48 ± 6 (76)
Eulemur fulvus	46 ± 6 (60)	86 ± 9 (60)	51 ± 9 (60)
Saguinus oedipus	42 ± 3 (46)	90 ± 5 (52)	49 ± 6 (25)
Saimiri sciureus	43 ± 5 (60)	84 ± 6 (60)	56 ± 8 (60)

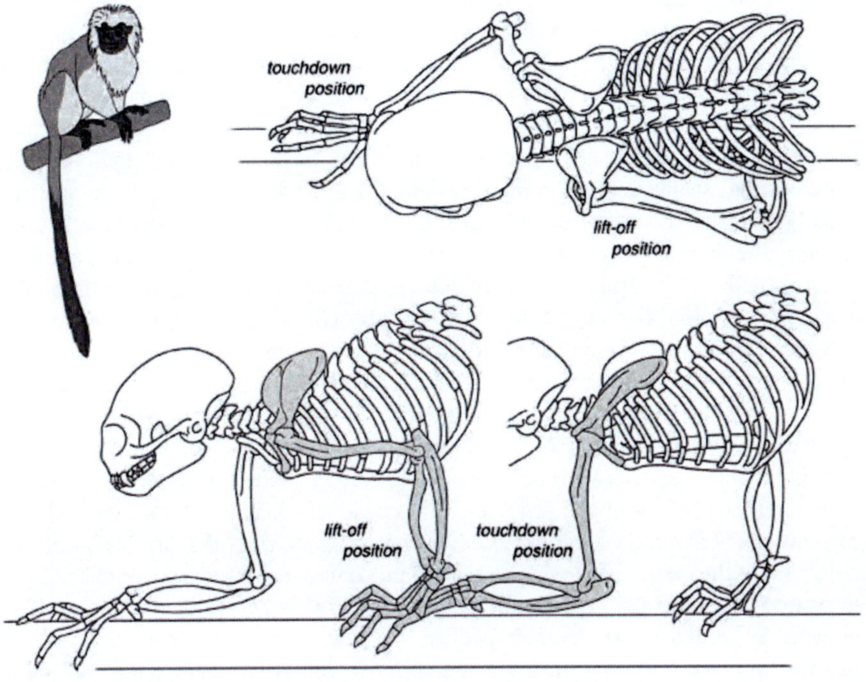

Fig. 6.2 Forelimb position of the cotton-top tamarin *Saguinus oedipus* at touchdown and lift-off shown in lateral and dorsal perspective

The overall amplitude of scapular cranio-caudal rotation (Fig. 6.1a, angle α) is very similar among the four arboreal primates considered in this study (Table 6.2). Figure 6.2 illustrates the touchdown and lift-off posture of the forelimb in the cotton-top tamarin, *Saguinus oedipus*. It shows that the scapula is oriented at an oblique angle of about 45° to the horizontal plane at touchdown of the forelimb. The following caudal rotation of the scapula is continuous and ends in an almost vertical position of the shoulder blade at the end of the contact phase. Caudal rotation of the scapula produces a significant backward and downward translation of the glenohumeral joint. Knowing about this motion principle, glenohumeral translation is detectable without X-ray analyses and can also be observed in field studies (Schön Ybarra and Schön 1987). The glenohumeral joint lies close to the neck of the animal at the beginning of a step cycle, but more ventrally and beside the rib cage at the end of the support phase. Interestingly, touchdown and lift-off angles of the scapula hardly differ among the four primates (Table 6.2). They are also very similar to scapular angles of other noncursorial mammals, such as tree shrews, rats, and opossums (Jenkins 1971; English 1978a; Fischer 1994; Schilling and Fischer 1999; Fischer et al. 2002), indicating that neither phylogeny nor locomotor habitat has a significant influence on the angular excursion of the scapula (Fischer et al. 2002). Accordingly, the first common principle of scapula motion is that *the amplitude of cranio-caudal rotation is highly invariant in mammals* up to at least cat size or lemur size. Caudal rotation starts at an angle of 40° to 45° and ends in a vertical position of the shoulder-blade. Studies about asymmetrical gaits (gallop, half-bound) reported that the cranio-caudal excursion range of the scapula does not change significantly with gait transition (Rocha-Barbosa et al. 1996; Schilling and Fischer 1999; Fischer et al. 2002). Therefore, we can predict to find the observed cranio-caudal scapular rotation in all primates that prefer to walk and run quadrupedally on arboreal substrates and that show a similar body size as the investigated species. Predictions would refer to members of the following primate families: Cheirogaleidae, Lemuridae, Daubentoniidae, the more quadrupedal forms of the Galagonidae (genus *Otolemur*), the Aotidae, Calltrichidae, Cebidae, and Pitheciidae. Whether Loridae could also be included here is uncertain because several morphological peculiarities of these primates indicate that there are very particular (see later). Also uncertain is how the cranio-caudal scapular excursion range is in cercopithecine Old World monkeys. Cercopithecinae comprise arboreal, terrestrial, and intermediate species (Rowe 1996; Fleagle 1999). Most of them are larger in size than the aforementioned primates. Whitehead and Larson (1994) reported that the cercopithecine vervet monkeys (*Chlorocebus aethiops*) display a significant reduction of the cranial excursion of the scapula compared to other mammals. They measured touchdown angles of only about 50° to 55°, whereas the lift-off angle is between 85° and 90° like in other species. Such a restriction of the cranial rotation of the scapula occurs also in larger cursorial mammals such as dogs and goats (Goslow et al. 1981; Fischer et al. 2002). Therefore, it seems likely to assume that the terrestrial cercopithecines (baboons, patas monkeys) correspond to the vervets in their cranio-caudal scapular excursion. Because the last common ancestor of the Cercopithecidae was rather a terrestrial or semiterrestrial form than an arboreal primate (Napier 1967; Rollinson and Martin 1981; McCrossin et al. 1998; Fleagle 1999; Hartwig 2002) it could be that the (secondary) arboreal cercopithecines (many guenons and macaques) retained the limited cranio-caudal scapular rotation as they did with several other limb features (more extended limb posture, intra- and interlimb proportions) (Schultz 1970; Rollinson and Martin 1981; Schmitt 1998; Schmidt 2008). Field studies could help to provide a comparative background to clarify this open question. The position of the glenohumeral joint relative to the neck could be used to indicate whether cranial scapular rotation is limited (the joint would be located well beneath the neck) or not (the joint would reach the level of the neck). So, the overall similarity of scapular excursion in quadrupedal mammals, being influenced solely by body size, facilitates estimation of the amplitude of scapular rotation even without detailed cineradiographic analyses. This implies that scapular motion can be assessed relatively easily and accurately in field studies of locomotor behavior. Schön and Schön-Ybarra (1987) demonstrated on the example of red howlers how this can be done. They identified morphological adaptations to climbing by relating anatomical details of the limb bones to there field observations of positional behavior. With respect to the shoulder movements, the authors carefully documented the position of the glenohumeral joint in their samples of observational recordings. Then, with the bones at hand, observations were rendered onto the skeletal structures. This study demonstrates that such a deductive analysis requires a profound knowledge about the motions that can possibly occur in the shoulder region of a primate. With the present chapter we attempt to provide a useful contribution to this knowledge by illustrating that scapular motion during locomotion is not as diverse as the morphology of the shoulder region may implicate.

Scapular Motion and Its Importance for Three-Dimensional Forelimb Excursions

Cranio-caudal rotation of the scapula is that component of the whole motion that is projected onto the parasagittal plane and that serves for forelimb pro- and retraction. However, when the scapula is moving along the thoracic wall, it follows the curvature of the ribs keeping an almost parallel position to the chest. Accordingly, the shoulder blade is parasagittal only at the end of the support phase when it is in vertical orientation relative to the horizontal plane (Fig. 6.2). At the beginning of the step cycle (at touchdown), the scapula lies more dorsally onto the rounded thorax. In this position, the glenoid is more lateral than the vertebral border. Hence, the scapula is abducted (Fig. 6.1b). In addition, the caudal border is more lateral than the cranial border: the scapula is rotated inwards (Fig. 6.1c). Abduction and inwards rotation always occur simultaneously. These two components of scapular motion, however, have been observed only in claviculate mammals (Jenkins 1974; Schmidt and Fischer 2000) elucidating the intimate relationship between the presence of a clavicle and three-dimensional scapular motions. But, what is the functional context of these motions or are they still an unnecessary consequence of having a clavicle? Mammals that rely to a considerable degree on three-dimensional forelimb excursion for climbing, flying, or digging have all retained clavicles. In contrast, mammals that rely on rapid or long-distance walking and running have lost their clavicles during the course of their evolution. Jenkins (1974) tested experimentally the function of the clavicle. He excised the clavicles in laboratory rats and compared the forelimb excursion of these rats with those of intact animals. Jenkins found that in normal rats the glenoid of the scapula moved in an arc (in dorsoventral perspective), with the clavicle functioning as strut (see Fig. 6.1c), but the shoulder of clavectomized rats collapsed medially and the limb moved only in a parasagittal plane.

Hence, the clavicle plays an important role in guiding the scapula on its path along the thoracic wall. Arboreal primates normally have to place their limbs in a line underneath the body because they frequently move on branches with diameters smaller than the individual's trunk. Such a placement requires adducted forearms. Because elbow joint excursions are restricted to a single plane, forearm adduction is produced via rotation of the total forelimb about its longitudinal axis at which both the forearm and the upper arm just follow the inwards rotation of the scapula. Scapular motion in arboreal quadrupedal primates is three-dimensional to a much higher degree than in tree shrews or laboratory rats (Schmidt and Fischer 2000), indicating that this mobility is an important prerequisite for the locomotion on small-diameter substrates. Both the rounded shape of the thorax and the long and curved clavicles of primates facilitate the three-dimensional mobility of the shoulder blade. In contrast, the mobility of the glenohumeral joint is more limited with respect to sideward directed excursions, meaning that the humerus always moves in the scapular plane. This coupled motion of scapula and humerus is likely the ancestral pattern for primates. It occurs in mouse lemurs, brown lemurs, and cotton-top tamarins, but also in tree shrews and rats. Squirrel monkeys, however, deviate from this pattern. The scapula of this species is performing the described three-dimensional motion but the humerus does not follow. The humerus retains a parasagittal position throughout the entire contact phase and moves in a flat arc when the limb is swinging forward. Again, Schön Ybarra and Schön (1987) were able to deduce a similar kind of shoulder motion from their field study on red howlers. By correlating their observations with a detailed anatomical analysis of the glenohumeral joint, they concluded that scapula and humerus are aligned if the forelimb is protracted (and the glenohumeral joint is extended). In this position, arm abduction and adduction is achieved by dorsolateral scapular displacement on the rib cage. When the forelimb is retracted (and the glenohumeral joint is flexed), the humerus can be adducted independently from the position of the scapula. Hence, we can propose a second common principle of shoulder motion: *the humerus cannot abduct against the plane of the scapula.* Whenever we observe—in the field or in the laboratory—that a primate moves its arm sideward, this movement results from scapular abduction and inward rotation but not from sideward directed movements of the humerus in the glenohumeral joint. This principle applies to all primate species that display highly three-dimensional scapular excursions. It does not apply to the Atelidae and Hominoidea. These primates show a fundamental reorganization of the shoulder girdle with a permanent dorsal position of the scapula. Interestingly, adduction of the humerus against the scapula is possible in some primates, but not in others.

Different Patterns of Shoulder Motion and Their Evolutionary Aspects

From these observations on arboreal quadrupedal primates we can conclude that the overall shoulder mobility in primates is affected by *scapular mobility* (together with the mobility of the clavicle), on one hand, and *glenohumeral mobility* on the other hand. It seems reasonable to assume that these two elements of the total shoulder mobility evolved, to some extent, independently. The propulsive component of scapular motion—the cranio-caudal rotation—is highly invariant in arboreal quadrupedal primates and corresponds to the motion observed in other mammals. However, the two other components—tilting and inward rotation—are more variable with respect to the way they produce three-dimensional forelimb motions with or without support through the glenohumeral joint. Based on the different patterns of shoulder motion described so far, we can hypothesize that forelimb mobility of arboreal primates is first enhanced by scapular mobility, as reflected in the large excursions of the shoulder blade of the brown lemur, *Eulemur fulvus* (Schmidt and Fischer 2000). At a later stage in primate evolution, the independence or "emancipation" of the humerus from the scapular plane, and thus glenohumeral mobility, gained more importance. Good explanations for this evolutionary step are still lacking. The cases of squirrel monkeys and red howlers may indicate that enhanced glenohumeral mobility occurs not only when suspensory locomotor behaviors evolved but may instead be a prerequisite for these derived locomotor modes.

Climbing might be an interesting "candidate" when we search for an locomotor mode that might require a greater independence of humeral motion from scapular motion. Schön Ybarra and Schön (1987) stated that climbing is not merely walking on sloped substrates. They emphasized that climbing differs from walking by its irregularity of the available supports, and thus, by noncyclic and nonsymmetrical limb motion. Therefore, it seems reasonable to assume that such a locomotor mode might have a driving effect on the evolution of glenohumeral mobility and its independence from scapular mobility. However, to test this hypothesis we need support from field studies and answers to the question about the preferences of climbing locomotion over walking in a broad comparative approach. Based on the available information (e.g., Fleagle and Mittermeier 1980; Garber 1980; Fontaine 1990) it seems that the suspected evolutionary shift from a predominance of walking and running to a predominance of climbing occurred during the evolution of the Anthropoidea. Whether the particular mode of climbing in the strepsirhine Loridae has also produced a greater independence between humeral and scapular excursion is still unknown but will probably be clarified in the near future. A cineradiographic study of the shoulder movements in pygmy slow loris (*Nycticebus pygmaeus*) is now in preparation.

Figure 6.3 depicts the known patterns of shoulder motion in primates and applies them to a current phylogenetic system of primates. Together with the previous X-ray studies on spider monkeys (Jenkins et al. 1978) and vervets (Whitehead and Larson 1994), we can define four different patterns of combined scapular and humeral motion. The various degrees of glenohumeral flexion and extension in primates are not considered here, but group-specific differences are well documented by Larson and her co-workers (Larson et al. 2000). The character states present in some closely related mammalian orders (within the Superorder Euarchontoglires) are also demonstrated to show the direction of character transformation. As already

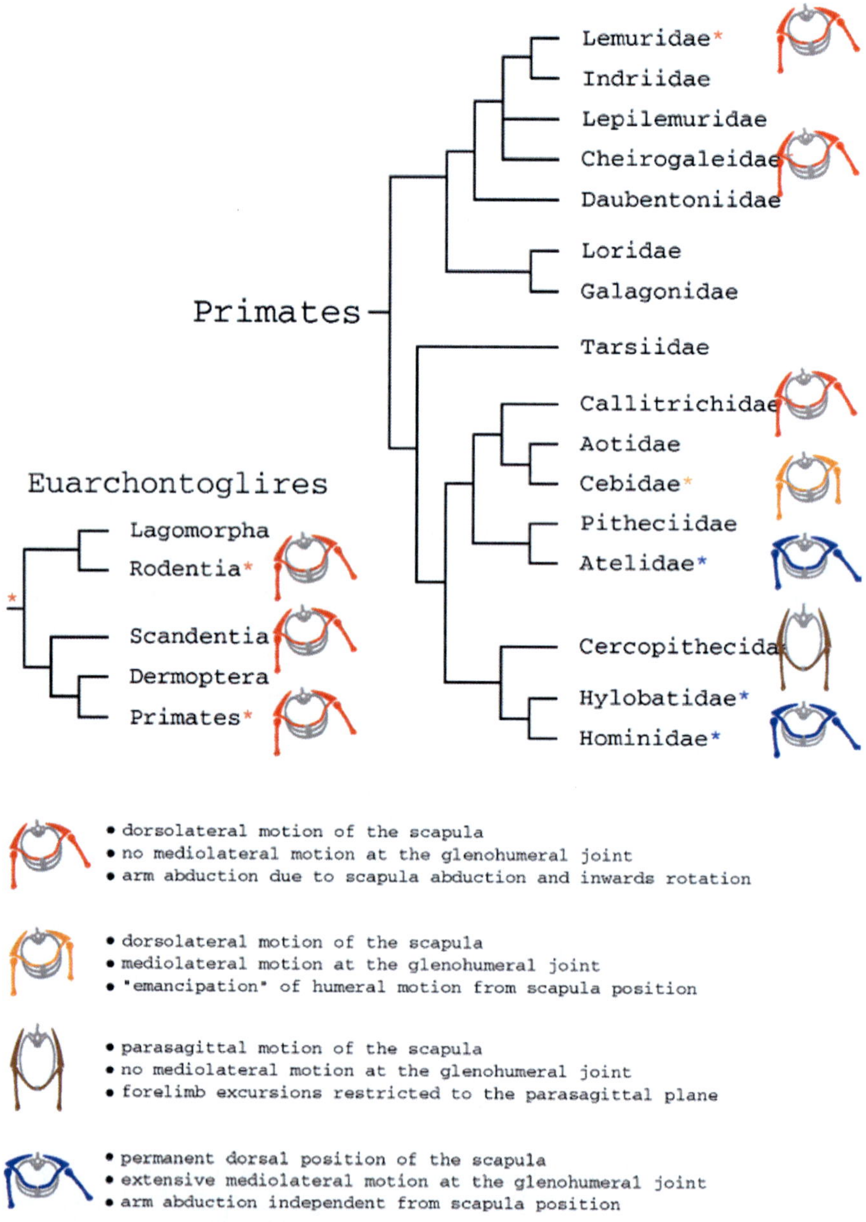

Fig. 6.3 Motion patterns of scapula and humerus applied to a current phylogenetic system of the order Primates (Geissmann 2003). Character states of closely related mammalian groups (Euarchontoglires) are shown to illustrate the direction of character transformation

mentioned in the preceding text, it is very likely that closely related primate species, e.g., members of the same family, display very similar patterns of shoulder motion provided their locomotor behavior is similar.

This is probably not the case for Old World monkeys (Cercopithecidae) for which differences in forelimb mobility can be expected based on their different locomotor behavior (Ashton and Oxnard 1964b; Napier 1967; Fleagle 1976; Morbeck 1979; Rollinson and Martin 1981; Gebo and Chapman 1995; Isler and Grüter 2005). Cercopithecine vervets not only display a rather limited cranio-caudal scapular motion, but the motion of the shoulder blade is also strictly restricted to the parasagittal plane. Several authors (e.g., Rollinson and Martin 1981) have related the reduced mobility of the forelimb of terrestrial and semiterrestrial cercopithecines to the greater need for postural stability and reduced bone stresses, comparable to locomotor adaptations of cursorial mammals. Field studies on colobine Old World monkeys (Fleagle 1976; Morbeck 1979; Gebo and Chapman 1995), however, indicate that these highly arboreal primates display a much greater forelimb mobility than their cercopithecine relatives. Because many colobines prefer a climbing mode of progression, some of them even use suspensory postures, it seems likely that their scapular motion corresponds to that of red howlers. Unfortunately, colobine monkeys are underrepresented in morphofunctional studies, although field studies provided interesting information about their locomotor behavior (e.g., Fleagle 1976; Mittermeier and Fleagle 1976; Morbeck 1979; Gebo and Chapman 1995; Isler and Grüter 2005). Schmidt (2008) has shown that colobine forelimb proportions differ significantly from those of arboreal and terrestrial quadrupedal primates. The proportions are shifted toward a significant reduction of the relative length of the scapula. However, the consequences of this proportional shift for the mobility of the shoulder and the forelimb are not clear so far.

The observed pattern of shoulder motion in *Ateles geoffroyi* (Jenkins et al. 1978) likely corresponds to that of Hylobatidae and Hominidae with respect to the permanent dorsal position of the scapula and free mobility of the glenohumeral joint. Figure 6.3 also illustrates that there are several primate families for which shoulder motion with respect to its three-dimensional excursions cannot be deduced from similarities in locomotor behavior with other species. So, we have no idea about what happens in the shoulder region of vertical clingers and leapers (Indriidae, Galagonidae, Tarsiidae) during locomotion. In terms of the quadrupedal climbers and walkers (Loridae, Pitheciidae, and Colobinae) we could speculate that they are more similar to squirrel monkeys and red howlers, but we should not exclude the possibility that more variations of shoulder motion exist that we now have identified.

Morphological Correlates of Scapular Mobility

One reason for expecting more variation in possible shoulder motions is given by the great morphological variability of the shoulder girdle elements. Figure 6.4

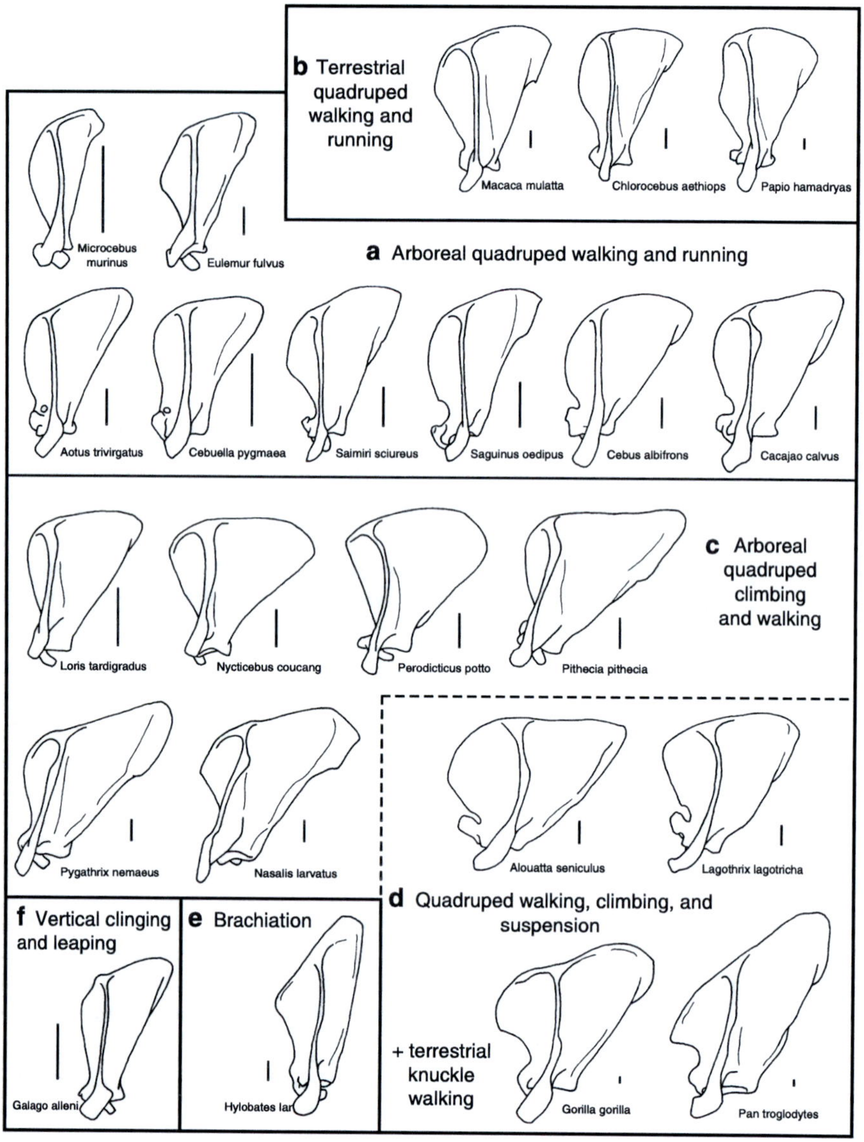

Fig. 6.4 Variety of scapula shape in primates related to preferred locomotor modes. The scale bar beside the drawings corresponds to 1 cm. (**a**) Arboreal quadruped walking and running. (**b**) Terrestrial quadruped walking and running. (**c**) Arboreal quadruped climbing and walking. (**d**) Quadruped walking climbing, and suspension. (**e**) Brachiation. (**f**) Vertical clinging and leaping

depicts a sample of scapula forms for several primate species illustrating the great variability of shapes but also the similarities among species with similar locomotor preferences. We see that quadrupedal primates have long and narrow shoulder blades. This applies both to the arboreal (Fig. 6.4a) and to the terrestrial quadrupeds (Fig. 6.4b). Species that engage more in climbing have broader scapulae, mainly due to a progressive enlargement of the infraspinous fossa (Fig. 6.4c). The supraspinous fossa even seems to decrease in its area. In addition, the spine of the scapula is no longer perpendicular to the blade, but more inclined towards the cranial border. Species for which suspensory postures belong to their locomotor repertoire differ from quadrupedal forms by a further enlargement of both the supraspinous and infraspinous fossa (Fig. 6.4d). The acromion is also large and reaches well beyond the glenoid cavity. The shape of the scapula of the brachiating gibbon (Fig. 6.4e) is very particular and shows no similarity to the other species considered herein. In contrast, the scapula of the vertical clinging and leaping *Galago* almost looks like the shoulder blade of arboreal quadrupedal lemurs (Fig. 6.4f). But, we also see that even within the used groupings of preferred locomotor modes the variation in shape is considerable. Several authors attempted to capture this variability into a framework of morphometric parameters (Miller 1932; Inman et al. 1944; Davis 1949; Smith and Savage 1956; Ashton and Oxnard 1963, 1964a; Erikson 1963; Oxnard 1963; Müller 1967; Roberts 1974; Corruccini and Ciochon 1976; Fleagle 1976; Kimes et al. 1981; Shea 1986; Taylor 1997; Krause 2008). Pioneering work was done during the 1960s and 1970s by Ashton, Oxnard, and their colleagues, who introduced multivariate discriminant statistics to sort the variability into groupings (Ashton and Oxnard 1964a; Oxnard 1967; Ashton et al. 1965a, 1971). The groupings corresponded well to the just established locomotor categories (Napier and Walker 1967), but a closer correspondence to shoulder function has not been attained. All aforementioned researchers acknowledged the broad diversity of scapular shape among primates but were not able to relate it to the functional diversity of the shoulder simply because of the limited knowledge on shoulder movements. Ashton et al. (1965b) even suggested that certain features of scapular shape had no functional significance in locomotion and can therefore be used for classificatory purposes. Although we know much more about the variability of shoulder motions in primates today, a reliable correlation between form and function is still difficult to find. The comparison of scapular dimensions across the primate families shows that the length of the scapula relative to body size is fairly homogeneous (Fig. 6.5a; Table 6.3). There are no significant differences among the families that would indicate a close correlation between relative scapula length and a certain pattern of scapular movement with the exception of the gibbons. The Hylobatidae—highly specialized to overhead suspension—have the relative longest scapulae among all groups herein. The scapula of the gibbons is lying inclined onto the dorsum of the rib cage and the glenoid faces cranially (Larson 1988). The shortest scapulae relative to body size occur in the Colobinae and Hominidae. Such a relative short scapula also allows for a dorsal position but combined with a more laterally facing glenoid.

The ratio between the length and the width of the scapula (the scapular index), however, shows a much broader variation with significant differences among certain families (Fig. 6.5b, Table 6.3). The lowest index can be assigned to pronograde climbers like lorises, atelids (*Alouatta seniculus*, *Lagothrix lagotricha*), and colobines.

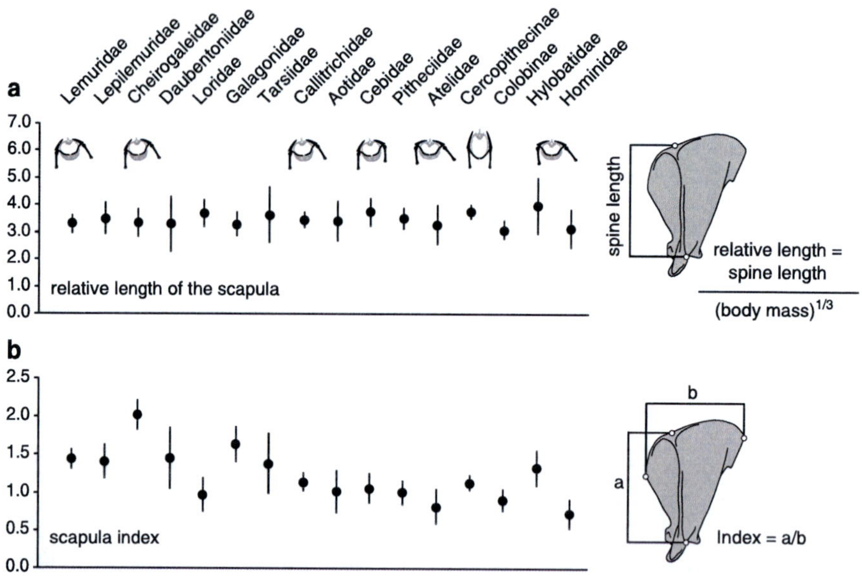

Fig. 6.5 Comparison of scapular dimensions across primate families. Mean values and their comparison intervals are shown for the length of the scapula relative to body mass (**a**) and for the scapular index (length/width) (**b**)

Table 6.3 Scapular dimensions across primate families

	Scapular length/(body mass)$^{1/3}$	Scapular length/scapular width
	Mean ± Co.I.[a] (*n*)	Mean ± Co.I.[a] (*n*)
Lemuridae	3.32 ± 0.35 (9)	1.45 ± 0.13 (9)
Lepilemuridae	3.51 ± 0.61 (3)	1.41 ± 0.23 (3)
Cheirogaleidae	3.35 ± 0.52 (4)	2.02 ± 0.20 (4)
Daubentoniidae	3.32 ± 1.05 (1)	1.45 ± 0.40 (1)
Loridae	3.70 ± 0.52 (4)	0.97 ± 0.23 (3)
Galagonidae	3.31 ± 0.47 (5)	1.64 ± 0.23 (3)
Tarsiidae	3.65 ± 1.05 (1)	1.38 ± 0.40 (1)
Callitrichidae	3.47 ± 0.32 (11)	1.15 ± 0.13 (10)
Aotidae	3.43 ± 0.74 (2)	1.01 ± 0.29 (2)
Cebidae	3.77 ± 0.52 (4)	1.06 ± 0.20 (4)
Pitheciidae	3.53 ± 0.40 (7)	1.00 ± 0.17 (6)
Atelidae	3.30 ± 0.74 (2)	0.82 ± 0.23 (2)
Cercopithecinae	3.76 ± 0.27 (15)	1.13 ± 0.10 (15)
Colobinae	3.11 ± 0.37 (8)	0.91 ± 0.15 (7)
Hylobatidae	4.02 ± 1.05 (1)	1.33 ± 0.23 (1)
Hominidae	3.17 ± 0.74 (3)	0.73 ± 0.20 (3)

[a]Co.I. = Comparison index calculated by GT2 method (Sokal and Rohlf 1995)

The difficulty in explaining scapular shape, and its diversity among primates, arises from the fact that scapular form is a compromise of various determinants (Roberts 1974; Larson 1993). The scapula provides a framework for the musculature that acts on the glenohumeral joint, for the muscles that move and stabilize scapular motion itself, and for muscles that transmit the bodyweight onto the forelimbs. Many studies have related the shape of the scapula to its mechanical function, either as a means of providing leverage or as a means of absorbing the stress produced by its associated musculature (Inman et al. 1944; Davis 1949; Ashton and Oxnard 1963, 1964b; Müller 1967; Roberts 1974). Roberts (1974) has attempted to relate the shape of the scapula to its function as a dynamic structure when he considered the motion and its spatial limitations along the thoracic wall. For example, he proposed that the superior angle of the blade is rounded off in a smooth curve in species that display a large extent of scapular rotation. This is well visible in the arboreal quadrupedal primates and corresponds to the observed scapular motion. The progressive enlargement of the infraspinous fossa has often been associated with climbing and a related need for an increasing leverage for the teres major muscle (Inman et al. 1944; Davis 1949; Roberts 1974). However, beside the study of Schön Ybarra and Schön (1987) on *Alouatta seniculus*, we have no information on how the scapula is moving during climbing. Moreover, it would be very interesting to know whether Loridae, Pitheciidae, and Colobinae display a different mode of scapular motion related to the differences in shape between these primates and red howlers (see Fig. 6.4). The aim of understanding the functional morphology of the primate shoulder is complicated further by the fact that the range of scapular motion is geometrically constrained, not only by the dimensions and shape of the scapula itself but also by the shape and dimensions of clavicle and thorax (Jenkins et al. 1978; Larson 1993; Preuschoft et al. 2003; Chan 2007). Recently, Chan (2007) considered the complex geometrical relationships among the clavicle, the thorax, and the scapula in describing scapular position in preserved cadavers of a variety of primate species. His findings that arboreal quadrupedal primates display a more dorsally positioned scapula than terrestrial quadrupeds corresponds to the observations made via X-ray analyses. Cineradiography revealed that the scapula of quadrupeds moves upon the thoracic wall in the region anterior to thoracic vertebrae 5 and 6. At this level, the thorax is rounded in shape in all quadrupeds without any clear separation between arboreal and terrestrial primates (Fig. 6.6a, Table 6.4). This has also been demonstrated in a recent paper of Kagaya and her co-workers (Kagaya et al. 2008). Consequently, it must be the length of the clavicle and its shape, e.g., curvature, that limits the degree of dorsolateral scapular excursions. However, the observed variation (Fig. 6.6b) cannot be sufficiently explained on the basis of the available knowledge of shoulder motion. Only in the Hylobatidae is the combination of a long clavicle and a dorsoventrally flattened thorax (Fig. 6.6c, Table 6.4) clearly related to a permanent dorsal scapular position (Takahashi 1990; Kagaya et al. 2008).

Table 6.4 Thorax and clavicula dimensions across primate families

	Thorax length/thorax width at the 4th rib	Clavicle length/(body mass)$^{1/3}$	Clavicle length/thorax width at the 1st rib
	Mean ± Co.I.[a] (*n*)	Mean ± Co.I.[a] (*n*)	Mean ± Co.I.[a] (*n*)
Lemuridae	0.90 ± 0.08 (7)	2.46 ± 0.29 (9)	0.91 ± 0.10 (7)
Lepilemuridae	0.85 ± 0.16 (2)	3.00 ± 0.51 (3)	1.04 ± 0.19 (2)
Cheirogaleidae	0.80 ± 0.16 (2)	2.79 ± 0.44 (4)	1.01 ± 0.19 (2)
Daubentoniidae	0.81 ± 0.22 (1)	2.95 ± 0.88 (1)	1.08 ± 0.28 (1)
Loridae	0.93 ± 0.11 (4)	3.49 ± 0.44 (4)	1.22 ± 0.14 (4)
Galagonidae	0.86 ± 0.10 (5)	2.89 ± 0.39 (5)	1.09 ± 0.14 (4)
Tarsiidae	0.79 ± 0.22 (1)	2.84 ± 0.88 (1)	0.86 ± 0.28 (1)
Callitrichidae	0.88 ± 0.07 (10)	3.14 ± 0.26 (12)	1.07 ± 0.09 (10)
Aotidae	1.01 ± 0.22 (1)	3.22 ± 0.62 (2)	1.14 ± 0.28 (1)
Cebidae	0.84 ± 0.16 (2)	3.29 ± 0.44 (4)	1.05 ± 0.16 (3)
Pitheciidae	0.92 ± 0.11 (4)	3.43 ± 0.33 (7)	1.13 ± 0.14 (4)
Atelidae	0.88 ± 0.16 (2)	3.66 ± 0.62 (2)	1.07 ± 0.19 (2)
Cercopithecinae	0.95 ± 0.06 (12)	2.89 ± 0.23 (15)	1.13 ± 0.08 (12)
Colobinae	0.92 ± 0.08 (8)	2.90 ± 0.31 (8)	1.09 ± 0.10 (8)
Hylobatidae	0.75 ± 0.22 (1)	5.17 ± 0.88 (1)	1.55 ± 0.28 (1)
Hominidae	0.91 ± 0.13 (3)	4.00 ± 0.62 (3)	1.31 ± 0.16 (3)

[a]Co.I. = Comparison index calculated by GT2 method (Sokal and Rohlf 1995)

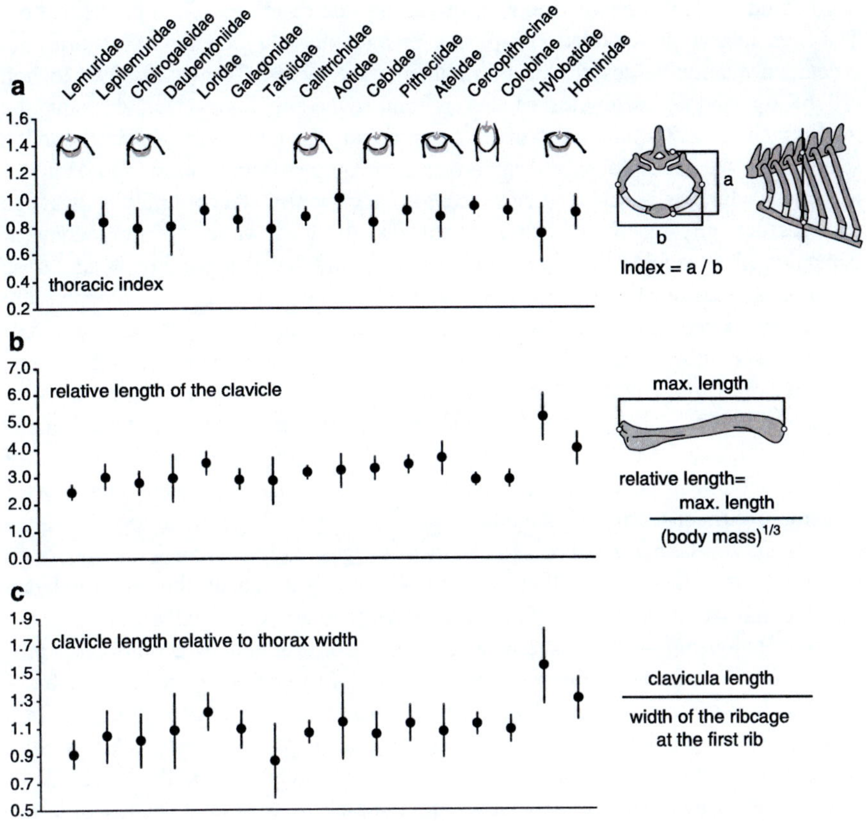

Fig. 6.6 Comparison of thorax and clavicula dimensions across primate families. Mean values and their comparison intervals are shown for the thoracic index (length/width) (**a**); for the length of the clavicle relative to body mass (**b**), and for the ratio between clavicle length and thorax width (**c**)

Conclusion

During the past century a huge amount of information on the morphology of the primate shoulder has been accumulated, including insights into the variation in shape and dimensions of scapula, clavicula, thorax, and proximal humerus. Many studies have dealt with investigating the bony morphology and many others with investigating the function of shoulder muscles. A detailed revision of these studies lies outside the scope of this chapter, and we refer the reader to the literature for a full overview of this topic (e.g., Miller 1932; Inman et al. 1944; Ashton and Oxnard 1963, 1964a; Erikson 1963; Roberts 1974; Corruccini and Ciochon 1976; Tuttle and Basmajian 1978; Larson and Stern 1986, 1989; Rose 1989; Larson 1993). However, when investigating the literature, it becomes obvious that there is still a contrast between the huge amount of information about form and the scarce information about function. The latter means not the relationship between anatomy and locomotor categories but a detailed qualitative description of shoulder movements. Corruccini and Ciochon (1976) argued that locomotor categories tend to be typological and discount the importance of total behavioral profile and potential. The authors emphasized that the structure of the musculoskeletal apparatus correlates with the total range of all its activities, which is today a widely accepted perspective. Accordingly, this range includes not only locomotor activities but also the great variety of nonlocomotor behaviors like reaching for (food) objects, self- and social grooming. It might be reasonable to assume that nonlocomotor activities put an important selective pressure on the evolution of the locomotor system, especially in primates with their great repertoire of social behavior. This might explain why morphometric features that determine the maximum range of shoulder motion, such as clavicle length, scapular length, and thoracic shape, are obviously limited in their predictive power of the locomotor motion of the shoulder (at least in quadrupeds). The available studies of shoulder motion in primates document the important role of scapular mobility for forelimb excursions, whereas glenohumeral mobility is often limited to secure postural stability. Whether the observed excursion ranges really document the maximum mobility of the shoulder is rather unlikely, which might explain the only moderate correlation between variation in form and known function. For a better understanding of this relationship—which is crucial for the reconstruction of evolutionary processes—it goes without saying that we need more information about shoulder function. Laboratory studies should not ignore scapular excursions in describing forelimb movements. Cineradiographic studies revealed that scapular motion is very similar across a great variety of species. The first common principle is that the amplitude of cranio-caudal rotation is highly invariant in mammals up to at least cat size or lemur size. Caudal rotation starts at an angle of 40° to 45° and ends in a vertical position of the shoulder blade. Larger quadrupedal primates may display a reduced amount of cranial rotation. Therefore, with the available studies at hand, for many primates it is possible to deduce scapular motion from video recordings. We have shown that the position of the glenohumeral joint relative to the neck or to the thoracic well indicates the amplitude of cranio-caudal scapular rotation. A second principle is that the humerus cannot abduct against the plane of the scapula. In quadrupedal primates, sideward directed arm movements result from scapular abduction and inward rotation but not from sideward directed movements of the humerus in the glenohumeral joint. The limitation of the glenohumeral mobility during locomotion is clearly related to the needs for muscular stabilization of this joint. Whether a greater mobility occurs during nonlocomotory forelimb activities is not known. Future field studies could provide the functional morphologist with a description of the repertoire of nonlocomotor activities that require high shoulder mobility. Certainly, locomotion is not the only task of the postcranial musculoskeletal system. Adaptations to nonlocomotory activities could also have had a driving effect on the evolution of the shoulder. Assuming that the reconstruction of primate evolution is the common ultimate objective of all our laboratory-based and field-based studies, we are convinced that a better scientific exchange between both strategies will produce a novel quality of research about the evolution of the primate locomotor apparatus.

Acknowledgments We thank Evie Vereecke and Kristiaan D'Août for inviting us to participate in the symposium "Primate Locomotion: Linking *In Situ* and *Ex Situ* Research" held at the 2008 meeting of the International Primatological Society in Edinburgh and for their motivation and great effort in organizing this book volume.

We thank Dieter Haarhaus (Institut für den Wissenschaftlichen Film, Göttingen) and his team for their patience and competence in cineradiography. Danja Voges kindly provided the X-ray films of the cotton-top tamarins. We also thank the German Primate Research Centre (Göttingen), Gettdorf Zoo, and the Institute of Zoology at the Veterinary University Hannover for kindly providing our institute with animals. We also thank the curators of the Phylogenetisches Museum Jena, Museum für Naturkunde Berlin, and Naturhistorisches Museum Bern for access to skeletal material in their care.

This research was supported by the Deutsche Forschungsgemeinschaft (Innovationskolleg Bewegungssysteme, INK 22/B1-1).

References

Arms A, Voges D, Fischer MS, Preuschoft H (2002) Arboreal locomotion in small New World monkeys. Z Morph Anthrop 83:243–263.

Ashton EH, Oxnard CE (1963) The musculature of the primate shoulder. Trans Zool Soc Lond 29:553–650.

Ashton EH, Oxnard CE (1964a) Functional adaptations in the primate shoulder girdle. Proc Zool Soc Lond 142:49–66.

Ashton EH, Oxnard CE (1964b) Locomotor patterns in primates. Proc Zool Soc Lond 142:1–28.

Ashton EH, Healy MJR, Oxnard CE, Spence TF (1965a) The combination of locomotor features of the primate shoulder girdle by canonical analysis. J Zool, Lond 147:406–429.

Ashton EH, Oxnard CE, Spence TF (1965b) Scapular shape and primate classification. Proc Zool Soc Lond 145:125–142.

Ashton EH, Flinn RM, Oxnard CE, Spence TF (1971) The functional and classificatory significance of combined metrical features of the primate shoulder girdle. J Zool, Lond 163:319–350.

Cartmill M (1972) Arboreal adaptations and the origin of the order Primates. In: Tuttle R (ed), Functional and Evolutionary Biology of Primates, Aldine Atherton, Chicago, pp. 97–122.

Cartmill M (1974) Rethinking primate origins. Science 184:436–443.

Chan LK (2007) Scapular position in primates. Folia Primatol 78:19–35.

Corruccini RS, Ciochon RL (1976) Morphometric affinities of the human shoulder. Am J Phys Anthropol 45:19–38.

Davis DD (1949) The shoulder architecture of bears and other carnivores. Fieldiana Zool 31:285–305.

English AW (1978a) Functional analysis of the shoulder girdle of cats during locomotion. J Morph 156:279–292.

English AW (1978b) An electromyographic analysis of forelimb muscles during overground stepping in the cat. J Exp Biol 76:105–122.

Erikson GE (1963) Brachiation in New World monkeys and in anthropoid apes. Symp Zool Soc Lond 10:135–164.

Fischer MS (1994) Crouched posture and high fulcrum. A principle in the locomotion of small mammals: the example of the rock hyrax (*Procavia capensis*) (Mammalia: Hyracoidea). J Hum Evol 26:501–524.

Fischer MS, Schilling N, Schmidt M, Haarhaus D, Witte H (2002) Basic limb kinematics of small therian mammals. J Exp Biol 205:1315–1338.

Fleagle JG (1976) Locomotor behavior and skeletal anatomy of sympatric Malaysian leaf-monkeys (*Presbytis obscura* and *Presbytis melalophos*). Yrbk Phys Anthropol 20:440–453.

Fleagle JG, Mittermeier RA (1980) Locomotor behavior, body size, and comparative ecology of seven Surinam monkeys. Am J Phys Anthropol 52:301–314.

Fleagle JG (1999) Primate Adaptation and Evolution. Academic Press, San Diego.

Fontaine R (1990) Positional behavior in *Saimiri boliviensis* and *Ateles geoffroyi*. Am J Phys Anthropol 82:485–508.

Garber PA (1980) Locomotor behavior and feeding ecology of the Panamanian tamarin (*Saguinus oedipus geoffroyi*, Callitrichidae, Primates). Int J Primatol 1:185–201.

Gebo DL (2004) A shrew-sized origin for primates. Yrbk Phys Anthropol 47:40–62.

Gebo DL, Chapman CA (1995) Positional behavior in five sympatric Old World monkeys. Am J Phys Anthropol 97:49–76.

Geissmann T (2003) Vergleichende Primatologie. Springer Verlag, Berlin.

Goodman M, Porter CA, Czelusniak J, Page SL, Schneider H, Shoshani G, Gunnell G, Groves CP (1998) Toward a phylogenetic classification of primates based on DNA evidence complemented by fossil evidence. Mol Phylogenet Evol 9:585–598.

Goslow GE Jr, Seeherman HJ, Taylor CR, McCutchin MN, Heglund NC (1981) Electrical activity and relative length changes of dog limb muscles as a function of speed and gait. J Exp Biol 94:15–42.

Grzimek B (1987) Enzyklopädie der Säugetiere. Kindler Verlag, München.

Hartwig WC (2002) The Primate Fossil Record. Cambridge University Press, Cambridge.

Inman VT, Saunders M, Abbott LC (1944) Observations on the function of the shoulder joint. J Bone Joint Surg 26:1–30.

Isler K, Grüter CC (2005) Arboreal locomotion in wild black-and-white snub-nosed monkeys (*Rhinopithecus bieti*). Folia Primatol 77:195–211.

Jenkins FA Jr (1971) Limb posture and locomotion in the Virginia opossum (*Didelphis marsupialis*) and in other non-cursorial mammals. J Zool Lond 165:303–315.

Jenkins FA Jr (1974) The movement of the shoulder in claviculate and aclaviculate mammals. J Morph 144:71–83.

Jenkins FA Jr, Dombrowski PJ, Gordon EP (1978) Analysis of the shoulder in brachiating spider monkeys. Am J Phys Anthropol 48:65–76.

Jenkins FA Jr, Weijs WA (1979) The functional anatomy of the shoulder in the Virginia opossum (*Didelphis virginiana*). J Zool, Lond 188:379–410.

Kagaya M, Ogihara N, Nakatsukasa M (2008) Morphological study of the anthropoid thoracic cage: scaling of thoracic width and an analysis of rib curvature. Primates 49:89–99.

Keith A (1923) Man's posture: its evolution and disorders. Br Med J 1:451–454, 499–502, 545–548, 587–590, 624–626, 669–672.

Kimes KR, Siegel MI, Sadler DL (1981) Musculoskeletal scapular correlates of plantigrade and acrobatic positional activities in *Papio cynocephalus anubis* and *Macaca fascicularis*. Am J Phys Anthropol 55:463–472.

Krause C (2008) Die Beweglichkeit des Schultergelenks arborikoler und terrikoler Primaten. Ph.D thesis. Friedrich Schiller Universität Jena.

Kuznetsov AN (1985) Comparative functional analysis of the fore- and hind limbs in mammals. Zool J, Moscow 64:1862–1867.

Larson SG (1988) Subscapularis function in gibbons and chimpanzees: implications for interpretation of humeral head torsion in hominoids. Am J Phys Anthropol 76:449–462.

Larson SG, Stern JT Jr (1986) EMG of scapulohumeral muscles in the chimpanzee during reaching and "arboreal" locomotion. Am J Anat 176:171–190.

Larson SG, Stern JT Jr (1989) The role of propulsive muscles of the shoulder during quadrupedalism in vervet monkeys (*Cercopithecus aethiops*): implications for neural control of locomotion in primates. J Motor Behav 21:457–472.

Larson SG (1993) Functional morphology of the shoulder in primates. In: Gebo DL (ed), Postcranial Adaptation in Nonhuman Primates. Northern Illinois University Press, DeKalb.

Larson SG, Schmitt D, Lemelin P, Hamrick M (2000) Uniqueness of primate forelimb posture during quadrupedal locomotion. Am J Phys Anthropol 112:87–101.

Martin RD (1972) Adaptive radiation and behaviour of the malagasy lemurs. Philos Trans Roy Soc Lond B 264:295–352.

McCrossin ML, Benefit BR, Gitau SN, Palmer AK, Blue KT (1998) Fossil evidence for the origins of terrestriality among Old World higher primates. In: Strasser E, Fleagle JG, Rosenberger A, McHenry H (eds), Primate Locomotion. Recent Advances. Plenum Press, New York, London, pp. 353–396.

Miller RA (1932) Evolution of the pectoral girdle and forelimb in the primates. Am J Phys Anthropol 17:1–56.

Mittermeier RA, Fleagle JG (1976) The locomotor and postural repertoires of *Ateles geoffroyi* and *Colobus guereza*, and a reevaluation of the locomotor category semibrachiation. Am J Phys Anthropol 45:235–256.

Morbeck ME (1979) Forelimb use and positional adaptation in *Colobus guereza*: integration of behavioral, ecological, and anatomical data. In: Morbeck ME, Preuschoft H, Gomberg N (eds), Environment, Behavior and Morphology: Dynamic Interactions in Primates. Gustav Fischer, New York, pp. 95–117.

Müller HJ (1967) Form und Funktion der Scapula. Vergleichend analytische Studien bei Carnivoren und Ungulaten. Z Anat Entw 126:205–263.

Napier JR (1967) Evolutionary aspects of primate locomotion. Am J Phys Anthropol 27:333–342.

Napier JR, Walker AC (1967) Vertical clinging and leaping – a newly recognized category of locomotor behaviour of primates. Folia Primatol 6:204–219.

Oxnard CE (1963) Locomotor adaptation of the primate forelimb. Symp Zool Soc Lond 10:165–182.

Oxnard CE (1967) The functional morphology of the primate shoulder as revealed by comparative anatomical, osteometric and discriminant function technique. Am J Phys Anthropol 26:219–240.

Preuschoft H, Schmidt M, Hayama S, Okada M (2003) The influence of three-dimensional movements of the forelimb on the shape of the thorax and its importance for erect body posture. Cour Forschungsinstitut Senckenberg 243:9–24.

Roberts D (1974) Structure and function of the primate scapula. In: Jenkins FA (ed), Primate Locomotion. Academic Press, New York, pp. 171–200.

Rocha-Barbosa O, Renous S, Gasc J-P (1996) Comparison of the fore and hind limbs kinematics in the symmetrical and asymmetrical gaits of a caviomorph rodent, the domestic guinea pig, *Cavia porcellus* (Linné, 1758) (Rodentia, Caviidae). Ann Sci Nat Zool Paris 17:149–165.

Rollinson J, Martin RD (1981) Comparative aspects of primate locomotion, with special reference to arboreal cercopithecines. Symp Zool Soc Lond 48:377–427.

Rose MD (1973) Quadrupedalism in primates. Primates 14:337–358.

Rose MD (1989) New postcranial specimens of catarrhines from the Middle Miocene Chinji Formation, Pakistan: descriptions and a discussion of proximal humeral functional morphology in anthropoids. J Hum Evol 18:131–162.

Rowe N (1996) The pictorial guide to the living primates. Pogonias Press, Charlestown.

Schilling N, Fischer MS (1999) Kinematic analysis of treadmill locomotion of tree shrews, *Tupaia glis* (Scandentia: Tupaiidae). Z Säugetierk 64:129–153.

Schmidt M, Fischer MS (2000) Cineradiographic study of forelimb movements during quadrupedal walking in the brown lemur (*Eulemur fulvus*, Primates: Lemuridae). Am J Phys Anthropol 111:245–262.

Schmidt M (2005) Quadrupedal locomotion in squirrel monkeys (Cebidae: *Saimiri sciureus*): a cineradiographic study of limb kinematics and related substrate reaction forces. Am J Phys Anthropol 128:359–370.

Schmidt M (2008) Forelimb proportions and kinematics: how are small primates different from other small mammals? J Exp Biol 211:3775–3789.

Schmitt D (1998) Forelimb mechanics during arboreal and terrestrial quadrupedalism in Old World monkeys. In: Strasser E, Fleagle JG, Rosenberger A, McHenry H (eds), Primate Locomotion. Recent Advances. Plenum Press, New York, London, pp. 175–200.

Schön Ybarra MA, Schön MA (1987) Positional behavior and limb bone adaptations in red howling monkeys (*Alouatta seniculus*). Folia Primatol 49:70–89.

Schultz AH (1970) The comparative uniformity of the Cercopithecoidea. In: Napier JR, Napier PH (eds), Old World Primates: Evolution, Systematics and Behavior. Academic Press, New York, pp. 39–51.

Shea BT (1986) Scapula form and locomotion in chimpanzee evolution. Am J Phys Anthropol 70:475–488.

Smith MJ, Savage RJ (1956) Some locomotory adaptations in mammals. J Linn Soc Lond, Zool 42:603–622.

Sokal RR, Rohlf FJ (1995) Biometry: The Principles and Practice of Statistics in Biological Research, 3rd ed. W H Freeman and Company, San Francisco.

Sussman RW (1991) Primate origins and the evolution of angiosperms. Am J Primatol 23: 209–223.

Takahashi LK (1990) Morphological basis of arm swinging: multivariate analyses of the forelimb of *Hylobates* and *Ateles*. Folia Primatol 54:70–85.

Taylor AB (1997) Scapula form and biomechanics in gorillas. J Hum Evol 33:529–553.

Tuttle RH, Basmajian JV (1978) Electromyography of pongid shoulder muscles. II. Deltoid, rhomboid and "rotator cuff". Am J Phys Anthropol 49:47–56.

Walker AC (1979) Prosimian locomotor behavior. In: Doyle GA, Martin RD (eds), The study of Prosimian Behaviour. Academic Press, New York, pp. 543–565.

Ward SC, Sussman RW (1979). Correlates between locomotor anatomy and behavior in two sympatric species of *Lemur*. Am J Phys Anthropol 50:575–590.

Whitehead PF, Larson SG (1994) Shoulder motion during quadrupedal walking in *Cercopithecus aethiops*: intergration of cineradiography and electromyographic data. J Hum Evol 26:525–544.

Chapter 7
The Influence of Load Carrying on Gait Parameters in Humans and Apes: Implications for the Evolution of Human Bipedalism

Jo Watson, Rachel Payne, Andrew Chamberlain, R. Jones, and William Sellers

Abstract We provide a comparison of loaded and unloaded carrying gait parameters in humans, common chimpanzees (*Pan troglodytes*), bonobos (*Pan paniscus*), Western lowland gorillas (*Gorilla gorilla gorilla*), and Bornean and Sumatran orang-utans (*Pongo pygmaeus* and *Pongo abelii*). We consider whether freeing the hands to carry loads could have been a precursor to the evolution of bipedal walking in humans. We attempt to bridge the gap between *in situ* and *ex situ* research by collecting data from three locations: the field, the laboratory, and zoos. We use the results of the field work to design the laboratory experiments and then consider the results from the laboratory and zoo together. We observed infant carrying preferences in humans under field conditions and measured gait parameters during loaded and unloaded walking in humans in the laboratory. The results showed that human males and females prefer carrying infants asymmetrically, and carrying a load has no significant influence on spatiotemporal gait parameters. We measured gait parameters in nonhuman apes during free-ranging loaded and unloaded locomotion in zoos. The results showed that carrying a load had little influence on gait parameters.

Keywords Hominins • Infant carrying • Locomotion • Spatiotemporal parameters

Abbreviations

Bi	bipedal
BM	body mass
C	contralateral
DG	diagonality
D	dorsal

W. Sellers (✉)
University of Manchester, UK
e-mail: wsellers@man.ac.uk

K. D'Aoû t and E.E. Vereecke (eds.), *Primate Locomotion: Linking Field and Laboratory Research*, Developments in Primatology: Progress and Prospects,
DOI 10.1007/978-1-4419-1420-0_7,

M	mouth
I	ipsilateral
RHS	right hand side
Tri	tripedal
V	ventral
Quad	quadrupedal

Introduction

Bipedalism has long been recognized as one of the earliest and perhaps the most influential structural and behavioral adaptation that distinguishes modern humans from our primate relatives. Numerous hypotheses have been proposed to explain the adaptive advantages and selective pressures that gave rise to the adoption of habitual bipedalism in early hominin development, yet the origins of the evolutionary transition to bipedality are poorly understood. Hypotheses for the evolution of bipedality include various foraging strategies, tool use, thermoregulation, and predator avoidance. A common element in several of these hypotheses is the requirement for the hands to be free (Hewes 1961; Videan and McGrew 2002), e.g., the postural feeding hypothesis necessitates the hands to be free to pick fruit from high branches (Hunt 1994), and the provisioning model requires pair-bonded males to carry food back to dependent females (Lovejoy 1981). Hand use is required for tool transportation (Gould and Lewontin 1979; Videan and McGrew 2002), central place foraging, and infant carriage. Certainly early hominins would have needed to carry their dependent infants actively after a reduction in prehensile capacity in the hominin foot, which would have prevented infants from clinging for long durations in the way that infant apes and monkeys necessarily do to be transported safely.

Most nonhuman primates spend their days in stable groups, and it is the group as a whole that ranges, foraging through the environment. This is particularly apparent in the commonly used models for ancestral hominid behavior: chimpanzees and baboons (Elton 2006). Central place foraging where individuals collect food from remote locations and bring them back to central location is commonly found in birds and nesting mammals (Stephens and Krebs 1986) but very rarely seen in nonhuman anthropoid primates (Chapman et al. 1989). A change in foraging strategy would have resulted in a large shift in the basic daily activity pattern, particularly in terms of day range and time spent carrying, and thus would have potentially been of great importance to the energy budgets of early hominins. The magnitude of the benefits of this major change in foraging strategy, coupled with greater reliance on bipedalism and carrying, could have given the necessary impetus that resulted in anatomical adaptation for bipedality.

Modern human carrying has been extensively analyzed in both an ergonomic and comparative ethnographic context (Goldman and Iampietro 1962; Soule and Goldman 1969; Keren et al. 1981; Pierrynowsky et al. 1981; Cook and Neumann 1987; Holewijn 1990; Bhambhani and Maikala 2000; Wall-Scheffler et al. 2007; Watson et al. 2008), but there are few comprehensive, quantitative data on the frequencies of different methods of infant carrying (Bruser 1981). This information not only can give insight into the methods of infant carrying possibly used by early humans but is also essential for the experimental design of evolutionary load-carrying studies.

In the first of the three studies described here we observed how modern humans carry infants without the aid of adjuncts like slings or baby backpacks. Most laboratory-based load carriage studies conducted on humans have used a single method of load carriage, usually backpacks, and have concentrated on the energetic cost of load carrying (Soule and Goldman 1969; Cymerman et al. 1981; Legg 1985; Holewijn 1990; Quesada et al. 2000; Abe et al. 2004; Stuempfle et al. 2004; Bastien et al. 2005). Few load-carrying studies report gait parameters, and in those that have, there are conflicting results on the effect of load carriage. Peak ground reaction forces increase in proportion to the load carried, but it has been found that lower limb loading and unloading rates do not rise in relation to load (Tilbury-Davis and Hooper 1999). Ghori and Luckwill (1985) found that there was a significant decrease in stance phase duration with loads of 20% body mass (BM) carried in a single hand. This is surprising because the time averaged vertical force must equal the body weight, so that any decrease in stance phase duration must lead to a concomitant increase in limb force, which would increase energetic cost and could potentially lead to injury. No significant differences were found in the stance phase with back-loads of up to 50% BM (Ghori and Luckwill 1985). Taylor et al. (1980) found that in rats, dogs, humans, and horses there was no change in stride frequency or contact time when carrying loads between 7% and 27% body mass on the back.

In the second of the three studies described here, we collected human gait parameters in the laboratory to detect differences between loaded and unloaded locomotion. We aimed to identify changes resulting from large alterations in the way that a mass is carried, particularly the modes of carrying thought to have been used by early hominins. We chose a range of items (a weighted vest, dumbbells, and a mannequin infant) as loads. Each load represented a different method of load carriage. We used the weighted vest as a method of evenly distributing the mass of the load, and the dumbbells provided an excellent method of carrying a large mass in one hand. The mannequin infant was included because infant carrying would have become an important challenge for early hominins once pedal grasping capabilities were lost. Carrying an infant is an increasingly costly behavior during the period of nutritional dependence; however, in female baboons, it is estimated to be more energetically efficient to carry offspring than to allow the inefficient infant to travel independently (Altmann and Samuels 1992). Altmann and Samuels (1992) found that the faster a female baboon traveled, the more likely she was to carry her infant. This highlights that the decision to carry is not necessarily influenced by instantaneous energetics but also by the potential to enhance the mother's future reproductive success. In modern humans, and presumably other infant-carrying mammals, there is a theoretical break-even point, beyond which mothers do not carry their children. It has been argued that to maximize fitness, the cost of carrying must be less than the combined cost of the child and its mother walking independently (Kramer 1998). This argument is based on the assumption that energetic economy is paramount and does not account for the reality that energetically unfavorable carrying will occur in situations where speed or safety are important (Altmann and Samuels 1992).

In the third of the three experiments described here, we collected gait parameter data from nonhuman great apes at zoos during loaded and unloaded locomotion. Carrying ability has been studied in nonhuman primates in relation to tool use (Whiten et al. 1999; Ottoni and Mannu 2001; Whiten and Boesch 2001; Mercader et al. 2002), however, no attempt has been made to research the link between primate load carrying and gait parameters. Among primates the great apes are the most morphologically similar species to early humans, and in the absence of experimental kinematic data collection from early humans, apes are preferred as a referential model. We do not suggest that they demonstrate early hominin kinematics, and we appreciate that great ape locomotion is just as much the result of natural selection as is human bipedality. However, studies on load carrying in the nonhuman great apes do provide information on gait selection, footfall sequences, and mode of load carriage in habitual quadrupeds.

The experiments are divided into three sections that comprise field-, lab-, and zoo-based experimental work.

Human Carrying Observations

We collected observational data from adults carrying infants to determine which is the most commonly used method of infant carrying without the aid of adjuncts like slings or baby backpacks. There are many different methods of carrying infants, but surprisingly little information is available on the frequency of different methods of infant carrying (Bruser 1981). Many studies mention the use of slings and other devices for carrying infants (Denham 1974; Lawrence et al. 1985; Wall-Scheffler et al. 2007), but as it is unlikely that the earliest hominins had the technological capacity to manufacture a passive carrying device like a sling, we concentrated on methods that required active transport.

Human Gait Parameters

We then used the "field observations" in the experimental design of the human lab experiment. We collected human gait parameters, duty factor, stride length, and stride frequency during steady speed treadmill walking under laboratory conditions during loaded and unloaded locomotion to determine the influence of load. Duty factor is the percentage of the entire stride for which the limb is load bearing. A high duty factor indicates a long stance time. A long stance time increases the time course of force development and so reduces peak vertical limb force when the foot is on the ground. Limb force determines the load experienced by musculoskeletal structures and the cost of locomotion (Biewener 1990; Kram and Taylor 1990). We tested the hypothesis that the addition of load leads to increased foot contact time during the stance phase in order to minimise limb forces and thus reduce the energetic cost of carrying loads. Such a mechanism to reduce energetic costs would have facilitated bipedal load carrying were it present in early hominins.

Ape Gait Parameters

We investigated how habitually quadrupedal apes approached the challenges of carrying different loads and how carrying these loads influenced their gait parameters. We collected ape gait parameters, duty factor, stride length, speed, and stride frequency during free-ranging loaded and unloaded locomotion. The maximum force experienced by an individual limb during ground contact has been shown to constrain maximum running speed in humans and has been proposed as a trigger for gait transition in both bipeds and quadrupeds (Nilsson and Thorstensson 1989; Farley and Taylor 1991; Hreljac 1993; Weyand et al. 2001; Wickler et al. 2003). We may therefore expect the addition of load to increase limb force and as a result trigger a change in footfall sequence. We tested the hypothesis that load carrying leads to a change in footfall sequence by presenting the gait parameter data in Hildebrand gait diagrams (Hildebrand 1965, 1980). These diagrams express footfalls as points on a bivariate plot. The two dimensions on the bivariate plot are duty factor and diagonality. Diagonality is defined as the percentage of the stride interval that the footfall of a forefoot lags behind the footfall of the hind foot on the same side of the body.

In the three studies described here we are interested in determining the relationship between load carrying and gait parameters in the full range of human and nonhuman great ape species. Studying the differences in gait parameters between loaded and unloaded locomotion enables us to determine whether certain modes of carrying perturb gait more than others. Assuming that large perturbations are more energetically costly than steady locomotion, kinematic data collection could provide valuable insights into the energetic costs of carrying loads.

By identifying the methods by which modern humans choose to carry infants we can gain an insight into the possible methods of infant carrying employed by early hominins and thus assess the feasibility of infant carrying as a precursor to the evolution of bipedality.

Methods

Human Carrying Observations

We collected observational data on infant carrying without adjuncts at UK zoos and shopping centers. We collected frequency data manually using a tick sheet when a particular mode of carrying was observed. We identified the sex of the carrier, the approximate age of juveniles (<1, 1–3, 3–5, 5–7, >7 years), and the method of carrying (ventral facing forward, ventral facing backwards, left or right hip, left or right side above hip, back, shoulders, cradling). In all instances, the individuals were unaware that they were being observed and were walking normally. We collected data only from walking sequences where five simultaneous strides were taken while using a single mode of carrying. To identify differences between the observed frequencies of carrying modes, we performed χ^2 tests.

A χ^2 test for associations was initially performed to identify differences between male and female carrying preferences. We were primarily interested in female carrying preferences for use in the experimental laboratory work and so a χ^2 test for differences was then performed on the female preference data only. We are interested in females because we consider early hominin females more likely to have been infant carriers than males, and the morphology of modern human females is more representative of early hominin fossil remains than the morphology of a modern human male.

We then performed a χ^2 test for associations to identify alterations in mode of carrying between the different age groups of children.

Human Gait Parameters

We collected gait parameter data during speed-constant, level walking in seven physically fit females of child-bearing age. The loads all weighed 10 kg and the average mass carried was about 18% of the participants' mean BM. The loads were an adjustable, weighted vest (Reebok Ironwear), a 5-kg dumbbell in each hand, a 10-kg dumbbell in a single hand, and a weighted emergency-rescue-training mannequin (Ruth Lee, Model RL10, www.ruthlee.co.uk). The weighted vest contains flexible rubber weights in small pockets distributed evenly over the front and back. The mass was adjusted further by adding strips of malleable lead to the pockets. The masses of the dumbbells and the mannequin were adjusted by adding heavy gauge copper wire. The handheld masses were cast-iron dumbbell plates with threaded bars and collars. The 10-kg dumbbell was carried in a single hand, but participants were allowed to swap hands during the trial. They were not, however, allowed to support the mass with two hands at any time. The emergency-rescue-training mannequin is manufactured with approximately realistic mass distribution and size. The load mass chosen corresponds to the weight and size of a toddler. The mannequin wore an all-in-one suit that ensured it was comfortable to carry. Based on the results of the human carrying observational experiment, the mannequin was carried on the hip selected by the participant and remained on the same hip for the duration of data collection. Although there is no direct evidence to suggest that early hominins carried infants asymmetrically, this position was chosen as a method popularly used by modern females. Each walking task was carried out on a treadmill (Vison T9250, Wisconsin) at a constant speed (3.7 km/h, 1.028 m/s) chosen as the speed at which the net cost of locomotion for human walking is minimum (Sellers et al. 2005). We collected gait parameter data using 10 motion analysis cameras (ProReflex, Qualysis, Sweden) and passive retroreflective markers. We collected data at 240 Hz for a 10-s period during each carrying condition. We randomized the order of carrying conditions and gave participants a minimum of 5 min to become accustomed to walking on the treadmill. Data collection commenced after this period of acclimation and continued for 10 s for each condition. Participants rested for a minimum of 5 min between tasks. We performed the analysis in Visual 3D (C-Motion Version 3.79). Anatomical markers were placed on repeatable bony landmarks at the proximal and distal ends of the shank segment. Markers were also placed on the calcaneus, 1st, 2nd, and 5th metatarsals to define the foot segment. Heel strike was defined as the lowest point in the trajectory of the calcaneus. A stride was defined as heel strike to heel strike, and this was used to calculate duty factor, stride length and stride frequency for each stride. We used six strides per participant for statistical analysis.

We statistically tested the variables across all modes of load carrying for a general effect of load using a one way repeated measures analysis of variance. Where a significant effect of load was found a Bonferroni *post hoc* multiple comparison test (SPSS 13.0) was used to identify where the effect lay.

We performed paired *t*-tests on data from the right and left sides of the body identified no lateral asymmetry within participants, and so used data from the right hand side (RHS) of the body for analysis during unloaded walking, walking with the weighted vest, and with 5-kg dumbbells carried in both hands. We used data from the left and right sides of the body to give information on the ipsilateral and contralateral sides when the mannequin and 10-kg dumbbells were carried.

Ape Gait Parameters

We collected kinematic data from common chimpanzees (*Pan troglodytes*), bonobos (*Pan paniscus*), Western lowland gorillas (*Gorilla gorilla gorilla*), and Bornean and Sumatran orang-utans (*Pongo pygmaeus* and *Pongo abelii*) from four UK zoos (Chester Zoo, Twycross Zoo, Porte Lympne Wild Animal Park, and Paignton Zoo) and one Belgian zoo (Wild Animal Park Planckendael) under free-ranging conditions (Table 7.1). We placed items in the enclosures whenever possible to encourage carrying behavior. The light loads (<1 kg) available for carrying were primarily fruit but occasionally included blocks of frozen juice, boxes, bottles, sacks, and leafy branches. When infants were present they were carried by mature individuals. At Paignton, Porte Lympne, and Planckendael, the gorillas and bonobos were given wooden logs with drilled holes filled with honey, peanut butter, and raisins (mass range 5.0–7.1 kg). The apes poked thin sticks in the holes to access the food. The logs were the most effective at encouraging carrying behavior and constituted excellent enrichment items. The honey soaked into the wood and so the apes continued sucking, biting and picking at the wood long after other enrichment items

Table 7.1 Locations and focal individuals

Zoo	Species	No. of individuals	Mass range (kg)	Number of hind limb strides analyzed
Chester	Common chimpanzee	30 individuals	50–70	33
Chester	Sumatran orang-utan	4 adults 2 infants	50–90	24
Planckendael	Bonobo	6 adults 3 infants	10–45	145
Twycross	Bornean orang-utan	3 adults 1 infant	10–120	7
Twycross	Common chimpanzee	5 adults	50–70	42
Twycross	Bonobo	10 adults 2 infants	10–60	10
Twycross	Western lowland gorilla	3 adults	150–200	17
Paignton	Western lowland gorilla	2 adults	145–149	39
Porte Lympne	Western lowland gorilla	3 adults	150–200	25

had lost their appeal. In addition to this, the logs often filled with ants once they had been left on the ground for a short time. The apes used sticks to "fish" inside the holes which also helped to maintain interest and encouraged carrying behavior. At Planckendael, the bonobos were also given ice filled basketballs (mass 1–3 kg) and sealed drain pipes filled with gravel and walnuts (mass 2.5–3 kg). In compliance with the zoos' ethical guidelines, there was no physical contact with the apes during the study; they did not undergo any training and were unmarked.

We videotaped the apes in lateral view whilst walking and running in their enclosures. The subjects could adopt their preferred gait at any time. We collected gait parameter data via a standard video camera (Sony, NTSC, 59.96 fields per second). We measured reference points in the enclosures at Planckendael and Twycross to enable calculation of stride length and speed. At Planckendael, a 1 m × 2 m grid was marked with chalk on the concrete ground (Fig. 7.1). We selected walking sequences from the video footage to determine duty factor, stride frequency, stride length, and stride time. Owing to long grass sometimes covering the feet, we digitized the center of rotation of the ankle joint at the beginning and end of a stride and used it to calculate stride length. We calculated stride length only when the ape was walking directly next to measured reference points and was defined as the distance between foot contact to the subsequent foot contact in the same limb. We performed digitization using Didge (2.2.0) (http://biology.creighton.edu/faculty/cullum/Didge/index.html). Selected sequences contained walking that was perpendicular to the camera, over reasonably flat ground and at a steady speed. We determined the latter by digitizing the position of the center of rotation of the hip and plotting the x coordinate against time. We fitted a linear regression line to the displacement time data. A straight line plot indicated constant velocity, the data were discarded if the R^2 value fell below 0.97 (D'Août et al. 2002). To determine stride frequency and duty factor, the video was played back field by field to enable stance time and swing time, determined from the time plots between subsequent foot contacts, to be identified in the hind limbs. We defined stride frequency as 1/stride time. We defined duty factor as the percentage of an entire stride cycle for which a limb was in contact with the ground. We calculated diagonality for all

Fig. 7.1 Data collection area in the bonobo enclosure at Wild Animal Park Planckendael

quadrupedal and tripedal locomotion. We defined diagonality (DG) as the percentage of the cycle period by which the left hind footfall precedes the left fore footfall.

We performed statistical analyses on the walking trials (defined as strides with a duty factor of 50% or more). A one-way analysis of variance (ANOVA) (SPSS 13.0) was used to determine whether there was a significant change in gait parameters between the quadrupedal unloaded walking condition and any other condition in bonobos, gorillas, chimpanzees and orang-utans. Where we found a significant effect of load, we used a Bonferroni *post hoc* test to determine which condition differed from the quadrupedal unloaded condition.

Data Normalization

To account for the size differences between the different humans and apes, we normalized the data using hind limb length (Hof 1996; Aerts et al. 2000). Shank length has previously been shown to be less susceptible to digitization errors (Aerts at al. 2000) and was measured from knee to heel. Dimensionless walking speed = $v/(\sqrt{(gl)})$, dimensionless frequency = $f/\sqrt{(g/l)}$, and dimensionless stride length = stride length/l, where v = walking speed (m/s), g = gravitational acceleration (9.81 m/s^2), l = shank length (m). We measured the shank lengths of the bonobos from Planckendael (range 19–26 cm) directly from the video data, as the chalk grid on the ground enabled length measurements. For each individual, we measured the shank length and averaged it over several unloaded walking trials. We used these individual leg measurements to normalize the bonobo data. In the gorilla, we measured shank lengths in two of the gorillas and used the average of these (24 cm) to normalize all strides. Lower limb length was not available from the video data for chimpanzees, and so we took lower hind limb lengths (28 cm) as an average of values from Schoonaert et al. (2007). In the orang-utan hind limb length was unavailable from the video data, so we took lower limb length (25 cm) as an average of values from Isler et al. (2006).

We measured human shank lengths were measured for each participant and normalized the data for each participant by its own leg length (range 34–39 cm).

Results

Human Carrying Observations

We recorded 155 individual observations of infant carrying in males and females during normal walking using one of the seven different carrying methods. There was a significant difference in the selection of infant carrying mode between males and females: χ^2 (6) = 41.290, $p < 0.001$. The majority of females (53%) chose to carry infants on the hip, whereas the highest proportion of males (44%) chose to carry infants on one side but above the hip (Fig. 7.2). A separate χ^2 test to determine the frequency of a particular type of infant carrying in females found a significant effect of carrying mode: χ^2 (5) = 61.102, $p < 0.001$, indicating that females prefer to carry infants on the hip significantly more than other any other mode of carriage. Males often carried infants on the shoulders, but females tended to avoid carrying infants on the shoulders altogether.

We were particularly interested in physically fit females of child bearing age as we consider this group most likely to resemble the early hominins faced with the challenges of infant carrying. We performed a χ^2 test on the female data to determine whether the age of the infant influenced the manner in which it is carried.

There was no association between the age of the infant and carrying method: χ^2 (15) = 16.156, $p = 0.372$. This indicates that regardless of age (and therefore weight of the infant) females prefer to carry asymmetrically (Fig. 7.3).

Human Gait Parameters

In humans (Table 7.2), there were no significant differences in duty factor, dimensionless frequency, or dimensionless stride length between unloaded walking and the carrying conditions. The absolute values of stride frequency and stride length are shown in Table 7.2. We collected the human data at constant speed during treadmill

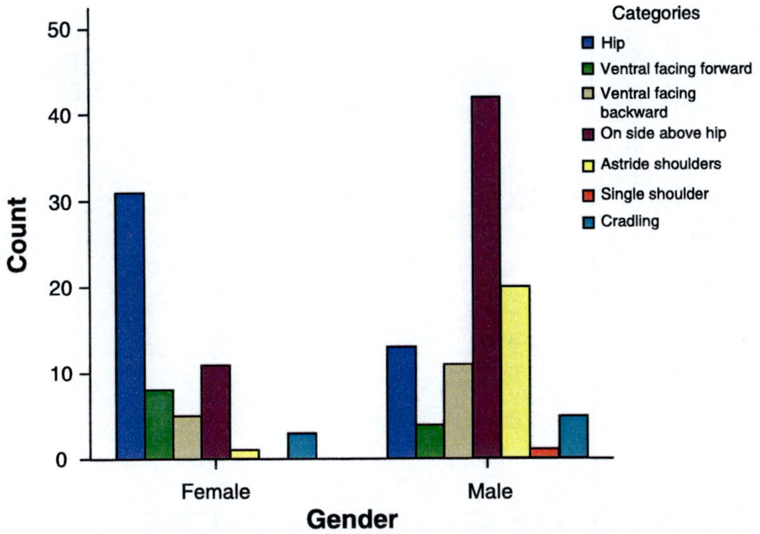

Fig. 7.2 Frequencies of infant carrying modes in males and females. The majority of females preferred to carrying infants on the hip. The majority of males preferred to carry infants on one side above the hip

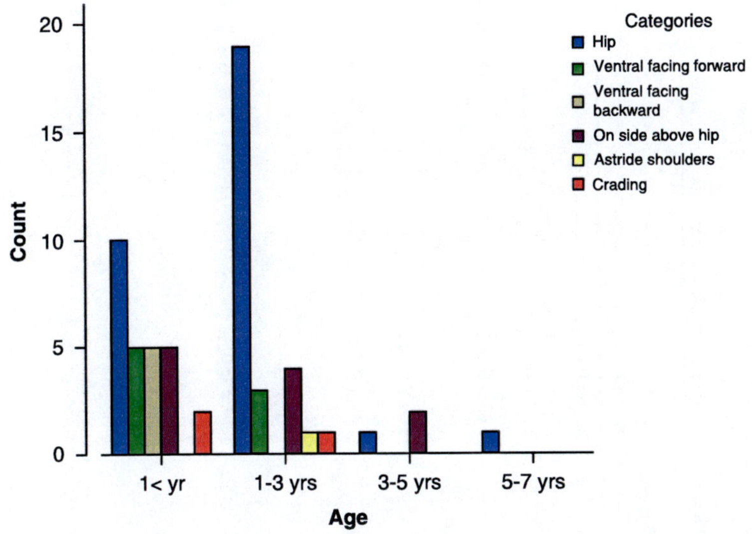

Fig. 7.3 Frequencies of infant age related carrying in females. Regardless of the age of the infants females preferred to carry asymmetrically

Table 7.2 Human gait parameters showing values for duty factor, dimensionless and absolute stride frequency, and stride length ± SE

Human	UL	Vest	5-kg dumbells	10-kg dumbbell ipsilateral hand	10-kg dumbbell contralat hand	Mannequin ipsilateral hip	Mannequin contralat hip
Duty factor	0.57 ± 0.01	0.56 ± 0.02	0.56 ± 0.01	0.56 ± 0.02	0.55 ± 0.01	0.56 ± 0.02	0.57 ± 0.01
n	6	6	6	6	6	6	6
p		ns	ns	ns	ns	ns	ns
Dimensionless stride frequency	0.17 ± 0.003	0.18 ± 0.004	0.18 ± 0.005	0.18 ± 0.005	0.18 ± 0.005	0.17 ± 0.007	0.18 ± 0.003
n	6	6	6	6	6	6	6
p		ns	ns	ns	ns	ns	ns
Dimensionless stride length	3.15 ± 0.08	3.13 ± 0.07	3.08 ± 0.08	3.08 ± 0.06	3.10 ± 0.07	3.20 ± 0.14	3.11 ± 0.07
n	6	6	6	6	6	6	6
p		ns	ns	ns	ns	ns	ns
Stride length (m)	1.14 ± 0.03	1.13 ± 0.03	0.93 ± 0.03	1.11 ± 0.03	1.12 ± 0.03	1.15 ± 0.05	1.12 ± 0.03
n	6	6	6	6	6	6	6
Stride frequency (Hz)	0.91 ± 0.02	0.91 ± 0.02	1.11 ± 0.03	0.93 ± 0.03	0.92 ± 0.03	0.90 ± 0.04	0.92 ± 0.02
n	6	6	6	6	6	6	6

UL = unloaded, n = number of strides analyzed, ns = not significant locomotion, so speed is not presented here. There were no significant differences in duty factor after the addition of load, which does not support our hypothesis that duty factor increases in order to minimise limb force during the stance phase.

Ape Gait Parameters

The observations on the manner in which apes chose to carry loads were noted here for interest but were not statistically tested. Loads were carried by the apes in many combinations of ways, dorsally, ventrally, and using the fore- and hind limbs during bipedal, tripedal, or quadrupedal locomotion. Initial observations indicated that all apes appeared to be able to carry light loads quadrupedally without any obvious perturbation of the normal gait cycle. Pieces of fruit or leaves were easily held tucked into the palm of a hand or carried in the mouth. Infants of all species were generally carried ventrally, but we also observed dorsal carrying. All four species of great ape occasionally employed tripedal carrying when quadrupedal carrying was not possible due to the dimensions or weight of the load. With the exception of one particular gorilla from Port Lympne Zoo that regularly walked bipedally, it appeared that only bonobos regularly chose to carry bipedally.

There were few differences in gait parameters between loaded and unloaded locomotion. In several cases there were not sufficient data to perform statistics but in these situations the values did not represent a great departure from the unloaded quadrupedal condition.

Bonobos

In the bonobo hind limb (Table 7.3) there was a significant difference in duty factor between the quadrupedal unloaded condition and bipedal locomotion with a load carried in both hands and ventrally. In the hind limb there was a significant difference in dimensionless frequency in both the bipedal locomotion load conditions: (1) bipedal with loads carried in both hands and ventrally and (2) bipedal with a load carried in both hands. In the bonobo hind limb, there were no significant differences in dimensionless speed or dimensionless stride length between quadrupedal unloaded walking and any loaded condition. The absolute values of stride frequency, speed, and stride length are shown in Table 7.3.

Gorillas

In the gorilla hind limb (Table 7.4) there was a significant difference in duty factor between the unloaded quadrupedal condition and tripedal walking with a load in the ipsilateral forelimb. There were no significant differences in dimensionless frequency in the gorilla hind limb with load. Data for dimensionless speed and dimensionless

Table 7.3 Bonobo dimensionless gait parameters for the hind limb showing values for duty factor, dimensionless and absolute frequency, speed, and stride length ± SE

Bonobo	Quad UL	Bi I+C fore + V	Bi I+C fore	Quad V	Quad I fore + V/M/D	Tri C fore	Tri C fore + V/M/D	Tri I fore	Tri I fore + V
Duty factor	60.53 ± 1.16	67.88 ± 0.95	65.74 ± 1.65	61.16 ± 1.13	66.65 ± 4.50	63.54 ± 1.16	65.34 ± 1.11	62.49 ± 1.98	65.59 ± 1.13
n	30	19	10	17	4	20	17	13	20
p		<0.001	ns	ns	ns	ns	ns	ns	ns
Dimensionless frequency	0.12 ± 0.01	0.17 ± 0.01	0.19 ± 0.01	0.15 ± 0.01	0.10 ± 0.02	0.15 ± 0.01	0.14 ± 0.01	0.15 ± 0.01	0.13 ± 0.01
n	24	18	10	14	4	20	17	13	20
p		0.015	<0.001	ns	ns	ns	ns	ns	ns
Dimensionless speed	0.43 ± 0.05	0.28	0.44 ± 0.11	0.56 ± 0.04	—	0.58 ± 0.10	0.52 ± 0.07	0.47 ± 0.08	0.48 ± 0.01
n	11	1	4	5	—	4	5	3	2
p		—	ns	ns	—	ns	ns	Ns	—
Dimensionless stride length	4.29 ± 0.34	2.05	2.73 ± 0.58	4.58 ± 0.31	—	4.04 ± 0.24	3.83 ± 0.24	3.47 ± 0.29	4.10 ± 0.02
n	11	1	4	5	—	4	5	3	2
p		—	ns	ns	—	ns	ns	ns	—
Frequency (Hz)	0.80 ± 0.07	1.11 ± 0.09	1.27 ± 0.08	0.96 ± 0.08	0.66 ± 0.14	1.00 ± 0.03	0.90 ± 0.04	1.00 ± 0.06	1.03 ± 0.00
n	30	8	10	14	4	20	17	13	2
Speed (m/s)	0.65 ± 0.07	—	0.67 ± 0.17	0.83 ± 0.07	—	0.84 ± 0.17	0.77 ± 0.10	0.70 ± 0.12	0.75 ± 0.01
n	11	—	4	5	—	4	5	3	2
Stride length (m)	1.03 ± 0.06	—	0.63 ± 0.12	1.05 ± 0.07	—	0.86 ± 0.09	0.86 ± 0.05	0.80 ± 0.07	0.81 ± 0.03
n	11	—	4	5	—	4	5	3	20

Quad = quadrupedal; Tri = tripedal; Bi = bipedal, UL = unloaded; M = mouth; I = ipsilateral; C = contralateral; D = dorsal; V = ventral. The symmetrical load carrying conditions were grouped (V/D/M) to enable sufficient data for statistical analysis. ns indicates not significantly different from the Quadrupedal UL condition.

Table 7.4 Gorilla hind limb gait parameters showing values for duty factor, dimensionless and absolute stride frequency, speed, and stride length ± SE

Gorilla	Quad UL	Quad M/D/V	Tri I fore	Bi I fore
Duty factor	63.44 ± 0.80	60.26 ± 1.32	58.08 ± 3.26	63.34 ± 1.22
n	47	5	5	11
p		ns	0.004	ns
Freq dimensionless	0.47 ± 0.02	0.47 ± 0.01	0.79 ± 0.14	0.58 ± 0.02
n	47	5	6	11
p		ns	ns	ns
Dimensionlessspeed	0.33 ± 0.03	—	—	—
n	5	—	—	—
p		—	—	—
Dimensionless stride length	3.65 ± 0.13	—	—	—
n	5	—	—	—
p		—	—	—
Stride frequency (Hz)	0.72 ± 0.04	0.72 ± 0.02	1.19 ± 0.27	0.92 ± 0.03
n	47	5	5	11
Stride length (m)	0.87 ± 0.03			
n	5			
Speed (m/s)	0.51 ± 0.04			
n	5			

Table key as in Table 7.3.

stride length were available only in the unloaded condition for the gorilla. The absolute values of stride frequency, stride length and speed are shown in Table 7.4.

Chimpanzees

Speed and stride length data were unavailable for the chimpanzee, as it was not possible to measure reference markers inside the enclosures. In the chimpanzee hind limb (Table 7.5) there were no significant differences in duty factor between the quadrupedal unloaded condition and any of the other conditions, including unloaded bipedal walking. There was, however, a significant difference in dimensionless frequency between the quadrupedal unloaded condition and tripedal walking with a load in the contralateral forelimb. There was also a significant difference in dimensionless frequency between quadrupedal unloaded walking and bipedal unloaded walking. The absolute values of stride frequency are shown in Table 7.5.

Orang-utans

Speed and stride length data were unavailable for orang-utans. In the orang-utan hind limb (Table 7.6) there were no significant differences in duty factor between

Table 7.5 Chimpanzee hind limb gait parameters showing values for duty factor and dimensionless and absolute stride frequency ± SE

Chimpanzee	Quad UL	Quad M/V/D	Quad C fore	Tri C fore	Bi I fore	Bi UL
Duty factor	62.07 ± 1.12	65.85 ± 2.06	59.51 ± 0.94	67.77 ± 1.01	69.80 ± 2.62	63.53 ± 2.45
n	36.00	11.00	3.00	10.00	3.00	7.00
p		Ns	ns	ns	Ns	Ns
Dimensionless frequency	0.13 ± 0.004	0.13 ± 0.004	0.12 ± 0.002	0.19 ± 0.023	0.14 ± 0.018	0.17 ± 0.019
n	36.00	11.00	3.00	10.00	3.00	7.00
p		ns	ns	<0.001	ns	0.024
Frequency	0.77 ± 0.02	0.79 ± 0.02	0.70 ± 0.01	1.10 ± 0.14	0.83 ± 0.11	1.03 ± 0.11
n	36	11	3	10	3	7

Table key as in Table 7.3.

Table 7.6 Orang-utan hind limb gait parameters showing values for duty factor, dimensionless and absolute stride frequency ± SE

Orangutan	Quad UL	Quad M/V/D	Quad C hind	Quad I hind	Tri I fore + D
Duty factor	60.74 ± 0.57	64.15 ± 1.67	57.90 ± 1.85	59.22 ± 1.80	69.45 ± 1.70
n	3	15	3	6	4
p		Ns	ns	ns	ns
Frequency dimensionless	0.09 ± 0.006	0.09 ± 0.003	0.09 ± 0.007	0.09 ± 0.005	0.12 ± 0.005
n	3	15	3	6	4
p		ns	ns	ns	0.01
Frequency	0.55 ± 0.04	0.55 ± 0.02	0.54 ± 0.04	0.58 ± 0.03	0.76 ± 0.03
n	3	15	3	6	4

Table key as in Table 7.3.

the quadrupedal unloaded walking and any of the load conditions. There was a difference in dimensionless frequency between the unloaded quad condition and tripedal walking with an infant dorsal and load in the ipsilateral forelimb. The absolute values of stride frequency are shown in Table 7.6b.

Footfall Sequences

Calculation of duty factor and diagonality enabled production of Hildebrand's bivariate plot (Fig. 7.4), which shows all strides of unloaded quadrupedal locomotion from bonobos, gorillas, chimpanzees, and orang-utans as well as a range of additional, generally smaller primate data taken from Cartmill et al. (2002). This plot is designed to identify changes in footfall sequence between species (Fig. 7.5). Most of the data collected in these studies was at a walking gait, although some (with duty factors below 50%) was at a run. Statistical analysis has been performed on the walking data only. Our data fit well with the other primate data although an ANOVA with Bonferroni *post hoc* tests showed that there are some differences between the species. The other primate data are significantly different from all our ape data ($p < 0.005$). The bonobo data are significantly different from all other species ($p < 0.05$), but diagonalities in the chimpanzee and gorilla are not significantly different from each other.

At the walk, bonobos generally used a walking trot and occasionally an intermediate gait between walking trot and diagonal sequence walk. Generally gorillas

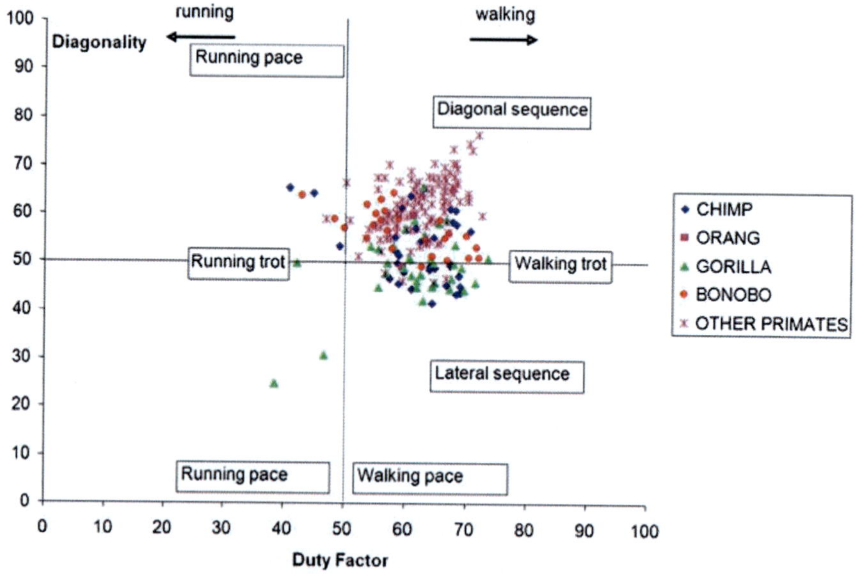

Fig. 7.4 Hildebrand diagram of quadrupedal unloaded strides for different primates

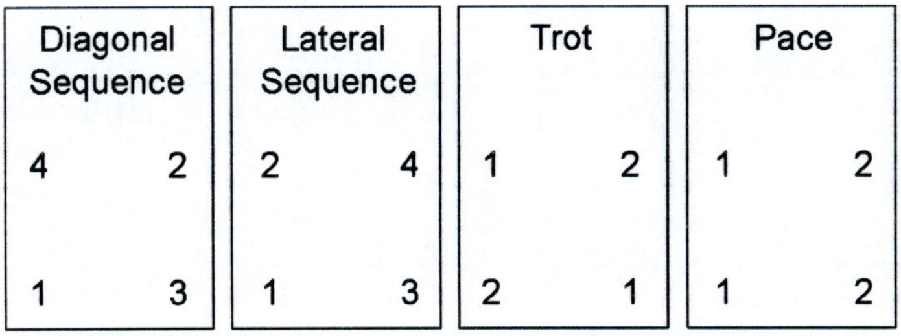

Fig. 7.5 Footfall sequences of different gait types

used a walking trot. We also observed intermediate gaits between a walking trot and diagonal sequence walk. The gorillas appeared to have lower diagonality values than the primate data from Cartmill et al. (2002), indicating that they use walking trots or lateral sequences in preference to diagonal sequences. The scatter of the chimpanzee data is centered on a diagonality of 50, indicating that when walking the chimps used a walking trot. In the two sequences of unloaded walking collected from the orang-utans, they also used a walking trot. We collected some running data from chimps, gorillas, and bonobos. Of the running, the gorillas paced or used an intermediate gait between a pace and a running trot. The chimps and bonobos utilized a running trot.

Bonobos

The strides of bonobo data (Fig. 7.6) under loaded conditions are largely clustered around the unloaded quadrupedal data and show little departure from the walking trot. None of the loaded conditions where sufficient data was available for statistically analyses elicited a significant departure in diagonality from the unloaded quadrupedal condition.

Gorillas

We collected a greater number of run sequences from gorillas than from the other apes. Tripedal locomotion with a load carried in the contralateral forelimb elicited, in one case, a change from a walking trot to a lateral sequence (Fig. 7.7). Statistics were performed on walking data only. In the gorilla there was no significant difference between unloaded walking and walking with a symmetrical load (mouth + dorsal, M+D).

Chimpanzees

Load carriage appears to have little influence on gait selection in the chimpanzee (Fig. 7.8). The quadrupedal ipsilateral + mouth (I+M) condition has a greater diagonality

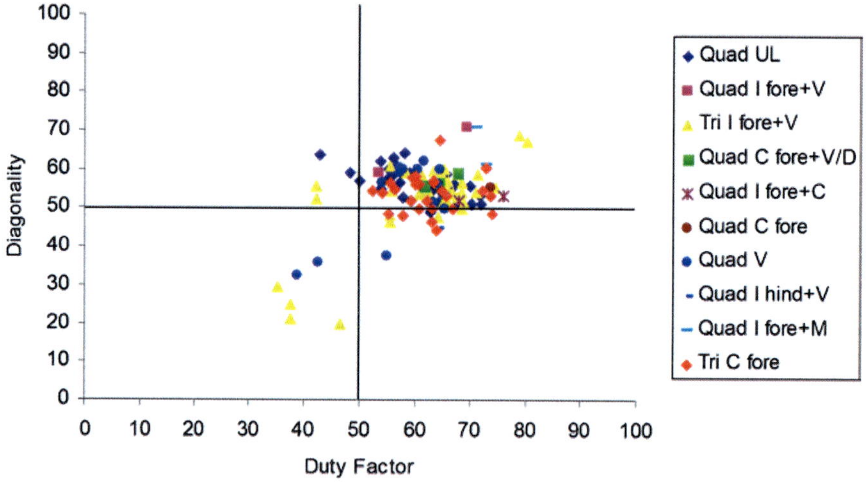

Fig. 7.6 Hildebrand diagram of all bonobo gaits

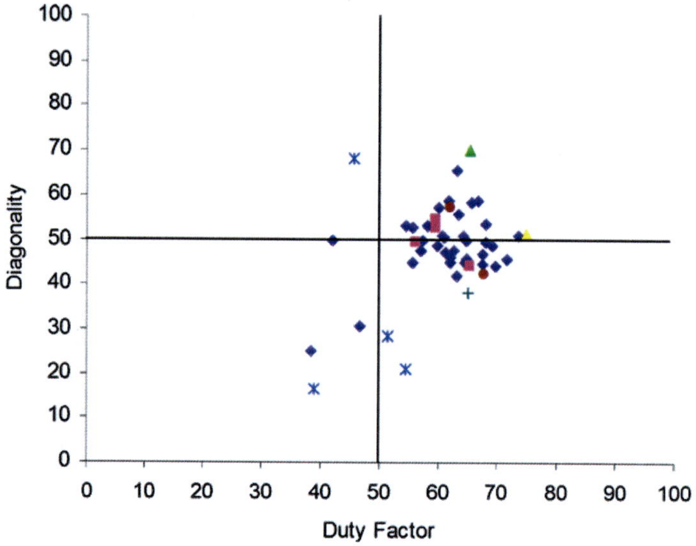

Fig. 7.7 Hildebrand diagram of all gorilla gaits

than some of the other data, indicating a diagonal sequence walk as opposed to a walking trot, but these data represent only two strides taken from a single individual which limits the reliability and prevents statistical analysis. Chimpanzees continued to use a walking trot footfall sequence when walking tripedally with a load. There were no significant differences in diagonality between the unloaded quadrupedal walking condition and quadrupedal walking with a symmetrical load (mouth + ventral + dorsal, M/V/D) load or with tripedal walking with a load in the contralateral forelimb.

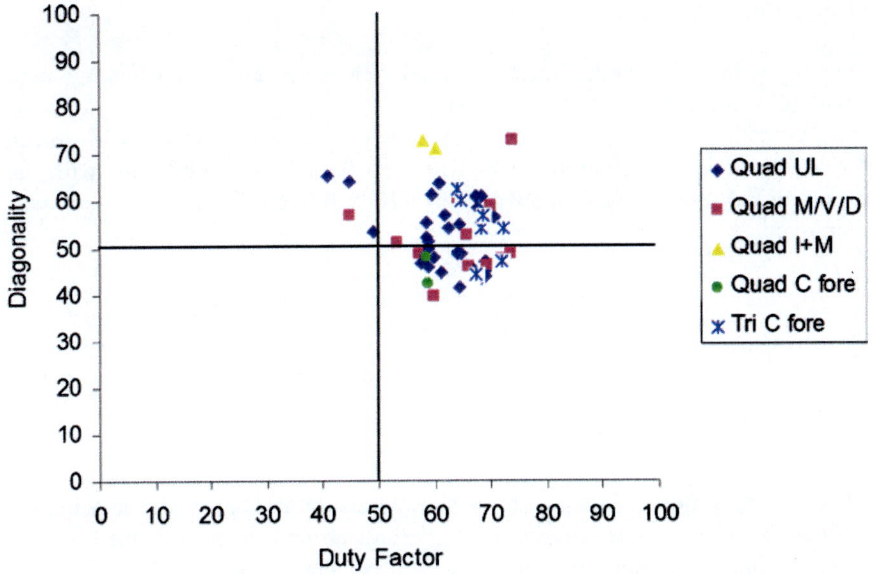

Fig. 7.8 Hildebrand diagram of all chimpanzee gaits

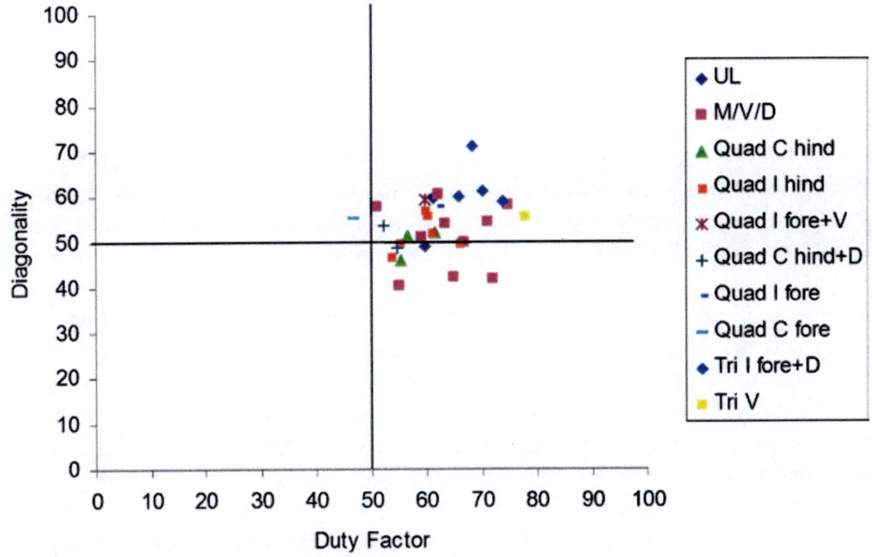

Fig. 7.9 Hildebrand diagram all orang-utan gaits

Orang-utans

Only two strides of unloaded quadrupedal walking were available in the orang-utan, so we did not perform statistical analyses. However, the diagram (Fig. 7.9) shows that load carriage has little influence on footfall sequence selection in the orang-utan. A walking trot gait was used for all load conditions with the exception of the tripedal sequences where a diagonal sequence walk was employed.

Discussion

Human Observations

Observations of human infant carrying showed that females prefer to carry infants on the hip and males also favor carrying infants on one side, above the hip. The most striking point about this result is that it has previously been found that the energetic cost of carrying infants on a hip is significantly greater that the cost of carrying an equivalent load evenly spread across the body (Watson et al. 2008). So, according to these studies, both males and females chose to carry infants in the most uneconomical manner. Previous studies on infant carrying have also found lateral preferences in child transport (Bruser 1981). The reasons for carrying infants asymmetrically are not clear, but carrying infants astride the hip is common in many cultures. In females the hip provides a natural "shelf" and the position is thought to provide social and sensory benefits for the infant (Jelliffe 1975). Certainly, carrying an infant on one side enables the infant to see where it is going as well as hold on and maintain its posture to aid the carrier. From the carrier's point of view, the asymmetric position enables either one arm to be free (Bruser 1981), or if necessary, enables the use of both forearms to lift and adjust the position of the infant compared to, e.g., carrying the infant on the back. The back was not a favored method despite the fact that it enables the infant to see where it is going and the symmetrical position would be expected to reduce the energetic cost. The unpopularity of this method may be due to the fact that the infant has to largely support itself as the arms of the carrier are not well adapted for strength when pointing backwards. If early hominins used the same method of infant carrying that we use today then infant carriage would have been an unlikely precursor to the evolution of bipedality owing to the high energetic cost. Certainly, the high energetic cost of carrying would have led to the rapid development of carrying aids; infants today are rarely carried for long distances without the use of a carrying device. We found that males carried infants on their shoulders, but this method is rarely used by females despite the energetic benefits of carrying the load symmetrically and the visual benefits to the child. This may be because many UK females do not have sufficient core strength to correct for the sway associated with having a mass positioned high above their own center of mass or it may be due to a lack of strength to feel comfortable lifting the child onto the shoulders.

Human Gait Parameters

We measured gait parameters during loaded and unloaded treadmill locomotion and found no significant differences in duty factor, stride frequency, or stride length. During the experiment, we performed simultaneous measurements of oxygen consumption to measure the difference in energetic cost between loaded and unloaded locomotion (Watson et al. 2008). Interestingly, we found a significant increase in the metabolic cost of locomotion after the addition of load and also found a significant increase in the cost of locomotion when the loads were carried asymmetrically compared to symmetrically. So, despite significant increases in the energetic cost of locomotion when loads are carried, there are no changes in gait parameters. It is therefore not possible to gain insight into the metabolic costs of load carrying from changes in gait parameters, as suggested in the introduction. The gait parameter data correspond with other kinematic data also collected during the same experiment (Watson et al. 2008) where joint angles did not alter following the addition of any of the loads. Other load-carrying studies have been inconclusive regarding changes in stride parameters. One study found a reduction in stride length when carrying a dummy infant in both arms compared to when the load was carried with one arm swinging or when the load was evenly distributed around the waist (Wall-Scheffler et al. 2007). Another study found that duty factor changed only in some conditions after the addition of load (Ghori and Luckwill 1985). We found no change in duty factor after the addition of load, which does not support our hypothesis that duty factor will increase to maximize stance time and thus minimize limb force and energetic cost. No researchers have reported an increase in duty factor after the addition of load. Assuming that energetic economy was paramount in early human bipeds, the fact that this possible energy-saving mechanism does not appear to exist could suggest that the transition to bipedality was not driven by load carrying and that load-carrying was employed only once bipedality was well established.

Ape Gait Parameters

In general, it was observed that quadrupedal locomotion was used for carrying in preference to other gaits. Tripedal locomotion appeared to be used when the load was too cumbersome to be carried quadrupedally and bipedal locomotion was used only rarely for carrying purposes. Further work, however, is required to confirm these observations. Few differences were observed in gait parameters (duty factor, stride frequency, speed, or stride length) in all species of ape between loaded and unloaded locomotion. The Hildebrand diagrams, generated to indicate changes in footfall sequence, showed no influence of load on gait in any of the apes. This does not support our hypothesis that the addition of load appreciably increases limb force which could trigger a gait transition; again, any changes caused by load carriage on kinematics are relatively small.

The differences that are observed in gait parameters may be due more to the nature of data collection in the non-laboratory environment than to differences in gait parameters between loaded and unloaded conditions. A limitation with length measurements in the field or zoo is that an approximation of length must be made based on markers a known distance apart in the field of view of the camera. If these markers are not in exactly the same plane as the measurements being taken then errors arise. This limitation extends to all length and velocity estimates used, however, the advantage in using dimensionless values for derived linear measurements is that these will be correct as long as the relative dimensions within the field of view are correct. Thus, absolute measures become less important but minimising parallax and ensuring perpendicularity are still essential.

Conclusion

Based on our findings, the most common manner in which human males and females carry infants is on one side, which shows infants are not carried in the most economical manner. Despite a significant increase in the metabolic cost of human locomotion when walking with a load, we found no significant differences in duty factor, stride frequency, or stride length between human loaded and unloaded locomotion. Similarly, in ape locomotion, few differences were observed in duty factor, stride frequency, stride length, or speed between loaded and unloaded gait parameters. The addition of load did not trigger changes in the footfall sequence. The high energetic cost of the infant carrying methods preferred by modern humans and the apparent lack of energy-saving mechanisms in human gait parameters suggests that load carrying would have been too energetically expensive to have been the driving force behind the evolution of bipedal walking in early hominins.

Acknowledgments We thanks the University of Salford for use of their facilities. We thanks Kristiaan D'Aoât, Anmin Liu, and Steve Horton for their help with data collection. We also thanks the keepers at Twycross Zoo, Paignton Zoo, Port Lympne Wild Animal Park, Chester Zoo, and Wild Animal Park Planckendael. This work was funded by NERC grant NE/C520447/1.

References

Abe D, Yanagawa K, Niihata S (2004) Effects of load carriage, load position, and walking speed on energy cost of walking. Appl Ergon 35:329.

Aerts P, Van Damme R, Van Elsacker L, Duchenne V (2000) Spatio-temporal gait characteristics of the hind-limb cycles during voluntary bipedal and quadrupedal walking in bonobos (*Pan paniscus*). Am J Phys Anthropol 111:503–517.

Altmann J, Samuels A (1992) Costs of maternal care: infant-carrying in baboons. Behav Ecol Sociobiol 29:391–398.
Bastien GJ, Willems PA, Schepens B, Heglund NC (2005) Effect of load and speed on the energetic cost of human walking. Eur J Appl Physiol 94:76–83.
Bhambhani Y, Maikala R (2000) Gender differences during treadmill walking with graded loads: biomechanical and physiological comparisons. Eur J Appl Physiol 81:71.
Biewener AA (1990) Biomechanics of mammalian terrestrial locomotion. Science 250:1097–1103.
Bruser E (1981) Child transport in Sri Lanka. Curr Anthropol 22(3):288–290.
Cartmill M, Lemelin P, Schmitt D (2002) Support polygons and symmetrical gaits in mammals. Zool J Linn Soc-Lond 136:401–420.
Chapman C A, Chapman L J, McLaughlin RL (1989) Multiple central place foraging by spider monkeys: travel consequences of using many sleeping sites. Oecologia 79:506–511.
Cook TM, Neumann DA (1987) The effect of load placement on the EMG activity of the low back muscles during load carrying by men and women. Ergonomics 30:1413–1423.
Cymerman A, Pandolf KB, Young AJ, Maher JT (1981) Energy expenditure during load carriage at high altitude. J Appl Physiol 51:14–18.
D'Août K, Aerts P, De Clercq D, De Meester K, Van Elsacker L (2002) Segment and joint angles of hind limb during bipedal and quadrupedal walking of the bonobo (*Pan paniscus*). Am J Phys Anthropol 119:37–51.
Denham WW (1974) Infant transport among the Alyawarra tribe of Central Australia. Oceania 44(4):253–277.
Elton S (2006) Forty years and still going strong: the use of hominin-cercopithecid comparisons in palaeoanthropology. J R Anthropol Inst 12(1):19–20.
Farley CT, Taylor CR (1991) A mechanical trigger for the trot-gallop transition in horses. Science 253:306–308.
Ghori GMU, Luckwill RG (1985) Responses of the lower limb to load carrying in walking man. Eur J App Physiol 54:145–150.
Goldman RF, Iampietro PF (1962) Energy cost of load carriage. J Appl Physiol 17:675–676.
Gould SJ, Lewontin RC (1979) The spandrels of San Marco and the Panglossian paradigm: a critique of the adaptationist programme. P Roy Soc B-Biol Sci 205:581–598.
Hewes G (1961) Food transport and the origin of hominid bipedalism. Am Anthropol 63:687–710.
Hildebrand M (1965) Symmetrical gaits of horses. Science 150:701–708.
Hildebrand M (1980) The adaptive significance of tetrapod gait selection. Am Zool 20:255–267.
Hof AL (1996) Scaling gait data to body size. Gait Posture 4:222–223.
Holewijn M (1990) Physiological strain due to load carrying. Eur J Appl Physiol 61:237–245.
Hreljac A (1993) Determinants of the gait transition speed during human locomotion: kinetic factors. Gait Posture 1:217–223.
Hunt KD (1994) The evolution of human bipedality ecology and functional morphology. J Hum Evol 26:183–202.
Isler K, Payne RC, Gunther MM, Thorpe SKS, Li Y, Savage R, Crompton RH (2006) Inertial properties of the hominoid limb segments. J Anat 209:201–218.
Jelliffe EF (1975) Recent trends in infant carrying. J Trop Pediatr Environ Child Health 21:93–108.
Keren G, Epstein Y, Magazanik A, Sohar E (1981) The energy cost of walking and running with and without a backpack load. Eur J Appl Physiol 46:317–324.
Kram R, Taylor CR (1990) Energetics of running: a new perspective. Nature 346: 265–267.
Kramer PA (1998) The costs of human locomotion: maternal investment in child transport. Am J Phys Anthropol 107:71–85.
Lawrence M, Singh J, Lawrence F, Whitehead RG (1985) The energy cost of common daily activities in African women: increased expenditure in pregnancy? Am J Clin Nutr 42:753–763.
Legg SJ (1985) Comparison of different methods of load carriage. Ergonomics 28:197–212.
Lovejoy CO (1981) The origin of man. Science 211:341–350.
Mercader J, Panger M, Boesch C (2002) Excavation of a chimpanzee stone tool site in the African rainforest. Science 296:1452–1455.

Nilsson J, Thorstensson A (1989) Ground reaction forces at different speeds of human walking and running. Acta Physiol Scand 136:217–227.
Ottoni EB, Mannu M (2001) Semifree-ranging tufted Capuchins (*Cebus paella*) spontaneously use tools to crack open nuts. Int J Primatol 22:3.
Pierrynowsky MR, Norman RW, Winter DA (1981) Metabolic measures to ascertain the optimal load to be carried by man. Ergonomics 24:393–399.
Quesada PM, Mengelkoch LJ, Hale RC, Simon SR (2000) Biomechanical and metabolic effects of varying backpack loading on simulated marching. Ergonomics 43:293–309.
Schoonaert K, D'Août K, Aerts P (2007) Morphometric and inertial properties in the body segments of chimpanzees (*Pan troglodytes*). J Anat 210:518–531.
Sellers WI, Cain GM, Wang W, Crompton RH (2005) Stride lengths, speed and energy costs in walking of *Australopithecus afarensis:* using evolutionary robotics to predict locomotion of early human ancestors. J R Soc Interface 2:431–441.
Soule RG, Goldman RF (1969) Energy cost of loads carried on the head, hands and feet. J Appl Physiol 27:687–690.
Stephens DW, Krebs JR (1986) Foraging Theory. Princeton University Press, Princeton, NJ.
Stuempfle K, Drury D, Wilson A (2004) Effect of load position on physiological and perceptual responses during load carriage with an internal frame backpack. Ergonomics 47:784.
Taylor RC, Heglund NC, McMahon TA, Looney TR (1980) Energetic cost of generating muscular force during running – a comparison of large and small animals. J Exp Biol 86:9–18.
Tilbury-Davis DC, Hooper RH (1999) The kinetic and kinematic effects of increasing load carriage upon the lower limb. Hum Mov Sci 18:693–700.
Videan EN, McGrew WC (2002) Bipedality in chimpanzee (*Pan troglodytes*) and bonobo (*Pan paniscus*): testing hypotheses on the evolution of bipedalism. Am J Phys Anthropol 118:184–190.
Wall-Scheffler CM, Geiger K, Steudel-Numbers KL (2007) Infant carrying: the role of increased locomotory costs in early tool development. Am J Phys Anthropol 133:841–846.
Watson JC, Payne RC, Chamberlain AT, Jones RK, Sellers WI (2008) The energetic costs of load-carrying and the evolution of bipedalism. J Hum Evol 54:675–683.
Weyand PG, Kelly M, Blackadar T, Darley JC, Oliver SR, Ohlenbusch NE, Joffe SW, Hoyt RW (2001) Ambulatory estimates of maximal aerobic power from foot–ground contact times and heart rates in running humans. J Appl Physiol 91:451–458.
Whiten A, Goodall J, McGrew W C, Nishida T, Reynolds V, Suguyama Y, Tutin CEG, Wrangham RW, Boesch C (1999) Cultures in chimpanzees. Nature 399:682–685.
Whiten A, Boesch C (2001) The cultures of chimpanzees. Sci Am 284:60–67.
Wickler SJ, Hoyt DF, Cogger EA, Meyers G (2003) The energetics of the trot-gallop transition. J Exp Biol 206:1557–1564.

Chapter 8
Field and Experimental Approaches to the Study of Locomotor Ontogeny in *Propithecus verreauxi*

Roshna E. Wunderlich, Richard R. Lawler, and Abigail E. Williams

Abstract In this chapter, we use field-behavioral, morphometric, and laboratory-based data to demonstrate complex links among morphology, performance, and fitness. Although *Propithecus verreauxi* become "ecological adults" at a very young age, skeletal growth of *Propithecus* is slow. This incongruity creates a challenge for a small, developing animal to move efficiently when traveling along the same pathways with larger adults. To explore the effects of this disparity, we quantified the relationships among postcranial morphology, behavior, and fitness in an ontogenetic sample of wild *Propithecus* and subsequently tested functional relationships in the laboratory. Juvenile *Propithecus* exhibit growth allometries and functional changes in locomotion related to decreasing emphasis on pedal grasping and increasing emphasis on thigh-powered leaping. Whereas adult *Propithecus* use their long, muscular thigh and leg segments to increase leaping distance and reduce collisional costs during galloping on the ground, juvenile *Propithecus* increase angular excursions and acceleration and use a hopping gait on the ground that reduces the number of collisions. We show how this juvenile locomotor strategy and other aspects of the "locomotor phenotype" are associated with fitness. Understanding how variation in morphology influences variation in performance throughout ontogeny and the consequences of these associations on fitness should be a major focus of both field and laboratory studies.

Keywords Biomechanics • Fitness • Locomotion • Ontogeny • *Propithecus*

Abbreviations

COM	center of mass
RMA regression	reduced major axis regression

R.E. Wunderlich (✉)
Department of Biology, James Medison University, Harrisonburg, USA
e-mail: wunderre@jmu.edu

K. D'Aoû t and E.E. Vereecke (eds.), *Primate Locomotion: Linking Field and Laboratory Research*, Developments in Primatology: Progress and Prospects,
DOI 10.1007/978-1-4419-1420-0_8,

Introduction

The quest to understand the patterns and processes of biological adaptation has formed the basis of much of modern evolutionary biology. Bock and von Wahlert (1965; Bock 1965) defined an evolutionary adaptation as a form-function complex whose "biological role" interacts with some selective forces. They recognized that energy conservation is one important aspect of performance (their effective fulfillment of biological role) and survivability. Nevertheless, studies of adaptation, especially in primates, tend to focus on design (form-function complex), performance, *or* selection rather than the integration of all three aspects of adaptation. Arnold (1983) formalized the ideas of Bock and von Wahlert (1965) and others and provided a framework to connect morphological design and fitness. He argued that because variation in morphology could be associated with variation in fitness through the critical intermediate variable, *performance*, one could measure the effect of a trait on some aspect of performance (the "performance gradient") and one could measure the effect of performance on fitness (the "fitness gradient"). Similar to the theme of this current volume, Arnold (1983) argued for the integration of laboratory and field studies; specifically, aspects of performance can best be measured in the laboratory, while aspects of fitness can best be measured in the field. Here, we apply the morphology-performance-fitness framework to locomotion in *Propithecus verreauxi*, the sifaka, but we do so from an ontogenetic perspective to provide a more comprehensive understanding and test of these relationships throughout an animal's life.

From an evolutionary perspective, organisms are life cycles (Rice 2002). Life cycles encapsulate the biologically important stages of a particular species. Selection is expected to construct organisms that maximize fitness at every stage in the life cycle, recognizing that both constraints and trade-offs will operate within and among stages (Stearns 1992). Individual animals flow through the life cycle with different propensities for survival, growth, and reproduction. As such, fitness is measured as one turn in the life cycle (Fig. 8.1) and it measures the average reproductive success

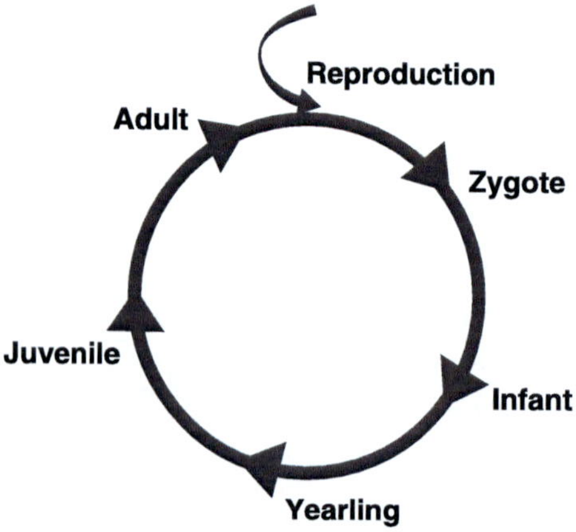

Fig. 8.1 Life cycle of a typical primate. Because fitness is measured as one complete turn in the life cycle, it is important to document patterns of selection acting on developmental stages throughout ontogeny

and/or population growth rate of the species in question. Given that fitness is measured across the life cycle, focusing only on one stage in the life cycle, such as adults, misses much of the evolutionary picture because each stage in the life cycle might have its own set of unique ecological demands and selection pressures.

When compared to other mammals, the primate life cycle is characterized by a substantially longer juvenile stage (Harvey and Clutton-Brock 1985; Pereira and Fairbanks 1993). Numerous theories have been proposed concerning the role of the extended juvenile period in primates as well as the risks associated with it (Poirier and Smith 1974; Clutton-Brock and Harvey 1979; Martin 1981, 1985; Janson and Van Schaik 1993; Joffe 1997; Deaner and Platt 2003; Leigh 2004; Walker et al. 2006). Although considerable work has addressed the evolutionary causes and consequences of the extended juvenile period on diet and foraging in primates, comparatively little work has considered the evolutionary causes and consequences for locomotor performance (Vilensky and Gankiewicz 1989; Dunbar and Badam 2000; Raichlen 2005a, b, 2006; Workman and Covert 2005; Herrel and Gibb 2006; Lawler 2006; Shapiro and Raichlen 2006). Compared to studies of adult animals, only a handful of locomotor field studies have explicitly sought to examine the ontogenetic bases of primate movement (Doran 1992, 1997; Wells and Turnquist 2001; Workman and Covert 2005; Lawler 2006).

Juvenile primates have to navigate the same ecological and social environment as adults, including keeping up with the social group, accessing food resources, and escaping predators. Yet, juveniles are less experienced with the environment in which they are moving, and they are also likely constrained by development of neuromuscular control of balance and locomotion, ongoing differentiation of tissues, and smaller overall body size (Hurov 1991; Carrier 1996; Wells and Turnquist 2001; Main and Biewener 2006, 2007). The juvenile period, therefore, is a time of great locomotor demand and great skeletal risk; due to these factors, selection on juvenile locomotor performance is probably very strong (Carrier 1996; Le Galliard et al. 2004). This effect is likely to be particularly enhanced in primates with a relatively long juvenile period that inhabit a three-dimensional arboreal environment.

In this chapter, we focus on locomotor ontogeny in *Propithecus verreauxi*, an indrid primate found exclusively in Madagascar. *Propithecus verreauxi* are group-living arboreal folivores that live in a highly seasonal environment in the dry and spiny forests of western Madagascar. The timing of dental development and weaning in *Propithecus* are closely tied so that juveniles can take advantage of transient food resources (Eaglen 1985; Godfrey et al. 2004). Juvenile *Propithecus* cope with seasonal food availability by having extremely fast dental growth, allowing them to become "ecological adults" at a very young age (Schwartz et al. 2002; Godfrey et al. 2004). *Propithecus* are born with their deciduous teeth fully erupted and are completely weaned by 6 months of age (Godfrey et al. 2004). Although it is typical for folivorous primates to exhibit more advanced dental development at the time of weaning (Janson and Van Schaik 1993; Leigh 1994), *Propithecus* is particularly precocious in this aspect compared to all other primates.

Propithecus are not precocious in other aspects of their development, and the evolutionary explanations for this pattern are explored in Godfrey et al. (2004) and Ravosa et al. (1993). At the time of weaning (ca. 6 months), juvenile *Propithecus* are still quite small (Fig. 8.2), and somatic growth proceeds slowly. By the time

Fig. 8.2 Adult female *Propithecus verreauxi coquereli* and a juvenile, six-months old. *Propithecus* manifest precocious dental development but their somatic and postcranial development is comparatively slow

juvenile *Propithecus verreauxi* are 8–9 months old, they are only about one-third of adult body mass. In fact, skeletal evidence suggests it takes 2–3 years for *Propithecus verreauxi* to approximate adult skeletal size and longer for the epiphyses to fuse completely (Godfrey et al. 2004). Upon sampling a large number of living individuals at Beza Mahafaly, Lawler (2006) found changes in body mass did not level off until age 8 in *Propithecus verreauxi verreauxi*. Indrids, in general, grow considerably more slowly than lemurids of similar body mass, yet indrids have considerably faster dental development. Juvenile *Propithecus*, therefore, must fulfill "adult-like" behaviors regarding group movements and foraging, but they do so with juvenile skeletal proportions and small body mass.

Given their slow somatic development, *Propithecus* are ideal for studying the consequences of small body size on juvenile locomotion. Not only are they under pressure to perform in a manner similar to adults relatively early in their life cycle, but their growth to adult size is postponed to relatively late in their life cycle. Moreover, their primary forms of locomotion (leaping, bipedal galloping) are associated with high potential joint loads, high energetic expense, and high risks of suboptimal performance. Juvenile *Propithecus* follow adults during travel, often leaping on the same sequence of substrates (Wunderlich and Lawler, unpublished data), and juveniles risk injury or death if they do not land on the substrate or keep up with the group. Both juveniles and adults use bipedalism on the ground. Bipedalism comprises 7–12% of the locomotor repertoire of wild *Propithecus verreauxi verreauxi* (Wunderlich and Lawler, unpublished data) and 13–26% of locomotor bouts in captive *Propithecus verreauxi coquereli* (Williams 2007). Bipedalism, although not intended to be a model for leaping, is kinematically similar to leaping in that it involves high hip and knee angular excursions, and stride length is facilitated by long hind limbs that allow long accelerations times. Whereas leaping distance is determined by the distance between trees, bipedal stride length can be more variable because of the substrate continuity and may therefore offer more opportunities for gait variations. Here we examine postcranial growth and locomotor dynamics in juvenile and adult *Propithecus* to ask how these ecologically precocious, yet postcranially small, individuals function such that they can keep up with adults. We then examine how postcranial traits associated with their habitual forms of locomotion affect fitness.

We measure performance variables using 1) ontogenetic data on limb growth in wild and captive *Propithecus*, 2) experimental data on kinematics and kinetics of locomotion in juvenile and adult *Propithecus*, and 3) field and captive behavioral data on juvenile and adult *Propithecus*. We measure fitness by merging genetic and demographic data with phenotypic measurements. In this way, selection can be measured in two stages: the relationship between the phenotype and performance and the effect of performance on fitness.

Methods

We draw from the morphology-performance-fitness framework to examine the locomotor behavior of *Propithecus verreauxi*. First, we examine aspects of postcranial morphology in wild and captive *Propithecus*. Ontogenetic series of limb measurements were taken on *Propithecus verreauxi verreauxi* at Beza Mahafaly reserve in southwest Madagascar and *Propithecus verreauxi coquereli* at the Duke University Lemur Center (Fig. 8.3a). Methods for measuring *Propithecus* are described in detail in Lawler (2006) and summarized in Table 8.1. Field measurements were taken at 1 year of age because of the limitations of capturing young animals. We measured a cross-sectional sample of 443 (103 resampled) *Propithecus verreauxi verreauxi* between the ages of 1 year and 30 years in the field. The use of captive individuals allows us to extend our growth series into the range between

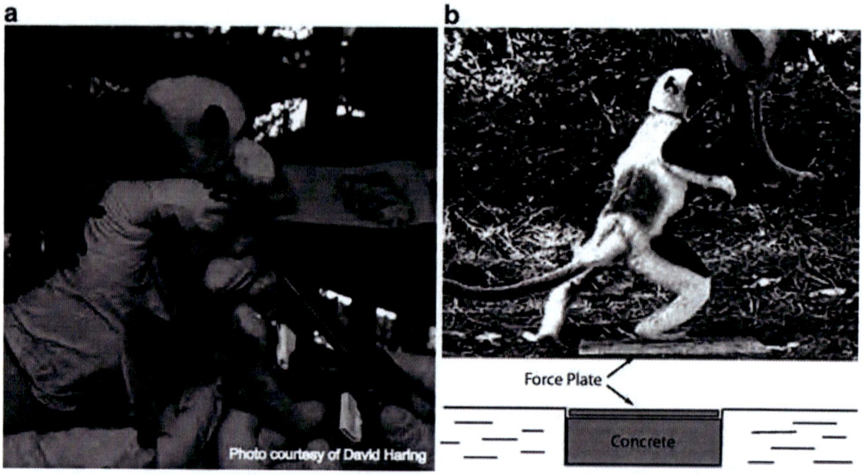

Fig. 8.3 Examples of some of the measurements used in this study. In (**a**), a newborn *Propithecus verreauxi coquereli* is measured using calipers. In (**b**), an adult sifaka bipedally traverses a Kistler force-plate buried flush with the ground on a concrete slab. Both individuals reside at the Duke Lemur Center

Table 8.1 Definition of morphometric traits used in this study

Trait	Description
Arm	Acromion process to lateral epicondyle of humerus
Forearm	Lateral epicondyle of humerus to radial styloid
Hand length	Base of thenar/hypothenar pad to tip of longest manual digit
Thigh	Greater trochanter to lateral epicondyle of tibia
Leg	Lateral epicondyle of tibia to lateral malleolus
Foot length	Back of calcaneus to tip of longest pedal digit
Circumferences	Circumference of arm, forearm, thigh, and leg were taken at the midpoint of each segment
Leg shape	First principal component of thigh length, tibia length and thigh circumference. Each linear measurement was first divided by the cube root of body mass for this trait
Body mass	Measured in kilograms

birth and 1 year of age. We measured six captive *Propithecus verreauxi coquereli* every 2 weeks for the first 6 months, every month for the second 6–9 months, and every 3 months in the second year. Because these two data sets are on two different subspecies of *Propithecus verreauxi* and because captive individuals tend to have higher *absolute* growth rates, we present each set of data separately. We performed reduced major axis (RMA) regressions on log-transformed data in JMP 7.0 (SAS Institute, Cary, NC). We used RMA because there is error associated with the data on both axes in these regressions (e.g., Martin et al. 2005). We used these analyses to examine patterns of allometric growth of the limbs.

We examined juvenile postcranial morphology in a functional context by examining one aspect of their locomotion, bipedalism, experimentally. Bipedalism is not intended to be a model for leaping. Bipedalism comprises about 10% of the locomotor repertoire of wild *Propithecus verreauxi* and more for captive individuals. It is therefore an important part of their locomotor repertoire. Further, studying bipedalism affords an opportunity to quantify kinematics and kinetics of locomotion in *Propithecus verreauxi* on a continuous substrate on which animals have more opportunities for gait variation. Methods for kinematic and kinetic analysis are described in detail elsewhere (Schmitt and Lemelin 2002; Kilkenny 2004; Wunderlich and Schaum 2007) and summarized here. We filmed subjects with lateral, frontal, and 30° cameras (60 Hz) while galloping bipedally along a path within their seminatural forested enclosure. In this setting, they are not limited by space and are moving along a natural dirt substrate. For some of these trials, as well as a number of trials conducted within the subjects' large indoor enclosure, a Kistler portable 9286A or a Kistler 9281B force plate was mounted on a cement block and buried along the runway such that it was flush with the runway (Fig. 8.3b). We encouraged subjects to move bipedally along the pathway by removing nearby vertical supports and using food rewards. We digitized anatomical landmarks (head, shoulder, elbow, wrist, hip, knee, ankle, foot) and filtered them using Peak Performance motion analysis software (Peak Performance Technologies, Centennial, CO), and calculated duty factor, maximum/minimum joint angles, joint angular excursion, and center of mass (COM) movement. All three components of raw force data were imported into MS Excel for analysis. Forces were sequentially integrated to obtain velocity and position (Cavagna 1975; Blickhan and Full 1993; Willems et al. 1995; Griffin et al. 2004; Bishop et al. 2008), and collisional mechanics were analyzed.

Performance measures of leaping and bipedalism used for this study include two primary factors: 1) distance traveled per stride (this is fixed when leaping between trees but not on the ground) and 2) energetic costs. The latter includes internal costs of each stride and collisional costs of the transition between strides. Collisional costs have recently been suggested to be a significant cost of locomotion (Ruina et al. 2005). A collision occurs when the limbs apply work to redirect the COM from generally downward to generally upward. We analyzed collisional mechanics by assessing the number of collisions per stride, the number of footfalls per collision, and the pseudo-elasticity of the collisions. We calculated the latter as the angle of the COM velocity to the substrate reaction resultant before and after the redirection of the COM (Ruina et al. 2005; Baumgartner et al., 2009).

We also assessed leaping and bipedal performance in both wild and captive *Propithecus* via behavioral measurements. We used locomotor bout sampling (Fleagle 1976) to collect data on juvenile (only 1-year-old individuals) and adult (6–25 years) individuals in the wild and in captivity. Definitions of locomotor behaviors are presented in Table 8.2. We calculated locomotor frequencies as well as distances per leap for juveniles and adults.

To estimate the effect of phenotypic differences on fitness, we estimated fitness surfaces via a combination of genetic, demographic, and morphometric data. Any time that one plots variation in some fitness measurement against variation in some trait, the function that unites these two variables is called a fitness function; when two traits are plotted against a fitness measurement, the function becomes a surface. Any number of regression techniques can be used to

Table 8.2 Descriptions of locomotor behaviors used in this study

Leap:	a Thigh-propelled, long-distance jump between vertical or oblique substrates. During take-off and landing the body is generally in an orthograde position (also see Demes et al., 1996)
Bipedal:	Movement using hind limbs only along a continuous substrate (usually the ground)

calculate the fitness function, including linear regression, polynomial regression, or any number of nonlinear techniques. We generated fitness surfaces by fitting a nonlinear neural network model to pairs of traits as well as a fitness measurement. Neural networks, like splines (Schluter 1988), provide a means to visualize the "basic shape" of the relationship between fitness and phenotypes. The shape of the fitness surface, in turn, can reveal what types of selection are acting on the traits in the analysis, e.g., a sloped surface often indicates directional selection, a hump-shaped surface indicates stabilizing selection, etc. These types of visualization techniques do not make any *a priori* assumptions about the form of the fitness surface but are a powerful method for providing an overall "picture" of which combinations of trait values confer the highest fitness (Schluter 1988; Schluter and Nychka 1994). We present two fitness surfaces. One surface reveals the relationship among hand length, foot length, and survival from age 1 to age 8. Survival data come from extensive field censusing collected on the Beza Mahafaly sifaka population (Richard et al. 2002). Both hand and foot length were corrected for age, using a least-squares regression, with the resulting residuals used in the analysis. The other surface shows the relationship between body mass, leg shape, and male fertility. We assessed male fertility using census data in conjunction with paternity analysis; fertility is measured as the number of offspring sired by a male divided by his reproductive lifespan. All males were adults, ages 5 and older. The fitness surfaces shown here are based on the statistically significant multivariate selection coefficients (Lande and Arnold 1983) that capture the relationship between some measure of fitness, i.e. survival or male fertility, and trait values. We analyzed several traits for their association with fitness; we present only fitness surfaces for traits found to be under strong selection (strong selection means the traits have significant p-values as measured via multivariate regression, and the fitness surface for these traits has a distinct, nonlinear shape). Further details of this methodology can be found in Lawler et al. (2005) and Lawler (2006).

Results and Discussion

Phenotype

In general, femur length exhibits strong positive allometry during the first year of life and grows isometrically in later years. Figure 8.4a illustrates the results of longitudinal growth measurements of six *Propithecus verreauxi coquereli*

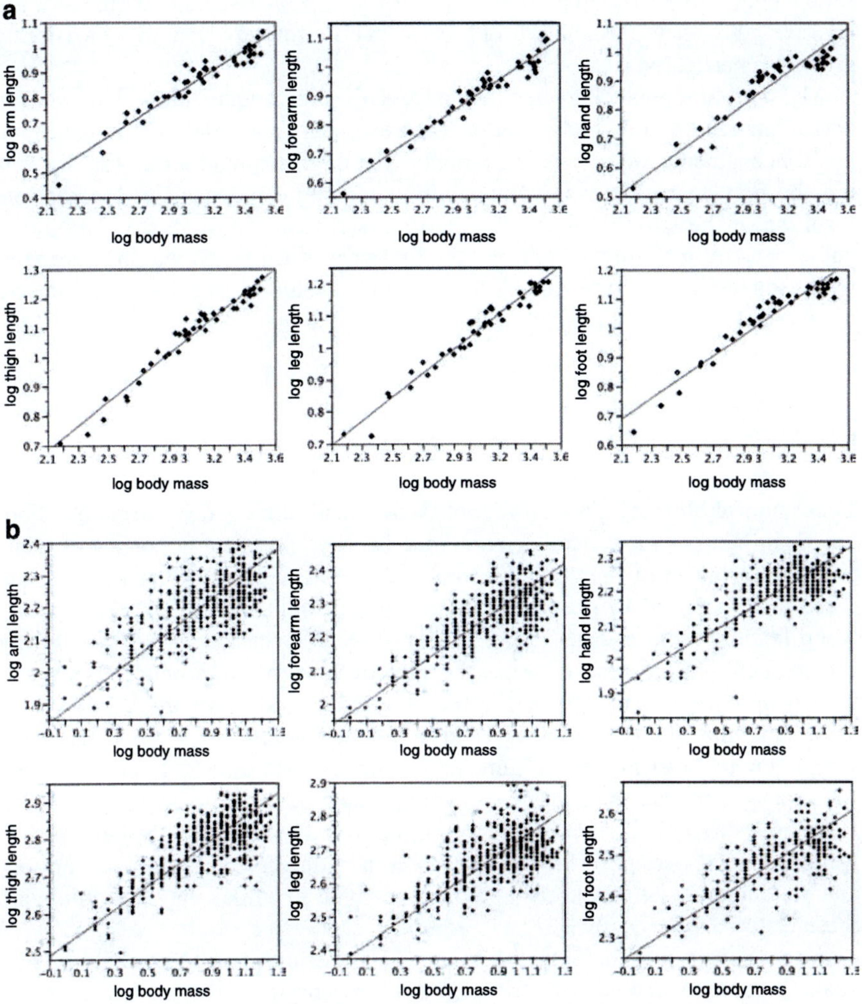

Fig. 8.4 Bivariate plots of limb segment lengths against body mass. In (**a**), the data come from captive *Propithecus verreauxi coquereli* ranging in age from 2 weeks to 18 months. During this time period, femur length exhibits strong positive allometry. In (**b**), the data come from wild *P. v. verreauxi* ranging in age from 1 year to 30 years old. During this time period, hand and foot length exhibit strong negative allometry

from the Duke University Lemur Center. These data are from a longitudinal sample of individuals 2 weeks to just over 1 year of age. The data from wild *Propithecus verreauxi verreauxi* represent a cross-sectional sample of individuals ages 1 year to 30 years (Fig. 8.4b). Tibia length, hand length, and foot length exhibit slight positive allometry during the first year of life, but hand and foot length exhibit negative allometry after year 1. Arm length exhibits slight positive allometry throughout ontogeny, while forearm length consistently grows isometrically.

The morphological characteristics of juveniles in comparison with adults can be summarized as relatively shorter thigh segments and relatively longer hand and foot segments in younger individuals. The thigh segment grows rapidly during the first year of life as young individuals begin independent locomotion, resulting in the strong positive allometry observed in thigh length. The hands and especially feet grow fast during the first year of life, such that they are relatively longer in yearlings compared to adults, resulting in negative allometry after year 1.

Function

These morphological data have behavioral and kinematic correlates. Foot length influences locomotor performance because foot length correlates positively with the span between the first and second digits. The space between these digits is used to grasp a branch during leaping and as a "catch-point" when landing from a leap (Fig. 8.5; Gebo 1985; Demes et al. 1996). During bipedal galloping, *Propithecus* use the span of their foot to produce a foot rollover from lateral to medial on the trail foot and medial to lateral on the lead foot. This may reduce the work of step-to-step transitions in much the same way as the rollover process in humans (Adamczyk 2006). The larger hand and foot spans can also enhance the grasping capabilities of young *Propithecus* (Lawler 2006) and may contribute to propulsive power during bipedal hopping. Longer feet also require higher foot clearance during terrestrial locomotion, and we demonstrate that kinematic differences in juveniles may accommodate these differences.

Although we present only kinematic and kinetic data on bipedalism here, sifaka bipedalism is kinematically similar to leaping in that it involves high hip angular excursions. Adult *Propithecus* use a unique form of bipedal galloping locomotion in which trail and lead limbs are sequenced, and the trunk is positioned 30° to the direction of travel (Fig. 8.6a). Juvenile *Propithecus*, however, use a bipedal hop on the ground (Fig. 8.6b). If we compare juvenile hopping to adult bipedal galloping, hopping strides tend to be longer and reach greater heights, ensuring foot clearance during the aerial phase as well as fewer contacts per distance traveled. Hopping strides involve much greater hip and knee angular excursions and much higher hip angular acceleration (Table 8.3, Fig. 8.7). Hopping may allow the use of two limbs to produce power in the absence of the increased time for acceleration afforded by the longer thighs of adults (see below); however, they have to produce higher peak vertical forces and greater impulse to accomplish the long and high bipedal hop.

Fig. 8.5 Photo showing how the span between the first and second digit on the foot plays a key role in grasping substrates as well as landing on vertical substrates. The span between the digits is relatively large in yearling sifaka

Fig. 8.6 Frame captures from video data. In (**a**), an adult *Propithecus verreauxi coquereli* is engaging in bipedal galloping. In (**b**), a juvenile *P. v. coquereli* is engaging in bipedal hopping

Performance

We evaluated locomotor performance based on stride length (leaping and bipedalism) and collisional energy loss (bipedalism). Wild and captive behavior studies indicate that leaping distance is similar in juveniles and adults (Table 8.3), but bipedal stride length is greater in juveniles than in adults. *Propithecus* tend to travel by following one another on the same substrates, so juveniles have to accomplish a similar level of leaping performance or choose another ("untested" and hence potentially riskier) route. Small-bodied adult prosimians use kinematically different leaping styles than larger-bodied prosimians (Demes et al. 1996) and sacrifice energetic efficiency by taking off at less-than-optimal angles to attain higher horizontal speeds (Crompton et al. 1993; Warren and Crompton 1998). Juveniles too have less time to accelerate because of their relatively shorter limbs, yet they have greater muscular cross-sectional area relative to body mass with which to produce greater force. We still do not know whether juvenile prosimians exhibit a leaping style that is kinematically different from adults or if juveniles simply leap with greater energetic cost. On the ground, however, hopping juveniles tend to take fewer, longer strides than galloping adults (Table 8.3). While this may increase the internal costs of the stride, it results in fewer collisions.

Baumgartner et al. (2009) demonstrate that galloping reduces costs of re-directing the COM. While galloping, *Propithecus* use pseudoelastic collisions, that is, the angle of the incoming (pre-collision) and outgoing (post-collision) velocity vectors are close to orthogonal (Fig. 8.8). Pseudo-elastic collisions, even without elastic recovery, reduce energetic costs by one-quarter relative to a purely absorbing collision (Ruina et al. 2005). Galloping also allows *Propithecus* to distribute each collision over two limb contacts and over the horizontal distance between them, again reducing the energetic expense of the collision (Fig. 8.8). Juvenile hopping does not distribute the collision over two limb contacts, resulting in greater costs for redirecting their COM (Ruina et al. 2005). It is unknown whether their larger joint excursions are indicative of storage and recovery of elastic energy.

Table 8.3 Kinematic and performance data for juvenile and adult *P.v.coquereli*

Variable	Juveniles	Adults
Hip Angular Excursion (degrees)	109	76
Knee Angular Excursion (degrees)	94	52
Stride Duration – Bipedalism (seconds)	0.67	0.74
Duty Factor – Bipedalism	0.31	0.44
Average distance per leap (meters)	1.34	1.28
Average number of strides or hops per bipedal series	1.8	3.5
Average distance traveled per bipedal series (meters)	1.6	2.4
Average length of bipedal stride or hop (meters)	0.9	0.7

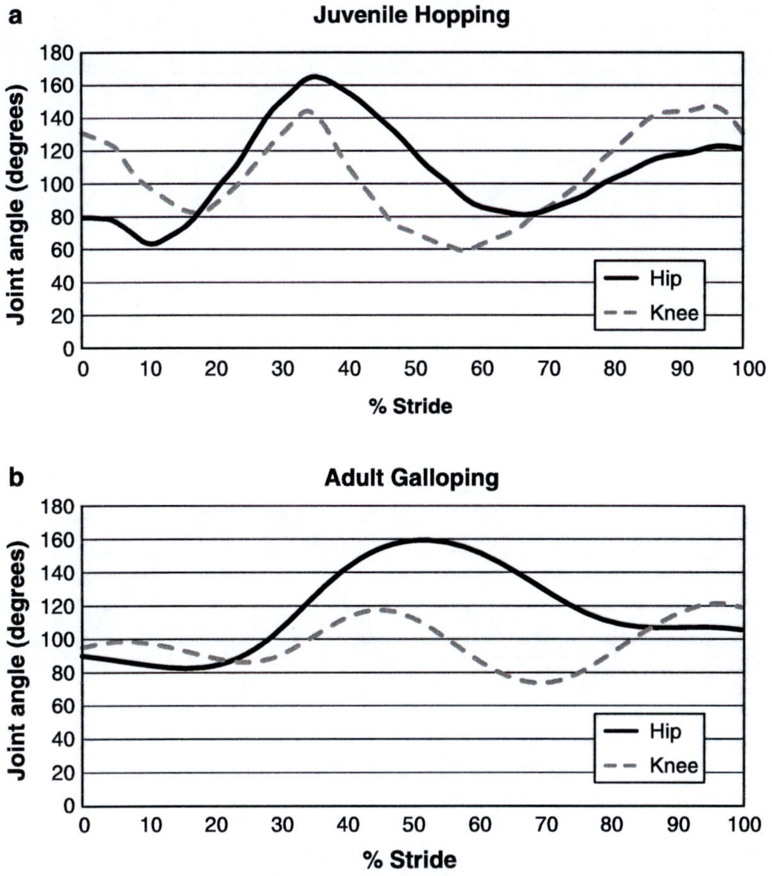

Fig. 8.7 Joint angles of hip and knee during a representative bipedal stride for (**a**) juveniles and (**b**) adults. Juveniles manifest larger joint angles during bipedal hopping relative to adult galloping

Fitness

Though some performance variables are best measured in the lab, estimating fitness and selection is probably best done using field data from a single population. This is because the unit of evolution is the population, and estimating the evolutionary consequences of variation in fitness requires large-sample data on individuals with known fates, phenotype, and kinship. Selection is a key component of the adaptive process, and it can be defined as the covariance between some aspect of fitness and some aspect of phenotype (Rice 2004). When selection acts on heritable traits, the distribution of the trait will change across generations and the population will adaptively evolve.

Using data from a long-term field study of wild *Propithecus*, we were able to estimate selection by collecting phenotypic measurements from individuals captured and released in the wild. In addition, we were able to estimate aspects of fitness using

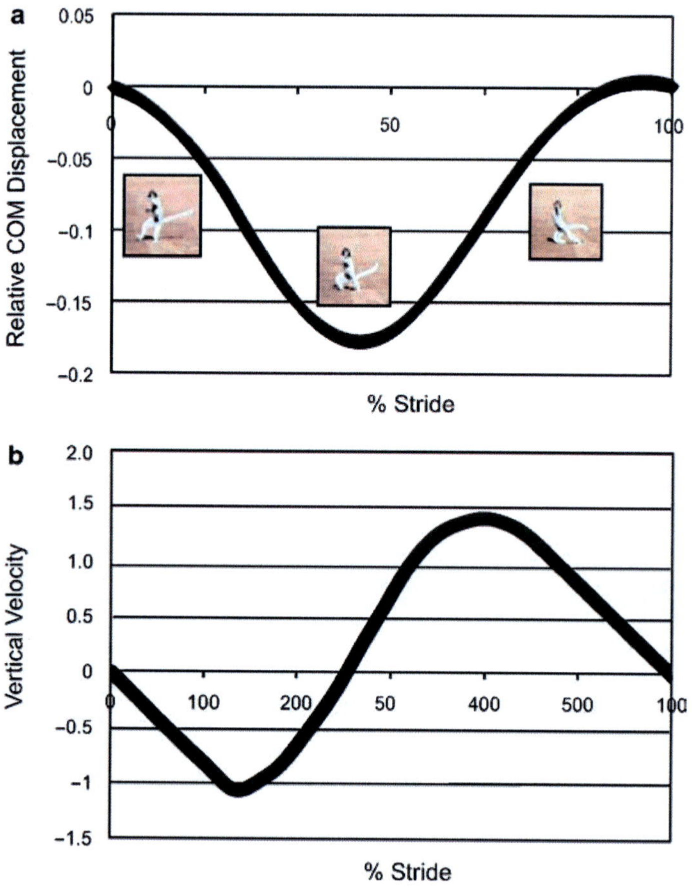

Fig. 8.8 (**a**) Center of mass (COM) and (**b**) vertical velocity plotted over time for a representative bipedal stride in *Propithecus verreauxi coquereli*. These individuals exhibit only one collision per stride, i.e., one change in direction of COM path, and one point per stride where vertical velocity shifts from negative to positive

information on the survival and reproduction of individual subjects. As mentioned previously, the fitness surfaces are generated by examining patterns of selection acting on postcranial traits. Three-dimensional fitness surfaces reveal the relationship between postcranial traits and their relationship with either survival or reproduction (Figs. 8.9, 8.10). One of the fitness surfaces we present (Fig. 8.10) does not concern ontogeny, per se, but we discuss it in the general framework of measuring selection in wild primate populations.

The fitness surface in Fig. 8.9 reveals the relationship between foot length and survivorship. This surface was generated by looking at all individuals that either survived past the age of 8 or died before this age. Foot length, but not hand length, shows a positive relationship with survivorship, indicating the action of positive directional selection. Individuals continue to gain body mass up until the age of 8, thus this surface reveals how foot length contributes to survival during

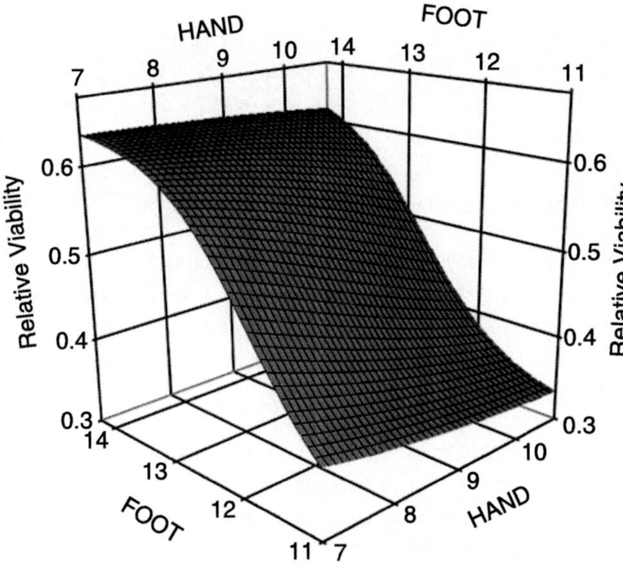

Fig. 8.9 Fitness surface for hand and foot length against relative survival in terms of living beyond age of 8 or dying before age of 8. The surface shows the positive relationship between foot length and survival, indicating positive directional selection on foot length. There is no selection on hand length

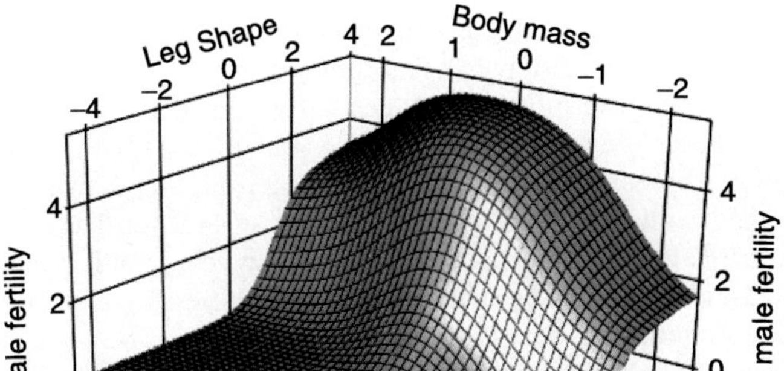

Fig. 8.10 Fitness surface for body mass and leg shape against relative male fertility (offspring sired per year/per male). The surface shows a concave relationship between body mass and fertility, indicating stabilizing selection. The surface also shows a positive relationship between leg shape and male fertility, indicating positive directional selection. Both male fertility and the two traits are standardized

a time when young individuals are still gaining body mass. The functional basis for this relationship was described earlier. Foot length increases grasping span and therefore provides a large "catch-point" when landing from leaps (Gebo 1985; Demes et al. 1996) and presumably during bipedalism. *Propithecus* achieve locomotor independence around 6 months but they must develop locomotor coordination throughout early ontogeny when their limbs and neuromuscular systems are still developing. Larger feet enable young *Propithecus* to grasp safely as well as leap between vertical substrates during the period of locomotor coordination (Lawler 2006). Large feet may also contribute to propulsion during bipedal hopping, although our data cannot speak to relative power generation across joints at this point. Our data reveal that young *Propithecus* have relatively large hands and feet, and we argue that this pattern is actively maintained by selection to allow *Propithecus*, particularly young *Propithecus*, to navigate safely between vertical substrates. However, relatively larger foot size may necessitate higher clearance during terrestrial locomotion, and the kinematically different bipedalism of young *Propithecus* affords this clearance. These data show how variation in morphology, specifically foot length, is associated with variation in fitness, in this case survivorship.

A fitness surface examining the relationship between male fertility, body mass, and leg shape illuminates the pattern of selection on male body mass and leg shape (Fig. 8.10). Selection favors a particular combination of traits with respect to successful reproduction. During the mating season, male *Propithecus* compete with each other for access to females. Mating competition takes on two primary forms: contact aggression and arboreal chases. The total pattern of selection acting on males suggests that traits related to locomotor contests, not aggression, are key determinants of fitness. Directional selection was not found to be operating on body mass or canine size (Lawler et al. 2005). Instead, traits pertaining to arboreal movement are under selection. Stabilizing selection acts on adult male body mass, favoring males that are not too large or too small, while directional selection acts on leg shape, favoring adult males with long legs and muscular thighs. Leg shape encapsulates limb length and thigh circumference and therefore muscle volume. Longer limbs can be used to accelerate for longer periods, thereby generating longer leaps. Thigh circumference represents cross-sectional area of the quadriceps femoris and hamstring muscle groups, major muscle groups used by *Propithecus* for leaping (Demes et al. 1998). Larger "thighed" males can generate more muscle force, and potentially more power, than smaller males. Thus successful males can use their strong, long legs and "streamlined" body mass to out-maneuver and/or out-last their sexual rivals during the mating season, ultimately leading to increased reproductive success. In this regard, intermediate body mass and leg shape are sexually selected traits (Lawler et al. 2005). This last set of results calls attention to an understudied area within locomotor studies: sexual selection. Within the context of sexual selection, locomotor traits pertaining to agility, maneuverability, and speed should be examined with respect to their influence on mate acquisition and mating competition.

Conclusions

Juvenile *Propithecus* exhibit growth allometries and functional changes in locomotion related to decreasing emphasis on manual and pedal grasping and increasing emphasis on thigh-powered leaping. Young *Propithecus* need to keep up with adults despite their small size, and locomotion such as vertical leaping comes with specific performance demands; changes in kinematics of locomotion associated with differences in postcranial shape are associated with performance. Whereas adult *Propithecus* use their long, muscular thigh and leg segments to increase leaping distance and reduce collisional costs on the ground, juvenile *Propithecus* increase angular excursions and acceleration, presumably to produce greater force at take-off. Using a hopping rather than galloping bipedal gait, they reduce the number of collisions on the ground rather than using multiple limb contacts to reduce energy loss as in adults.

As we have shown, morphological features can be related to fitness in wild *Propithecus*. In juvenile *Propithecus*, foot length experiences directional selection and ensures that the span between the first and second digit is large; this span facilitates grasping, leaping, and landing in growing *Propithecus* and may influence energy savings and propulsive power during bipedal hopping. In addition, directional selection targets leg shape in adult male *Propithecus*. These males also experience stabilizing selection on body mass. These traits are likely related to arboreal mating competition and indicate that locomotor contests rather than fighting are key determinants to male reproductive success. Although our field and lab analyses have, at times, focused on different behaviors, we have linked particular traits in growing *Propithecus* to performance and functional parameters and we have also ascertained their influence on fitness.

Future analyses need to examine more thoroughly the relationships among morphology, function and performance. These include the forces produced during locomotion, ontogenetic differences in leaping kinematics and kinetics, and the energetic consequences of juvenile design. The results of this study emphasize that both field and laboratory studies should design experiments that measure aspects of *performance* in order to link morphological variation to variation in fitness. Field primatologists often take a "standardized" set of morphometric measurements on wild animals (e.g., Richard et al. 2002; Kappeler and Schaffler 2008) in order to provide a "snapshot" of information on growth, size, health, and mass of each animal. However, these measurements were not initially defined with respect to functional or locomotor questions. To the extent that field studies include research on locomotion, we suggest that field primatologists move beyond collecting the standard set of measurements and consider taking measurements that are relevant to specific functional/biomechanical concepts or hypotheses. Similarly, laboratory studies should continue, when possible, to include realistic aspects of the species ecology and environment when studying performance variables in the lab. Not only should these studies incorporate aspects of the structural environment, e.g., branch compliance, but they should also pay attention to field studies that report locomotor differences between males and females as well as between age classes. Unlike field studies, laboratory studies have the power to isolate and analyze functional and biomechanical differences in locomotion among the biologically relevant stages that characterize primate ontogeny. We argue that an ontogenetic perspective is needed when studying locomotion. Given that selection likely operates differently on different developmental stages, it is important to document if developmental stages are associated with changes in locomotion. Understanding how variation in morphology influences variation in performance *throughout ontogeny* should be a major focus of both field and laboratory studies. Once we understand the ontogenetic associations between performance and morphology, we should strive to assess the fitness consequences of these linkages. Van Valen aptly observed the following: "Evolution is the control of development by ecology (1973: 488)." In this regard, the morphological configurations and locomotor behaviors we see in adult animals are ultimately the products of selection acting throughout the life cycle.

Acknowledgments Many thanks to Evie Vereecke and Kristiaan D'Août for organizing this symposium and volume and for inviting us to participate. Thank you also to Daniel Schmitt for useful comments on the manuscript and for presenting the talk upon which this manuscript is based. Various aspects of this project were supported by the Jeffress Memorial Trust, Sigma Xi, Yale University, Boston University and the National Science Foundation (DEB-9902146, DEB-0531988). We thank the Government of Madagascar and the Duke University Lemur Center for permission to conduct this research. Thank you to Alison Richard, Robert Dewar, Marion Schwartz, Joel Ratsirarson, Jeannin Ranaivonasy, Jacky Youssef, Enafa, Elahavelo, Emady Rigobert, and Ellis Edidy for assisting with various aspects of data collection in the field. Thank you also to Sarah Zehr, David Brewer, and Tracy Kivell for facilitating and assisting with data collection at the Duke University Lemur Center.

References

Adamczyk PG, Collins SH, and Kuo AD (2006). The advantages of a rolling foot in human walking. J Exp Biol 209(20):3953–3963.

Arnold SJ (1983) Morphology, performance, and fitness. Am Zool 23:347–361.

Baumgartner RE, Wunderlich RE, Schmitt D (2009) Collisional biomechanics during sifaka bipedalism. Proceedings of the Annual Meeting of the Society of Integrative and Comparative Biology P3.63.

Bishop KL, Pai AK, Schmitt D (2008) Whole body mechanics of stealthy walking in cats. PLoS ONE 3(11).

Blickhan R, Full RJ (1993) Mechanical work in terrestrial locomotion. In: Biewener AA (ed), Biomechanics: Structures and Systems. Oxford University Press, New York, pp 75–96.

Bock WJ (1965) The role of adaptive mechanisms in the origin of higher levels of organization. Systematic Zoology 14(4):272–287.

Bock WJ, von Wahlert G (1965) Adaptation and the form-function complex. Evolution 19(3):269–299.

Carrier DR (1996) Ontogenetic limits on locomotor performance. Physiol Zool 69(3):467–488.

Cavagna GA (1975) Force platforms as ergometers. *J Appl Physiol* 39:174–179.

Clutton-Brock TH, Harvey PH (1979) Comparison and adaptation. Proc R Soc Lond B 205(1161):547–565.

Crompton RH, Sellers WI, Gunther MM (1993) Energetic efficiency and ecology as selective factors in the saltatory adaptation of prosimian primates. Proc R Soc Lond B 253(1339):41–45.

Deaner RO, Platt ML (2003) Reflexive social attention in monkeys and humans. Curr Biol 13(18):1609.
Demes B, Jungers WL, Fleagle JG, Wunderlich RE, Richmond BG, Lemelin P (1996) Body size and leaping kinematics in Malagasy vertical clingers and leapers. J Hum Evol 31:385–399.
Demes B, Fleagle JG, Lemelin P (1998) Myological correlates of prosimian leaping. J Hum Evol 34: 385–399.
Doran DM (1992) The ontogeny of chimpanzee and pygmy chimpanzee locomotor behavior: a case study of paedomorphism and its behavioral correlates. J Hum Evol 23(2):139–157.
Doran DM (1997) Ontogeny of locomotion in mountain gorillas and chimpanzees. J Hum Evol 32: 323–344.
Dunbar DC, Badam GL (2000) Locomotion and posture during terminal branch feeding. Int J Primatol 21(4):649.
Eaglen RH (1985) Behavioral correlates of tooth eruption in Madagascar lemurs. Am J Phys Anthropol 66:307–315.
Fleagle JG (1976) Locomotion and posture of Malayan siamang and implications for hominoid evolution. Folia Primatol 26(4):245–269.
Gebo DL (1985) The nature of the primate grasping foot. Am J Phys Anthropol 67:269–277.
Godfrey LR, Samonds KE, Jungers WL, Sutherland MR, Irwin MT (2004) Ontogenetic Correlates of diet in Malagasy lemurs. Am J Phys Anthropol 123(3):250.
Griffin TM, Main RP, Farley CT (2004) Biomechanics of quadrupedal walking: how do four-legged animals achieve inverted pendulum-like movements? J Exp Biol 207(20):3545.
Harvey PH, Clutton-Brock TH (1985) Life history variation in primates. Evolution 39:559–581.
Herrel A, Gibb AC (2006) Ontogeny of performance in vertebrates. Phys Biochem Zool 79(1):1.
Hurov JR (1991) Rethinking primate locomotion: What can we learn from development? J Motor Behav 23:211.
Janson CH, Van Schaik CP (1993) Ecological risk aversion in juvenile primates: slow and steady wins the race. In: Perreira ME, Fairbanks LA (eds), Juvenile Primates. Oxford University Press, New York, pp. 57–76.
Joffe TH (1997) Social pressures have selected for an extended juvenile period in primates. J Hum Evol 32(6):593.
Kappeler PM, Schaffler L (2008) The lemur syndrome unresolved: extreme male reproductive skew in sifakas (*Propithecus verreauxi*), a sexually monomorphic primate with female dominance. Behav Ecol Sociobiol 62:1007–1015.
Kilkenny P (2004) Ground reaction forces during bipedalism in *Propithecus verreauxi*. Master's thesis, James Madison University.
Lande R, Arnold SJ (1983) The measurement of selection on correlated characters. Evolution 37: 1210–1226.
Lawler RR (2006) Sifaka positional behavior: ontogenetic and quantitative genetic approaches. Am J Phys Anthropol 131:261–271.
Lawler RR, Richard AF, Riley MA (2005) Intrasexual selection in Verreaux's sifaka (*Propithecus verreauxi verreauxi*). J Hum Evol 48:259–277.
Le Galliard JF, Clobert J, Ferrière R (2004) Physical performance and darwinian fitness in lizards. Nature 432(7016):502–505.
Leigh SR (1994) Ontogenetic correlates of diet in anthropoid primates. Am J Phys Anthropol 94(4):499–522.
Leigh SR (2004) Brain growth, life history, and cognition in primate and human evolution. Am J Primatol 62(3):139.
Main RP, Biewener AA (2006) In vivo bone strain and ontogenetic growth patterns in relation to life-history strategies and performance in two vertebrate taxa: goats and emu. Physiol Biochem Zool 79(1):57.
Main RP, Biewener AA (2007) Skeletal strain patterns and growth in the emu hindlimb during ontogeny. J Exp Biol 210(15):2676–2690.
Martin RD (1981) Relative brain size and basal metabolic rate in terrestrial vertebrates. Nature 293:57–60.

Martin RD and Harvey PH (1985) Brain size allometry: ontogeny and phylogeny. In: Jungers WL (ed), Size and Scaling in Primate Biology. Plenum Press, New York, pp. 147–173.
Martin RD, Genoud M, Hemelrijk CK (2005) Problems of allometric scaling analysis: examples from mammalian reproductive biology. J Exp Biol 208:1731–1747.
Pereira ME, Fairbanks LA (1993) Juvenile Primates. Oxford University Press, New York.
Poirier FE, Smith EO (1974) Socializing functions of primate play. Am Zool 14:275–287.
Raichlen DA (2005a) Effects of limb mass distribution on the ontogeny of quadrupedalism in infant baboons (*Papio cynocephalus*) and implications for the evolution of primate quadrupedalism. J Hum Evol 49(4):415.
Raichlen DA (2005b) Ontogeny of limb mass distribution in infant baboons (*Papio cynocephalus*). J Hum Evol 49(4):452.
Raichlen DA (2006) Effects of limb mass distribution on mechanical power outputs during quadrupedalism. J Exp Biol 209(4):633.
Ravosa MJ, Meyers DM, Glander KE (1993) Relative growth of the limbs and trunk in sifakas: Heterochronic, ecological, and functional considerations. Am J Phys Anthropol 92(4): 499.
Rice SH (2002) Lecture on Life History Evolution. Evolutionary Biology (E&Eb 225b), Yale University.
Rice SH (2004) Evolutionary Theory: Mathematical and Conceptual Foundations. Sinauer, Sunderland, MA.
Richard AF, Dewar RE, Schwartz M, Ratsirarson J (2002) Life in the slow lane? Demography and life histories of male and female sifaka (*Propithecus verreauxi verreauxi*). J Zool Lond 256: 421–436.
Ruina A, Bertram JEA, Srinivasan M (2005) A collisional model of the energetic cost of support work qualitatively explains leg sequencing in walking and galloping, pseudo-elastic leg behavior in running and the walk-to-run transition. J Theor Biol 237(2):170.
Schluter D (1988) Estimating the form of natural selection on a quantitative trait. Evolution 42:849–861.
Schluter D, Nychka D (1994) Exploring fitness surfaces. Am Nat 143:597–616.
Schmitt D, Lemelin P (2002) Origins of primate locomotion: gait mechanics of the woolly opossum. Am J Phys Anthropol 118(3):231–238.
Schwartz GT, Samonds KE, Godfrey LR, Jungers WL, Simons EL (2002) Dental microstructure and life history in subfossil Malagasy lemurs. Proc Natl Acad Sci USA 99(9):6124.
Shapiro LJ, Raichlen DA (2006) Limb proportions and the ontogeny of quadrupedal walking in infant baboons (*Papio cynocephalus*). J Zool 269(2):191.
Stearns SC (1992) The Evolution of Life Histories. Oxford University Press, Oxford.
Van Valen L (1973) Festschrift. Science 180: 488.
Vilensky J, Gankiewicz E (1989) Early development of locomotor behavior in vervet monkeys. Am J Primatol 17:11–25.
Walker R, Burger O, Wagner J, Von Rueden CR (2006) Evolution of brain size and juvenile periods in primates. J Hum Evol 51(5):480.
Warren RD, Crompton RH (1998) Diet, body size and the energy costs of locomotion in saltatory primates. Folia Primatol 69(S1):86–100.
Wells JP, Turnquist JE (2001) Ontogeny of locomotion in rhesus macaques (*Macaca mulatta*): II. Postural and locomotor behavior and habitat use in a free-ranging colony. Am J Phys Anthropol 115(1):80.
Willems PA, Cavagna GA, Heglund NC (1995) External, internal and total work in human locomotion. J Exp Biol 198:379–393.
Williams AE (2007) Ontogeny of locomotion in *Propithecus verreauxi*. Honors thesis, James Madison University.
Workman C, Covert HH (2005) Learning the ropes: the ontogeny of locomotion in red-shanked douc (*Pygathrix nemaeus*), Delacour's (*Trachypithecus delacouri*), and Hatinh langurs (*Trachypithecus hatinhensis*) I. Positional behavior. Am J Phys Anthropol 128(2):371.
Wunderlich RE, Schaum JC (2007) Kinematics of bipedalism in *Propithecus verreauxi*. J Zool 272(2).

Chapter 9
Comparisons of Limb Structural Properties in Free-ranging Chimpanzees from Kibale, Gombe, Mahale, and Taï Communities

Kristian J. Carlson, Richard W. Wrangham, Martin N. Muller, D. Rick Sumner, M.E. Morbeck, Toshisada Nishida, Atsushi Yamanaka, and Christophe Boesch

Abstract Structural characteristics of limb bones provide insight into how an animal dynamically loads its limbs during life. Cause-and-effect relationships between loading and the osteogenic response it elicits are complex. In spite of such complexities, cross-sectional geometric properties can be useful indicators of locomotor repertoires. Typical comparisons use primates that are distinguished by broad habitual locomotor differences, usually with samples garnered from several museum collections. Intraspecific variability is difficult to investigate in such samples because knowledge of their behavior or life histories, which are tools for interpreting intraspecific variability, is limited. Clearly, intraspecific variation both in morphology and behavior/life history exists. Here we expand an ongoing effort toward understanding intraspecific variation in limb structural properties by comparing free-ranging chimpanzees that have associated behavioral and life history data. Humeral and femoral data from 11 adult chimpanzees (*Pan troglodytes*) of Kibale National Park (Uganda) are compared to 29 adult chimpanzees from Gombe (Tanzania), Mahale Mountains (Tanzania), and Taï Forest (Côte d'Ivoire) National Park communities. Overall, limb structural morphology of Kibale chimpanzees most resembles limb structural morphology of Mahale chimpanzees. Shape ratios and percentage cortical areas of Kibale chimpanzees are most similar to non-Gombe chimpanzees, while Kibale structural properties, e.g., maximum rigidity, are most similar to non-Taï structural properties. Even after adding Kibale females, Taï females continue to stand out from females in other communities.

Keywords Cross-sectional geometry • Functional morphology • Locomotor behavior • *Pan troglodytes*

K.J. Carlson (✉)
Institute for Human Evolution, University of the Witwatersrand, Palaeosciences Centre, Johannesburg, South Africa
e-mail: kristian.carlson@wits.ac.za

K. D'Août and E.E. Vereecke (eds.), *Primate Locomotion: Linking Field and Laboratory Research*, Developments in Primatology: Progress and Prospects,
DOI 10.1007/978-1-4419-1420-0_9, © Springer Science+Business Media, LLC 2011

Abbreviations

ANOVA	analysis of variance
AP	anteroposterior
BM	body mass
%CA	percentage cortical area of cross sections
CA	cortical area
F	femur
FMSID	supero-inferior diameter of the femoral head
G	Gombe
H	humerus
HHMD	maximum diameter of the humeral head
I_{max}/I_{min}	maximum/minimum rigidity (principal moments of area)
I_x/I_y	second moments of area about anatomical planes
K	Kibale
KS	Kolmogorov-Smirnov
L	bone length
LSD	least significant difference
M	Mahale
ML	mediolateral
ROI	region of interest
sI_{max}	normalized I_{max} (maximal rigidity)
SD	standard deviation
T	Taï
TA	total cross-sectional area
VOI	volume of interest

Introduction

Functional morphologists rely on comparative approaches as well as experimental techniques in the laboratory, i.e., kinetics, kinematics, electromyography, strain analysis, to understand form-function relationships in the postcranium of animals. Shared or unique components of activity patterns provide a framework against which morphological commonalities or differences are evaluated. Often, comparative studies construct samples from specimens of museum collections (Ruff 2002, 2008; Green et al. 2007; Haeusler and McHenry 2007; Marchi 2007). Although museum specimens may be numerous and accessible—two necessary criteria for amassing large samples that rigorous statistical analyses favor—they also necessitate a seldom appreciated trade-off, namely, that though behavior and life history may vary among group individuals, this variation must be ignored to compare groups. Clearly, however, individuals within populations can vary substantially in behavioral or life history variables (Goodall 1986; Hunt 1992), which may in turn contribute to intragroup variability in morphological characteristics and reproductive fitness.

Chimpanzees (*Pan troglodytes*) offer a unique opportunity among animals to address functional morphology questions. Observational studies of free-ranging chimpanzee communities representing eastern (*Pan troglodytes schweinfurthii*) and western (*P. t. verus*) subspecies provide a detailed portrait of individual life histories (Goodall 1986; Nishida 1990; Morbeck 1999; Boesch and Boesch-Achermann 2000; Morbeck et al. 1991; Morbeck et al. 2002). Studies encompassing the last 45 years at several locations, e.g., Gombe Stream National Reserve (Tanzania), Kibale National Park (Uganda), Mahale Mountains National Park (Tanzania), and Taï Forest National Park (Côte d'Ivoire), document activity profiles of female and male chimpanzees of all ages in all sorts of situations or settings (Whitten et al. 1999). Skeletal collections at three of these locations, i.e., Gombe, Kibale, and Mahale, representing *Pan troglodytes schweinfurthii*, and one location (Taï) representing *P. t. verus*, have been accumulated over time, and thus, individual specimens frequently can be associated with contextual information, e.g., life history, activity, and habitat data. Such a sample is ideal for investigating form-function relationships in the primate postcranium, and in fact, is uniquely situated to investigate intraspecific variability in the primate postcranium. While intraspecific comparisons within *Pan* have noted variation in cranial features (Guy et al. 2003; Lockwood et al. 2004), intraspecific differences in the postcranium are less well-documented. Humeral and femoral lengths closely align Kibale and Taï chimpanzees (Carter et al. 2008; Zihlman et al. 2008), and exceed respective lengths of Mahale and Gombe chimpanzees, which appear to resemble one another more, particularly amongst the females (Gunji et al. 1998; Zihlman et al. 2008). In terms of overall body size, the western subspecies (*Pan troglodytes verus*) has been attributed higher sex-specific body weights relative to the eastern subspecies (*P. t. schweinfurthii*) (Smith and Jungers 1997), and Kibale and Mahale chimpanzees have higher estimates of body mass than Gombe chimpanzees (Carter et al. 2008).

Bone has an ability to readjust its diaphyseal structure over the course of the lifetime of an animal (Martin et al. 1998; Currey 2002). Cross-sectional geometric properties, as one means of quantifying *in vivo* adjustment, are frequently used in functional comparisons of human and nonhuman primate postcrania (Burr et al. 1982, 1989; Schaffler et al. 1985; Ruff 1987, 1989, 2002; Demes and Jungers 1989, 1993; Sumner et al. 1989; Demes et al. 1991; Ruff and Runestad 1992; Ohman 1993; Terranova 1995a, b; Sumner and Andriacchi 1996; Jungers et al. 1998; Polk et al. 2000; Stock and Pfeiffer 2001; Carlson 2002, 2005; Yamanaka et al. 2005; Carlson et al. 2006, 2008; Marchi, 2007). Efforts to quantify bone deformation during quadrupedal locomotion indicate the importance of using caution when inferring locomotor performance from cross-sectional properties alone (Demes et al. 1998, 2001; Lieberman et al. 2004). For example, the common assumption that tissue economy in diaphyseal cross sections should be optimized for resisting the observed bending loads does not always hold true (Demes et al. 1998, 2001; Pearson and Lieberman 2004; Ruff et al. 2006). When comparing limb loading during terrestrial quadrupedal locomotion and select modes of a primate arboreal locomotor repertoire, i.e., vertical climbing, brachiation, the latter are characterized by relatively greater variation in load orientations (Swartz et al. 1989; Demes et al. 2001). When using additional, but indirect measures to infer limb loading, e.g., substrate reaction forces, kinematics, and second moments of area, during even more locomotor behaviors, variability in loading regimes seems even greater than currently appreciated (Carlson et al. 2005; Demes et al. 2006; Carlson and Judex 2007; Demes and Carlson 2009). In addition to activity-induced deformations, other nonmechanical factors, e.g., genetics, hormones, and age, affect the bone modeling/remodeling process, which potentially could affect cross-sectional geometric properties independent of activity patterns (Martin et al. 1998; Turner et al. 2000; Wergedal et al. 2005; Xiong et al. 2006; Devlin and Lieberman 2007; Robling et al. 2007). Despite the list of cautionary notes, cross-sectional geometric properties retain value to paleoanthropologists and physical anthropologists as tools for inferring locomotor repertoires when animals cannot be observed visually, such as extinct taxa (Madar et al. 2002; Holt 2003; Marchi et al. 2006; Griffin 2008; Ruff 2008).

Shape variation in select regions of African ape femoral and humeral diaphyses is associated with reported frequencies of arboreal locomotion (Carlson 2002, 2005). More evenly distributed bone mass in a cross section, i.e., more circular shape, has been correlated with increased percentage of arboreal locomotion, while more elliptical cross sections have been correlated with increased percentage of terrestrial locomotion (Fig. 9.1). Frequencies of specific locomotor behaviors in a behavioral repertoire, however, have not been linked to diaphyseal shapes in a similarly straightforward fashion, whether using a museum collection sample (Carlson 2005) or a small sample of free-ranging chimpanzees (Carlson et al. 2006).

A comparison of Gombe, Mahale, and Taï chimpanzees tentatively linked differences in habitat characteristics between the three communities to structural differences in their diaphyseal morphology (Carlson et al. 2008). A terrain effect that has been observed in human lower limb structural properties (Ruff 1999) was also visible in some chimpanzee populations apart from any differences in locomotor mode frequencies. Qualitative variation within a single locomotor mode, such as habitat-induced variation in quadrupedalism, ultimately could impact deformation patterns experienced by limb elements. Mobility (cf. Carlson et al. 2007), which can be described as distance traveled plus the frequency of maneuvering around obstacles within a habitat, may reflect characters such as vegetation density or ground cover. Elevation changes and ruggedness of terrain are two additional factors that can distinguish habitats. Previous work has shown that mobility and terrain elevation/ruggedness are relevant to human patterns of lower limb loading (Burr et al. 1996; Ruff 1999). Such scenarios remain to be verified in nonhuman primate populations, but would seem to be present given the conservative nature of bone's response to loading amongst organisms (Martin et al. 1998; Currey 2002).

With a unique sample of primates we overcome ordinary challenges that can constrain functional morphologists, and in a sense, we bring the field into the laboratory. To reconfirm recently proposed form-function relationships in the chimpanzee postcranium, we add individuals from a fourth well-studied chimpanzee community to a previous comparison of three chimpanzee communities (Carlson et al. 2008). Do femoral and humeral diaphyseal shapes of Kibale chimpanzees differ from

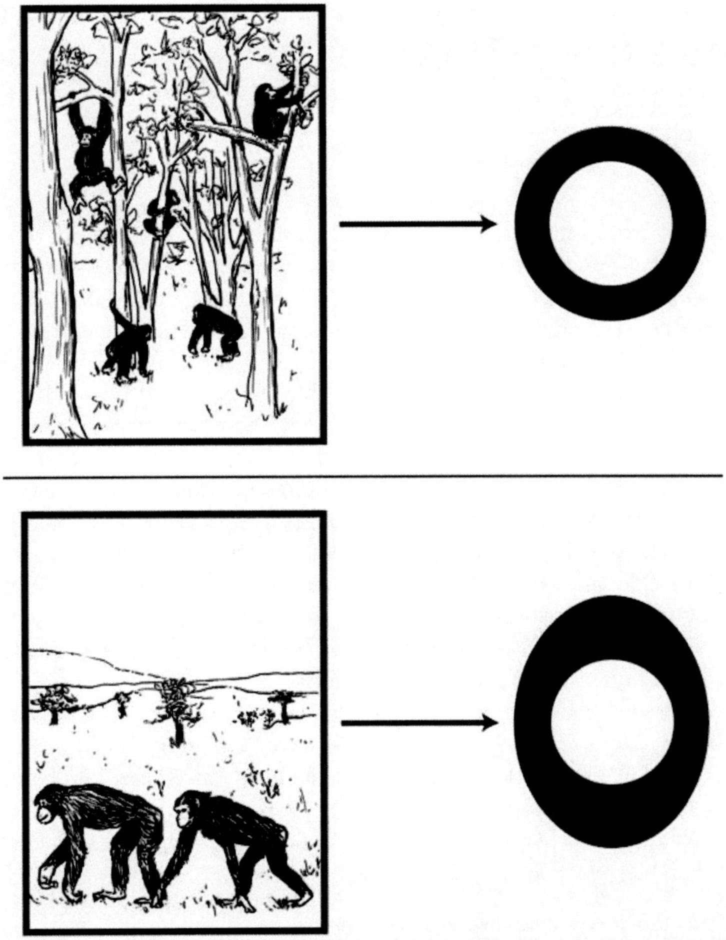

Fig. 9.1 Theoretical expectations for shape ratios of individuals occupying more closed habitat conditions, performing greater percentages of arboreal locomotion **(top row)**, and individuals occupying more open habitat conditions, performing lesser percentages of arboreal locomotion **(bottom row)**

shapes at analogous diaphyseal locations of Gombe, Mahale, or Taï chimpanzees? Do "strength" properties that estimate maximum bending rigidity distinguish Kibale chimpanzees from chimpanzees in the other communities? Do percent cortical areas (%CA) distinguish Kibale chimpanzees from chimpanzees in the other communities? Do observed morphological differences between Kibale chimpanzees and chimpanzees from other communities reinforce previous interpretations of morphological differences between Mahale, Gombe, and Taï chimpanzees that were attributed to habitat characteristics? Ultimately, if habitat characteristics can be used to differentiate loading patterns in human populations, they may similarly distinguish habituated chimpanzee communities, which would have important implications for reconstructing locomotor repertoires of early hominins.

Materials and Methods

To acquire cross-sectional geometric properties, we used serial computed tomography (CT) scans of humeri and femora representing 40 adult chimpanzees (*Pan troglodytes*) from Gombe (Tanzania), Kibale (Uganda), Mahale Mountains (Tanzania), and Taï Forest (Côte d'Ivoire) National Parks. The sample from Gombe, Mahale Mountains, and Taï Forest National Parks has been described previously (Carlson et al. 2006, 2008). Wherever possible we collected bilateral data from forelimbs and hind limbs of an individual. For each bone, we analyzed three regions of interest (ROIs): 35 (mid-distal), 50 (midshaft), and 65 (mid-proximal) percent diaphyseal lengths (Table 9.1). We excluded individuals that exhibited serious injuries or disabling diseases that we thought could have permanently altered locomotor repertoires. We retained several Taï individuals in the sample that died as a result of an Ebola epidemic (Boesch and Boesch-Achermann 2000), but because of the rapid onset of death associated with Ebola, we believe that individuals that died from an Ebola infection had insufficient time for any reduction in activity level to alter bone diaphyseal morphology significantly.

We followed an existing protocol for obtaining CT data from Kibale specimens (Carlson et al. 2006, 2008). Briefly stated, we saved CT images in DICOM format. We imported DICOM stacks corresponding to entire long bones into commercial software, Amira® 4.0 (Visage Imaging, Inc, Carlsbad, CA), segmented DICOM stacks to create isosurfaces, and then rendered volumes of interest (VOIs) using

Table 9.1 Specimens comprising community samples

	Females	Males	Unknown	Represented elements
Gombe[a]	5	4		7 left femora
				9 left humeri
Kibale	4	6	1	9 bilateral femora, 1 left femur
				9 bilateral humeri, 2 left humeri
Mahale[a]	3	1		4 bilateral femora
				3 bilateral humeri, 1 left humerus
Taï[a,b]	10	5	1	11 bilateral femora, 3 left femora, 1 right femur
				11 bilateral humeri, 2 left humeri

[a]From Carlson et al. (2008).

[b]An additional individual from Taï, Nipla, was included in the present sample. The additional Taï individual was listed in field notes as originating from a group to the south of the main group that contributed all other individuals. However, since the habitat conditions for the added individual were likely more similar to the habitat conditions for the other Taï chimpanzees relative to the habitat conditions for chimpanzees in any of the other three communities, it was reasonable to include this individual in the Taï sample.

thresholds that gave an accurate digital representation of bone surfaces. Selection of the appropriate threshold considered two criteria: eliminate artificial holes in surfaces and separate lower density objects, such as soft tissues, from bone surfaces. Variation in the appropriate thresholds for different bones was minor following these criteria. Once a VOI was rendered, it was aligned in virtual space using the same criteria as previous studies that aligned physical specimens in CT scanners (Ruff 2002; Carlson 2005; Carlson et al. 2006, 2008). Subsequent to identifying ROIs, we used cutting planes in the commercial software to virtually section the rendered VOIs. We used screen capture software programs, or options within the visualization software program, to record digital images of virtually "sectioned" surfaces. We calculated cross-sectional geometric properties of virtual "sections" using custom-written macros for Scion Image (release Beta 4.0.2; ported from NIH Image for Macintosh by Scion Corporation and freely available at http://www.scioncorp.com) and a modified version of the SLICE program (Nagurka and Hayes 1980).

The custom-written macro calculates standard cross-sectional geometric properties from a cross section, e.g., maximum rigidity (I_{max}). We treated cortical bone in cross sections as having homogeneous material properties, which is customary in analyses of cross-sectional properties (but see Bhatavadekar et al. 2006). We calculated shape ratios from principal moments of area (I_{max}/I_{min}) rather than second moments of area about anatomical planes (I_x/I_y) since the former ratio provides a more accurate reflection of overall shape (Carlson 2005). We calculated percentage cortical area (%CA) of cross sections as cortical area (CA) divided by total area (TA). Following Sumner et al. (1989), this measure provides a useful estimate of bone mass at a diaphyseal location.

When modeling bending deformation of a beam, bending is proportional to the product of the applied force and the length of the beam. When comparing a bone to a beam undergoing bending, body weight is a suitable substitute for applied force and bone length is a suitable substitute for the length of the beam. Chimpanzee body mass is known to vary according to several factors, including banana provisioning, community range size and density, seasonality, social rank in females, female reproductive cycles, and age (Pusey et al. 2005). Body mass estimates for several of the populations are available in the literature (Carter et al. 2008), however, we chose to use different estimates of body mass for two reasons: 1) estimating body mass at the level of the individual was more consistent with the stated goals of the present study, and 2) the predictors we used for estimating body mass provided the most accurate estimates of body mass in comparisons of several predictors (Carlson 2002). To estimate body mass for scaling femoral properties (see Eq. 9.1), we regressed body mass, BM, on supero-inferior diameter of the femoral head, FMSID (Ruff 2002; Carlson et al. 2006). Carlson (2002) estimated body masses for 25 African apes of known body mass using Eq. (9.1) and found that 80% had predicted values within 20% of their recorded body mass. We estimated body mass for each right and left femur separately. We use femoral mechanical length in scaling measures since this is a suitable estimate for the length of the bone as it undergoes bending forces (Ruff 2002; Carlson et al. 2006). We measure femoral mechanical length from each right and left femur separately in order to derive side-specific scaling factors. To compare individuals, we normalize I_{max} to the product of estimated body mass and bone length (Eq. 9.2), where sI_{max} = normalized I_{max}, bm = body mass, and L = bone length.

$$\log_{10}(BM) = 3.030(\log_{10} FMSID) + (-2.946) \tag{9.1}$$

$$sI_{max} = I_{max}((BM) \times (L))^{-1} \tag{9.2}$$

To estimate body mass for scaling humeral properties (see Eq. 9.3), we regressed body mass on maximum diameter of the humeral head, HHMD (Carlson 2002). Using Eq. (9.3), Carlson (2002) found that 87% of the 25 African apes had predicted body mass values within 20% of their recorded body mass. We estimated body mass for each right and left humerus separately. For humeral diaphyses, we use a maximum length of the humerus with the long axis of the diaphysis parallel to the longitudinal axis (Carlson 2002). We measured bone length from each right and left humerus separately in order to derive side-specific scaling factors for normalizing humeral I_{max} (Eq. 9.2).

$$\log_{10}(BM) = 2.824(\log_{10} HHMD) + (-2.896) \tag{9.3}$$

We report habitat characteristics of Kibale National Park (Chapman et al. 1997; Hunt and McGrew 2002) that are analogous to those reported in an earlier comparative study of chimpanzees from Gombe, Mahale Mountains, and Taï Forest National Parks (Carlson et al. 2008), and when possible, we update values for previously reported characteristics. We concentrate on specific habitat characteristics that we believe introduce qualitative differences into terrestrial quadrupedalism, such as estimates of obstacle frequency during quadrupedalism and the extent of locomotion over uneven terrain (Table 9.2).

We used Kolmogorov-Smirnov (KS) tests for normality to assess variable distributions against theoretical (normal) distributions. Since no variables departed significantly from normal distributions, parametric statistical analyses were justified. Accordingly, we used a series of one-way analyses of variance (ANOVA) to assess differences between Kibale, Gombe, Mahale, and Taï communities. In the event that groups differed significantly, we used a Levene test for homogeneity of variances to verify the assumption of equal group variances, which is a necessary assumption of the one-way ANOVA. The Levene test also dictated which *post hoc* analysis was used to determine which groups differed from one another. When we observed a nonsignificant Levene statistic, i.e., group variances did not significantly differ, we performed a Bonferroni *post hoc* analysis. In comparisons where there were three groups or fewer, we used Fisher's least significant difference (LSD) *post hoc* analyses rather than Bonferroni *post hoc* analysis because the conservative Bonferroni correction for multiple comparisons was deemed too restrictive. When we observed a significant Levene statistic, i.e., group variances significantly differed, we conducted a Tamhane's T2 *post*

Table 9.2 Habitat characteristics of habituated groups[a]

	Gombe	Kibale[b]	Mahale	Taï
Annual mean rainfall	1775 mm	1671 mm	1836 mm	1829 mm
Ground cover	More open woodland	Evergreen moist forest	M-group: closed forest, vine tangles	Tropical moist forest
Elevation range (above sea level)	772–1500 m	1000–1700 m	772–2462 m	ca. 120 m
Slope of terrain	16.5°	6.8°	8.3°	Slightly undulating

[a] Data sources described in Carlson et al. (2008), except annual mean rainfall data which has been updated to reflect data reported by Hunt and McGrew (2002). Hunt and McGrew (2002) broadly compare ecological parameters (e.g., annual mean rainfall) for numerous chimpanzee habitats, including, but not limited to the four habitats in the present study. Ultimately, we report values provided by Hunt and McGrew (2002) rather than those in Carlson et al. (2008) because the former favor more general comparisons and applications.
[b] Kibale values calculated from data reported in Chapman et al. (1997). Slope of terrain reported as the average of four study areas characterized as moderately undulating valleys.

hoc analysis. A Tamhane's T2 *post hoc* analysis is based on a *t*-test, and is preferable to other alternatives because it is conservative. In comparisons of sex-specific community trends, where sample sizes were often small, e.g., $n < 4$, we frequently chose a more conservative approach than one-way ANOVA by using the Kruskal-Wallis H test, a nonparametric test. The Kruskal-Wallis H test is a one-way ANOVA that compares groups by ranking data, and thus it does not assume normality.

We selected $p < 0.05$ as the level of statistical significance in all statistical testing. We used SPSS 15.0 (SPSS Inc., Chicago, IL) for statistical procedures.

Results

Kibale Chimpanzees versus Other Community Chimpanzees

Average shape ratios at diaphyseal ROIs for the four chimpanzee communities are reported in Table 9.3. Kibale chimpanzees exhibit significantly lower shape ratios than Taï chimpanzees at several locations: right F35 and F50 ROIs, as well as the right H50 ROI (Table 9.4). Kibale chimpanzees have significantly higher shape ratios at the right H35 ROI than either Taï or Mahale chimpanzees, and significantly higher shape ratios than Taï chimpanzees at the left H35 ROI (Table 9.4). While right humeri and femora of Gombe individuals were unavailable for comparisons, at the left H50 ROI, Kibale chimpanzees exhibited significantly higher shape ratios than Gombe chimpanzees.

Average maximum rigidities at diaphyseal ROIs are reported in Table 9.3. Normalized measures are available for only a subset of the communities, which means

Table 9.3 Community means (1 SD) for shape ratios (I_{max}/I_{min}), % CA, and normalized I_{max} (sI_{max})

ROI	Gombe (left)	Kibale (left)	Kibale (right)	Mahale (left)	Mahale (right)	Taï (left)	Taï (right)
F35 shape	1.38	1.42	1.40	1.31	1.30	1.50	1.53
	(0.19)	(0.12)	(0.11)	(0.05)	(0.04)	(0.08)	(0.13)
	7	10	9	4	4	14	12
F50 shape	1.44	1.37	1.33	1.30	1.35	1.49	1.56
	(0.16)	(0.15)	(0.13)	(0.08)	(0.10)	(0.08)	(0.17)
	7	10	9	4	4	14	12
F65 shape	1.42	1.42	1.40	1.40	1.38	1.43	1.52
	(0.11)	(0.17)	(0.13)	(0.14)	(0.14)	(0.14)	(0.11)
	7	10	9	4	4	14	12
H35 shape	1.26	1.40	1.46	1.24	1.20	1.13	1.17
	(0.10)	(0.21)	(0.22)	(0.02)	(0.03)	(0.08)	(0.10)
	9	11	9	4	3	13	11
H50 shape	1.12	1.26	1.21	1.19	1.15	1.35	1.38
	(0.05)	(0.15)	(0.13)	(0.05)	(0.02)	(0.12)	(0.11)
	9	11	9	4	3	13	11
H65 shape	1.11	1.20	1.18	1.16	1.11	1.26	1.29
	(0.05)	(0.09)	(0.07)	(0.10)	(0.02)	(0.13)	(0.17)
	9	11	9	4	3	13	11
F35 %CA	50.1	59.4	59.3	65.3	64.5	61.4	60.6
	(11.9)	(8.4)	(7.4)	(12.1)	(11.9)	(6.0)	(4.6)
	7	10	9	4	4	14	12
F50 %CA	53.0	65.2	64.9	72.2	71.2	67.2	66.8
	(12.4)	(9.0)	(8.7)	(13.1)	(13.4)	(5.7)	(4.4)
	7	10	9	4	4	14	12
F65 %CA	52.3	66.6	66.9	73.5	73.0	68.8	68.5
	(12.2)	(9.3)	(10.3)	(15.7)	(15.8)	(6.3)	(4.6)
	7	10	9	4	4	14	12

H35 %CA	51.5	65.3	67.6	69.5	67.1	64.4	65.2
	(13.8)	(7.3)	(5.8)	(13.9)	(15.9)	(8.1)	(7.1)
	9	11	9	4	3	13	11
H50 %CA	47.0	60.3	61.9	67.6	64.2	61.7	63.3
	(11.6)	(7.1)	(5.8)	(13.0)	(15.2)	(5.8)	(4.3)
	9	11	9	4	3	13	11
H65 %CA	43.5	58.1	59.2	62.9	59.6	61.1	61.9
	(9.3)	(6.4)	(6.3)	(13.8)	(16.5)	(7.3)	(6.1)
	9	11	9	4	3	13	11
F35 sI_{max}	—	1.18	1.18	1.05	1.09	1.37	1.35
		(0.20)	(0.18)	(0.12)	(0.10)	(0.32)	(0.16)
		10	9	4	4	6	7
F50 sI_{max}	—	1.20	1.18	1.07	1.15	1.41	1.41
		(0.18)	(0.16)	(0.14)	(0.14)	(0.33)	(0.16)
		10	9	4	4	6	7
F65 sI_{max}	—	1.22	1.21	1.12	1.14	1.34	1.37
		(0.20)	(0.18)	(0.14)	(0.08)	(0.24)	(0.13)
		10	9	4	4	6	7
H35 sI_{max}	—	0.77	0.79	—	—	0.78	0.78
		(0.15)	(0.17)			(0.09)	(0.11)
		11	9			7	5
H50 sI_{max}	—	0.83	0.81	—	—	0.94	0.95
		(0.12)	(0.12)			(0.18)	(0.21)
		11	9			7	5
H65 sI_{max}	—	0.85	0.85	—	—	0.95	0.93
		(0.13)	(0.11)			(0.18)	(0.21)
		11	9			7	5

F = femur, H = humerus; 35 = mid-distal diaphysis, 50 = midshaft, 65 = mid-proximal diaphysis. Cells contain mean values, 1 SD in parentheses, and n. Rows represent values of shape (I_{max}/I_{min}), %CA at a given ROI (e.g., F35), or normalized I_{max} (sI_{max}). Gombe individuals are represented by left elements only.
Gombe, Mahale, and Taï data from Carlson et al. (2008).

that not all four communities could be compared. Taï chimpanzees exhibit significantly greater normalized maximum rigidity than Kibale and Mahale chimpanzees at right F50 and F65 ROIs, and greater normalized maximum rigidity than Mahale chimpanzees at the right F35 ROI (Table 9.4). No significant community-level differences are observed at ROIs of the left femur or at ROIs of either humerus.

Average %CAs at diaphyseal ROIs for the four chimpanzee communities are reported in Table 9.3. Kibale chimpanzees do not differ significantly from Mahale or Taï chimpanzees at any femoral or humeral ROI. Kibale chimpanzees, similar to Mahale and Taï chimpanzees, have significantly higher %CA than Gombe chimpanzees at each left humeral ROI. Among femoral ROIs, Kibale chimpanzees exhibit significantly higher %CAs than Gombe chimpanzees at the left F65 ROI, while Mahale and Taï chimpanzees exhibit significantly higher %CAs than Gombe chimpanzees at left F50 and F65 ROIs. No significant community-wide differences in %CA were observed in right elements, for which Gombe individuals were not included. This is consistent with left ROIs, where the only significant differences in %CA occur between individuals in the Gombe community and individuals in the other three communities.

Sex-Specific Comparisons of Kibale Chimpanzees with Chimpanzees in Other Communities

To examine community-level comparisons more in-depth, average shape ratios and %CAs for females and males in the communities are reported in Table 9.5. Female chimpanzees from Kibale exhibit significantly greater shape ratios than Taï chimpanzees at left and right H35 ROIs (Table 9.6). Female Kibale chimpanzees also exhibit significantly lower shape ratios than Taï chimpanzees at the right H50 ROI (Table 9.6). Female chimpanzees from Kibale, Mahale, and Taï communities exhibit significantly greater %CA at each humeral ROI compared to female chimpanzees from Gombe (Table 9.7). Unlike Mahale and Taï female chimpanzees, female chimpanzees from Kibale do not differ from Gombe female chimpanzees in %CA at femoral ROIs (Table 9.7). Compared to female chimpanzees from all communities, female chimpanzees from Kibale differ least in average shape ratios or %CAs from Mahale females. Males from each of the four communities do not differ significantly in shape ratio or %CA at any femoral or humeral ROI.

Average normalized maximum rigidities are reported for females and males in Table 9.8. Normalizing measures are available for only a subset of the communities, which means that not all four communities can be compared. Female chimpanzees from Taï exhibit significantly greater normalized maximum rigidity than Kibale female chimpanzees at left H50 and H65 ROIs (Table 9.9). In a smaller sample of right humeri, the difference in normalized maximum rigidity between females from these two communities approaches, but does not reach statistical significance at each ROI ($p = 0.064$). Male chimpanzees from Kibale, Mahale, and Taï communities do not differ significantly from one another in normalized maximum rigidity at any femoral or humeral ROI.

Table 9.4 Community structural comparisons

I_{max}/I_{min}	n	F	p	Bonferroni *post hoc* analysis	I_{max}/I_{min}	n	F	p	LSD *post hoc* analysis
Left F35	35	3.560	0.025	T > M	Right F35	25	7.382	0.004	T>>M; T>K
Left F50	35	3.312	0.033	None	Right F50	25	6.901	0.005	T>>K; T>M
Left F65	35	0.047	0.986	—	Right F65	25	3.233	0.059	—
Left H35[a]	37	8.250	<0.001	K, M >> T; G > T	Right H35	23	9.244	0.001	K>>T; K>M
Left H50	37	8.066	<0.001	T >> G; K > G	Right H50	23	7.765	0.003	T>>K, M
Left H65[a]	37	4.175	0.013	T >> G	Right H65[a]	23	3.661	0.044	T>M
%CA					%CA				
Left F35	35	3.405	0.030	None	Right F35[a]	25	0.775	0.473	—
Left F50	35	5.080	0.006	M, T > G	Right F50[a]	25	0.887	0.426	—
Left F65	35	5.745	0.003	M, T >> G; K > G	Right F65[a]	25	0.632	0.541	—
Left H35	37	4.492	0.009	K, M, T > G	Right H35[a]	23	0.240	0.789	—
Left H50	37	7.419	0.001	M, T >> G; K > G	Right H50[a]	23	0.169	0.846	—
Left H65	37	9.465	< 0.001	K, M, T >> G	Right H65	23	0.312	0.735	—
sI_{max}[b]					sI_{max}[b]				
Left F35[a]	20	2.490	0.113	—	Right F35	20	3.614	0.049	T>M
Left F50[a]	20	2.948	0.080	—	Right F50	20	5.710	0.013	T>>K; T>M
Left F65	20	1.521	0.247	—	Right F65	20	3.717	0.046	T>K, M
Left H35	18	0.007	0.937	—	Right H35	14	0.014	0.907	—
Left H50	18	2.706	0.119	—	Right H50	14	2.656	0.129	—
Left H65	18	1.846	0.193	—	Right H65	14	0.713	0.415	—

$p < 0.05$; $p << 0.01$; G = Gombe, K = Kibale, M = Mahale, T = Taï; F = femur, H = humerus; 35 = mid-distal diaphysis, 50 = midshaft, 65 = mid-proximal diaphysis.

[a]Failed a Levene test of homogeneity of variances, thus when ANOVA was significant a Tamhane's T2 *post hoc* analysis was conducted.

[b]Femoral comparisons exclude Gombe, while humeral comparisons exclude Gombe and Mahale because of unavailable data for the scaling measurements.

Table 9.5 Sex-specific community means (1 SD) for shape ratios (I_{max}/I_{min}) and %CA

	Gombe (left)	Kibale (left)	Kibale (right)	Mahale (left)	Mahale (right)	Taï (left)	Taï (right)
Females							
F35 shape	1.38	1.48	1.50	1.30	1.29	1.48	1.53
	(0.21)	(0.07)	(0.11)	(0.05)	(0.04)	(0.09)	(0.14)
	5	3	3	3	3	8	8
F50 shape	1.45	1.45	1.47	1.29	1.33	1.47	1.53
	(0.17)	(0.12)	(0.13)	(0.09)	(0.10)	(0.09)	(0.14)
	5	3	3	3	3	8	8
F65 shape	1.43	1.46	1.46	1.42	1.40	1.45	1.51
	(0.09)	(0.05)	(0.07)	(0.16)	(0.16)	(0.15)	(0.09)
	5	3	3	3	3	8	8
H35 shape	1.24	1.36	1.39	1.25	1.21	1.14	1.15
	(0.09)	(0.09)	(0.09)	(0.02)	(0.04)	(0.09)	(0.09)
	5	4	2	3	2	9	8
H50 shape	1.13	1.24	1.18	1.19	1.16	1.38	1.40
	(0.07)	(0.16)	(0.20)	(0.06)	(0.01)	(0.12)	(0.09)
	5	4	2	3	2	9	8
H65 shape	1.08	1.16	1.12	1.15	1.10	1.29	1.35
	(0.04)	(0.09)	(0.01)	(0.12)	(0.03)	(0.14)	(0.15)
	5	4	2	3	2	9	8
F35 %CA	44.7	48.7	51.3	65.7	64.8	60.8	60.2
	(7.6)	(0.7)	(2.9)	(14.8)	(14.6)	(5.5)	(5.2)
	5	2	2	3	3	8	8
F50 %CA	48.1	55.1	57.2	71.7	70.3	66.4	65.7
	(9.8)	(7.5)	(9.7)	(15.9)	(16.3)	(5.3)	(4.8)
	5	2	2	3	3	8	8
F65 %CA	47.3	55.9	56.0	72.3	71.8	67.0	67.2
	(9.1)	(6.5)	(8.9)	(19.0)	(19.1)	(5.3)	(4.6)
	5	2	2	3	3	8	8

	(9.4)	(9.2)	(-)	(16.6)	(20.4)	(8.3)	(6.0)
	5	3	1	3	2	9	8
H50 %CA	39.9	54.6	56.6	67.0	61.6	60.9	62.8
	(8.0)	(6.6)	(-)	(15.8)	(20.5)	(6.3)	(3.9)
	5	3	1	3	2	9	8
H65 %CA	37.9	54.5	56.3	63.3	58.8	60.8	61.9
	(6.2)	(3.4)	(–)	(16.8)	(23.3)	(8.7)	(6.4)
	5	3	1	3	2	9	8
Males							
F35 shape	1.38	1.42	1.37	1.34	1.34	1.53	1.51
	(0.17)	(0.12)	(0.06)	(–)	(–)	(0.08)	(0.14)
	2	6	5	1	1	5	3
F50 shape	1.39	1.34	1.28	1.34	1.42	1.50	1.55
	(0.16)	(0.17)	(0.06)	(–)	(–)	(0.08)	(0.23)
	2	6	5	1	1	5	3
F65 shape	1.39	1.42	1.39	1.37	1.31	1.37	1.45
	(0.21)	(0.21)	(0.15)	(–)	(–)	(0.11)	(0.03)
	2	6	5	1	1	5	3
H35 shape	1.30	1.48	1.55	1.23	1.17	1.11	1.22
	(0.12)	(0.23)	(0.20)	(–)	(–)	(0.06)	(0.17)
	4	6	6	1	1	3	2
H50 shape	1.10	1.22	1.17	1.17	1.14	1.26	1.29
	(0.02)	(0.11)	(0.07)	(–)	(–)	(0.10)	(0.17)
	4	6	6	1	1	3	2
H65 shape	1.14	1.21	1.18	1.18	1.12	1.19	1.08
	(0.05)	(0.09)	(0.07)	(–)	(–)	(0.12)	(0.05)
	4	6	6	1	1	3	2

(continued)

Table 9.5 (continued)

	Gombe (left)	Kibale (left)	Kibale (right)	Mahale (left)	Mahale (right)	Taï (left)	Taï (right)
F35 %CA	63.6	63.3	62.4	63.9	63.6	63.4	63.0
	(10.7)	(7.7)	(6.1)	(–)	(–)	(7.2)	(2.3)
	2	4	4	1	1	5	3
F50 %CA	65.3	69.0	68.8	73.8	73.9	69.2	70.1
	(10.6)	(6.9)	(6.2)	(–)	(–)	(6.9)	(1.6)
	2	4	4	1	1	5	3
F65 %CA	65.0	70.4	71.7	76.9	76.7	72.2	72.6
	(11.1)	(6.2)	(6.0)	(–)	(–)	(7.7)	(3.2)
	2	4	4	1	1	5	3
H35 %CA	63.0	68.6	68.9	74.1	75.0	65.4	67.2
	(8.7)	(4.2)	(4.8)	(–)	(–)	(10.4)	(15.4)
	4	4	4	1	1	3	2
H50 %CA	55.9	63.2	63.4	69.2	69.5	63.3	63.6
	(9.1)	(5.5)	(4.9)	(–)	(–)	(6.1)	(8.2)
	4	4	4	1	1	3	2
H65 %CA	50.4	60.5	60.7	61.6	61.2	60.6	59.1
	(8.1)	(8.1)	(7.4)	(–)	(–)	(3.7)	(7.2)
	4	4	4	1	1	3	2

Consult Table 9.3 for definitions

Table 9.6 Community sex-specific comparisons in diaphyseal shape (I_{max}/I_{min})

	n	*F*	*p*	Bonferroni *post hoc* analysis		*n*	*F*	*p*	LSD *post hoc* analysis
Females (left)					Females (right)[a]				
F35	19	1.914	0.171	—	F35	14	4.270	0.042	T > M
F50	19	1.742	0.201	—	F50	14	2.600	0.119	—
F65	19	0.093	0.963	—	F65	14	1.201	0.338	—
H35	21	7.227	0.002	K >> T	H35	12	6.134	0.021	K >> T
H50	21	5.925	0.006	T >> G	H50	12	5.846	0.024	T > K, M
H65[b]	21	4.120	0.023	T > G	H65	12	4.462	0.045	T > M
Males (left)					Males (right)[a]				
F35	14	1.410	0.297	—	F35	9	2.424	0.169	—
F50	14	1.216	0.354	—	F50[b]	9	3.426	0.102	—
F65	14	0.089	0.964	—	F65	9	0.517	0.621	—
H35[b]	14	3.304	0.066	—	H35	9	3.231	0.112	—
H50	14	2.081	0.166	—	H50	9	1.408	0.315	—
H65	14	0.621	0.617	—	H65	9	2.080	0.206	—

$p < 0.05$; $p << 0.01$; G = Gombe, K = Kibale, M = Mahale, T = Taï; F = femur, H = humerus; 35 = mid-distal diaphysis, 50 = midshaft, 65 = mid-proximal diaphysis.

[a] Gombe individuals excluded from comparisons.

[b] Failed a Levene test of homogeneity of variances, thus when ANOVA was significant a Tamhane's T2 *post hoc* analysis was conducted.

Table 9.7 Community sex-specific comparisons in %CA

	n	*F*	*p*	Bonferroni *post hoc* analysis		*n*	*F*	*p*	LSD *post hoc* analysis
Females (left)					Females (right)[a]				
F35[b]	19	4.576	0.018	T > G	F35[b]	14	0.561	0.586	—
F50	19	4.858	0.015	M, T > G	F50[b]	14	0.469	0.637	—
F65[b]	19	4.906	0.014	T > G	F65[b]	14	0.443	0.653	—
H35	21	6.318	0.004	T >> G; K, M > G	H35[b]	12	0.371	0.700	—
H50	21	8.137	0.001	M, T >> G; K > G	H50[b]	12	0.017	0.983	—
H65	21	7.757	0.002	M, T >> G; K > G	H65[b]	12	0.078	0.926	—
Males (left)					Males (right)[a]				
F35	14	0.044	0.987	—	F35	9	0.099	0.907	—
F50	14	0.370	0.777	—	F50	9	0.657	0.552	—
F65	14	0.804	0.520	—	F65	9	0.502	0.629	—
H35	14	0.871	0.488	—	H35[b]	9	0.396	0.689	—
H50	14	1.690	0.232	—	H50	9	0.662	0.550	—
H65	14	2.119	0.161	—	H65	9	0.039	0.962	—

$p < 0.05$; $p << 0.01$; G = Gombe, K = Kibale, M = Mahale, T = Taï; F = femur, H = humerus; 35 = mid-distal diaphysis, 50 = midshaft, 65 = mid-proximal diaphysis.

[a] Gombe individuals excluded from comparisons.

[b] Failed a Levene test of homogeneity of variances, thus when ANOVA was significant a Tamhane's T2 *post hoc* analysis was conducted.

Table 9.8 Sex-specific community means (1 SD) for normalized (sI_{max})

	Kibale (left)	Kibale (right)	Mahale (left)	Mahale (right)	Taï (left)	Taï (right)
Females						
F35 sI_{max}	1.04	1.06	1.00	1.05	1.53	1.40
	(0.25)	(0.17)	(0.09)	(0.05)	(0.28)	(0.21)
	3	3	3	3	3	4
F50 sI_{max}	1.08	1.10	1.03	1.11	1.57	1.48
	(0.23)	(0.20)	(0.14)	(0.14)	(0.29)	(0.20)
	3	3	3	3	3	4
F65 sI_{max}	1.10	1.10	1.09	1.12	1.48	1.41
	(0.30)	(0.28)	(0.14)	(0.09)	(0.23)	(0.15)
	3	3	3	3	3	4
H35 sI_{max}	0.70	0.66	—	—	0.80	0.82
	(0.07)	(0.07)			(0.09)	(0.07)
	4	2			5	4
H50 sI_{max}	0.76	0.71	—	—	0.98	1.01
	(0.10)	(0.13)			(0.19)	(0.19)
	4	2			5	4
H65 sI_{max}	0.75	0.74	—	—	0.97	0.98
	(0.04)	(0.03)			(0.20)	(0.21)
	4	2			5	4
Males						
F35 sI_{max}	1.28	1.29	1.19	1.22	1.28	1.29
	(0.15)	(0.14)	(–)	(–)	(0.44)	(0.01)
	6	5	1	1	2	2
F50 sI_{max}	1.29	1.26	1.20	1.25	1.30	1.33
	(0.13)	(0.09)	(–)	(–)	(0.46)	(0.07)
	6	5	1	1	2	2
F65 sI_{max}	1.30	1.29	1.22	1.21	1.24	1.36
	(0.12)	(0.06)	(–)	(–)	(0.25)	(0.05)
	6	5	1	1	2	2
H35 sI_{max}	0.84	0.86	—	—	0.78	—
	(0.17)	(0.17)			(–)	
	6	6			1	
H50 sI_{max}	0.87	0.85	—	—	0.97	—
	(0.12)	(0.11)			(–)	
	6	6			1	
H65 sI_{max}	0.93	0.90	—	—	1.01	—
	(0.12)	(0.11)			(–)	
	6	6			1	

Consult Table 9.3 for definitions

Table 9.9 Kruskal–Wallis tests for community sex-specific comparisons[a] in normalized sI_{max}

	n	χ^2	p	Mean ranks		n	χ^2	p	Mean rank
Females(left)					Females(right)				
F35	9	4.622	0.099	—	F35	10	5.573	0.062	—
F50	9	4.356	0.113	—	F50	10	4.564	0.102	—
F65	9	4.356	0.113	—	F65	10	4.564	0.102	—
H35[b]	9	2.160	0.142	—	H35[b]	6	3.429	0.064	—
H50[b]	9	4.860	0.027	T > K	H50[b]	6	3.429	0.064	—
H65[b]	9	6.000	0.014	T > K	H65[b]	6	3.429	0.064	—
Males (left)					Males (right)				
F35	9	0.156	0.925	—	F35	8	1.200	0.549	—
F50	9	0.156	0.925	—	F50	8	1.917	0.384	—
F65	9	0.289	0.866	—	F65	8	4.200	0.122	—
H35[b]	7	0.250	0.617	—	H35	6	—	—	—
H50[b]	7	1.000	0.317	—	H50	6	—	—	—
H65[b]	7	0.250	0.617	—	H65	6	—	—	—

$p < 0.05$; K = Kibale, M = Mahale, T = Taï; F = femur, H = humerus; 35 = mid-distal diaphysis, 50 = midshaft, 65 = mid-proximal diaphysis.
[a]Gombe individuals excluded.
[b]Gombe and Mahale individuals excluded from comparisons.

Ranks of Community Structural Properties versus Habitat Differences

To assess one aspect of mobility, Carlson et al. (2008) used average annual rainfall and ground cover to estimate the density of obstacles, e.g., trees, in the habitats of Gombe, Mahale, and Taï chimpanzees. Taï Forest National Park was estimated to have the highest prevalence of obstacles, and chimpanzees from Taï often had higher shape ratios than chimpanzees from Gombe or Mahale (Carlson et al. 2008). Extending the comparison to include Kibale chimpanzees indicates that Kibale has the lowest average annual rainfall of the four communities, and presumably exceeds only Gombe in ground cover according to qualitative estimates (Table 9.10). Kibale chimpanzees, however, appear not to fit the proposed trend (Carlson et al. 2008) since they frequently have the highest or second highest shape ratio to Taï chimpanzees at each of the ROIs.

To assess a second aspect of mobility, Carlson et al. (2008) used elevation range and slope of terrain to estimate terrain elevation/ruggedness in the habitats of Gombe, Mahale, and Taï chimpanzees. Kibale National Park outranks Taï Forest National Park in these measures of habitat complexity, but falls below estimates from Gombe and Mahale. Kibale shape ratios at femoral and humeral ROIs, particularly at the former, usually exceed those of Gombe and Mahale individuals (Table 9.10). In other words, diaphyses of Kibale chimpanzees usually are less circular than chimpanzees from Gombe and Mahale, but more circular than

Table 9.10 Ranked shape ratios (I_{max}/I_{min}) versus ranked habitat characteristics

	Annual rainfall	Ground cover	Elevation range	Slope of terrain[a]
Community ranking	M > T > G > K	T > M ≈ K > G	M > G > K > T	G > M > K > T
Left femur shape	Female	35%: T = K > G > M	50%: T > K = G > M	65%: K > T > G > M
	Male	35%: T > K > G > M	50%: T > G > K = M	65%: K > G > T = M
Right femur shape	Female	35%: T > K > M	50%: T > K > M	65%: T > K > M
	Male	35%: T > K > M	50%: T > M > K	65%: T > K > M
Left humerus shape	Female	35%: K > M > G > T	50%: T > K > M > G	65%: T > K > M > G
	Male	35%: K > G > M > T	50%: T > K > M > G	65%: K > T > M > G
Right humerus shape	Female	35%: K > M > T	50%: T > K > M	65%: T > K > M
	Male	35%: K > T > M	50%: T > K > M	65%: K > M > T

G = Gombe, K = Kibale, M = Mahale, T = Taï.

[a]Quantitative data on slope of terrain from Taï Forest National Park were unavailable, but based on elevation ranges, it seems reasonable to characterize terrain at Taï as flattest among the sites.

chimpanzees from Taï. This supports the trend between terrain elevation/ruggedness and diaphyseal shape that was noted by Carlson et al. (2008).

Discussion

Communities differ significantly in shape at a majority of femoral and humeral diaphyseal locations. Kibale individuals differ specifically from Taï individuals at several locations, but except for the right humeral mid-distal diaphysis (H35) and the left humeral midshaft (H50), Kibale individuals do not differ significantly from Mahale or Gombe individuals at any location. Thus, addition of a fourth chimpanzee community to an earlier comparison of three chimpanzee communities that reported significantly greater shape ratios, i.e., more elliptical diaphyses, in Taï chimpanzees compared to other chimpanzees (Carlson et al. 2008) largely reinforces the same pattern: Taï chimpanzees, particularly females, tend to exhibit distinct diaphyseal shapes. Because diaphyseal circularity, e.g., more evenly distributed bone mass in multiple directions, is a stable structural solution in the face of multidirectional deformations (Biewener 2003), which according to a limited body of data (Swartz et al. 1989; Demes et al. 2001; Demes and Carlson 2009) probably characterizes arboreal locomotor behaviors compared to terrestrial locomotor behaviors, it is uncertain why Taï individuals exhibit more elliptical femoral diaphyses even after adding individuals from a fourth community (Kibale) to comparisons.

One possible explanation proposed by Carlson et al. (2008) focused on the fact that chimpanzees tend to exhibit greater mediolateral (ML) rigidity relative to anteroposterior (AP) rigidity throughout the femoral diaphysis, with a less consistent disparity present in the humeral diaphysis (Carlson 2002, 2005). In addition to potential community-level differences between the percentage of locomotion in arboreal and terrestrial settings, qualitative differences within terrestrial quadrupedalism at the communities might be possible to infer, if morphological trends are assessed in the context of ecological trends. Based on the effect that changes in direction have on mediolateral external forces (Demes et al. 2006) and bony morphology (Carlson and Judex 2007), Carlson et al. (2008) proposed that chimpanzees occupying more densely forested habitats may experience greater side-to-side forces during terrestrial quadrupedalism compared to chimpanzees occupying less densely forested habitats, which may further enhance the disparity between ML rigidity and AP rigidity in diaphyseal cross sections. The ultimate result in this case could be more elliptical diaphyses in chimpanzees inhabiting denser forested habitats. The inclusion of Kibale chimpanzees in the comparison of chimpanzee communities does not support this possibility in a straightforward manner. The forest at Kibale is comparatively more "open" than Taï and probably at least as open, if not more, than the forests at Gombe and Mahale according to estimates of average annual rainfall and descriptions of ground cover (Table 9.10). Yet, Kibale chimpanzees tend to possess more elliptical diaphyses than Gombe and Mahale chimpanzees, and occasionally more elliptical diaphyses than Taï chimpanzees. Additional comparative research is necessary to assess whether this ecomorphological relationship can be substantiated in free-ranging chimpanzees.

Describing forests at each of the four communities as “more dense” or “less dense” almost certainly oversimplifies complexities in forest structure. Densities of forest canopy and forest understory often are inversely related because the former can directly affect the amount of sunlight that is available to the latter (Gentry and Emmons 1987; Montgomery and Chazdon 2001; Dial et al. 2004). Further, forest understory can be a complex relationship between palm cover and small sapling cover (Harms et al. 2004), each of which may present different challenges when maneuvering through the understory. Thus, when considering a forest with more “closed” canopy conditions, e.g., Taï versus a forest with more “open” canopy conditions, e.g., Gombe, the understory conditions may be variable as well. In forests characterized by a dense understory, game trails and man-made transects cleared of vegetation, e.g., as can exist at habituated sites, may serve as alternate travel routes that would avoid much of the side-to-side terrestrial maneuvering that would be necessary in their absence. The frequency of maneuvering around obstacles during terrestrial quadrupedalism may not be easily predicted from simple measures of the degree to which the canopy is “open” versus “closed.” Further work toward identifying ecological variables that could be better estimators of maneuverability and quantifying the extent of maneuverability itself would be useful.

After adding Kibale chimpanzees to a previous comparisons of free-ranging chimpanzees (Carlson et al. 2008), Mahale chimpanzees exhibit the lowest or next to lowest average shape ratios in femoral and humeral ROIs, except among left humeral mid-distal diaphyses (H35) of females. Mahale is characterized by the greatest range of elevations compared to the other three communities, whereas Kibale exceeds only Taï (Table 9.10). Because chimpanzees generally have greater ML rigidity than AP rigidity at diaphyseal ROIs, particularly in the femur (Carlson 2002; Carlson et al. 2006), Carlson et al. (2008) suggested that an increase in circularity of Mahale femoral diaphyses is consistent with increased AP rigidity relative to ML rigidity. The addition to this comparison of a fourth habitat, Kibale, reinforces the possibility for a terrain effect in long bone diaphyseal structure of chimpanzees, which would appear to parallel a similar terrain effect in humans (Ruff 1999).

In comparing %CA at diaphyseal ROIs from chimpanzees of all four communities, i.e., left elements only, Kibale chimpanzees exhibit greater %CA than Gombe chimpanzees at four of the six ROIs. As noted in an earlier study (Carlson et al. 2008), Mahale and Taï chimpanzees exhibit greater %CA than Gombe individuals at five of these same ROIs (Table 9.4). In the present comparisons, Kibale chimpanzees appear similar to Mahale and Taï chimpanzees. The similarity between individuals from Kibale, Mahale, and Taï communities is illustrated further by comparing %CA in right diaphyses, where no significant group differences are observed after removing Gombe individuals from the comparisons (Table 9.4). Reduced %CA in concert with diaphyseal shape change is helpful for identifying reduced functional loading in the limbs (Carlson et al. 2008). Reduced functional loading of the limbs may be a result of alterations observed in chimpanzee locomotor repertoires with advancing age (Goodall 1986; Morbeck et al. 2002). Average ages for chimpanzees in the samples from Gombe, Mahale, and Tai are 33.3, 30.1, and 25.0 years, respectively (Carlson et al. 2008). Unfortunately, similarly precise age estimates for Kibale chimpanzees are unavailable (Carter et al. 2008). Based on the combination of trends in shape ratios and %CA comparisons, it is likely that chimpanzees comprising the Kibale sample had not experienced an age-induced reduction in activity levels, such as what seems to characterize many of the individuals comprising the Gombe sample.

Finally, overall similarity in structural properties of the chimpanzee femoral and humeral diaphyses between Kibale and Mahale chimpanzees, versus the relatively distinctive position of Taï chimpanzees among the four communities parallels current notions of genetic relatedness between chimpanzees in the four communities. The western chimpanzee (*Pan troglodytes verus*) diverged from the more closely related central (*P. t. troglodytes*) and eastern chimpanzees (*P. t. schweinfurthii*) approximately 0.84 million years ago (Becquet et al. 2007). A similar pattern can be found in that the western chimpanzee subspecies, e.g., Taï chimpanzees, has a higher reported body mass (Smith and Jungers 1997) than the eastern chimpanzee subspecies, whereas among *Pan troglodytes schweinfurthii* populations, Gombe chimpanzees are reportedly lower in estimated body mass than Kibale and Mahale chimpanzees (Carter et al. 2008). These trends are in contrast to greater reported similarity in humeral and femoral lengths of Kibale and Taï individuals relative to Gombe and Mahale individuals (Gunji et al. 1998; Carter et al. 2008; Zihlman et al. 2008). Clearly, genes are not the only contributing factor to variation in diaphyseal structural properties in the sample. Assessing cross-sectional geometric properties of the limb diaphyses of additional western chimpanzee populations could be illuminating in this regard. In addition, incorporating additional populations that would expand the range of habitats to include more extreme conditions such as dry, open habitats, e.g. Toro-Semliki Wildlife Reserve, Uganda, could be equally useful.

Acknowledgments We thank the L.S.B. Leakey Foundation for financially supporting this research. Support for earlier phases of this work also was graciously provided by the US National Science Foundation (DDIG, BCS-0002686). We acknowledge and thank the cooperation of the government of Uganda, the Uganda National Council for Science and Technology, the Uganda Wildlife Authority, and Makerere University. We wish to thank in particular, Dr. John Kasenene for facilitating access to the Kibale chimpanzee skeletal material. We also thank the Field Museum of Natural History, Chicago, IL for allowing access to specimens in their care. We thank Bill Stanley and Lawrence Heaney for their assistance during work conducted at the Field Museum. We thank the Kampala Imaging Centre for permitting use of their CT scanner in Kampala, and Craig Wambura for his skillful assistance in scanning specimens. We thank Rush University Medical Center, Department of Radiology for permitting use of their CT scanner in Chicago, and Chris Archer and Edward Jones for their valuable assistance in scanning specimens. We also thank the Institut für Radiologie, Kantonspital, Winterthur and the Universitätspital, Zürich for permitting the use of their CT scanners. Isuf Hoxha and Michael Krupa provided critical assistance with scanning specimens in Zürich. We also thank David Burr, Diane Doran-Sheehy, Jane Goodall, Harumoto Gunji, Kevin Hunt, and Tanya Smith for their assistance at various times during this research. Luci-Betti Nash kindly provided her artistic skill in drawing Figure 9.1. Lastly, we are deeply grateful to the symposium organizers and volume editors, Evie Vereecke and Kristiaan D'Août, both for their invitation to participate in the symposium, and for their editorial assistance in preparing the manuscript.

References

Becquet C, Patterson N, Stone AC, Przeworksi M, Reich D (2007) Genetic structure of chimpanzee populations. PLoS Genetics 3, e66.

Bhatavadekar NB, Daegling DJ, Rapoff AJ (2006) Application of an image-based weighted measure of skeletal loading stiffness to great ape mandibles. Am J Phys Anthropol 131:243–251.

Biewener AA (2003) Animal Locomotion. Oxford University Press, Oxford.

Boesch C, Boesch-Achermann H (2000) The Chimpanzees of the Taï Forest – Behavioural Ecology and Evolution. Oxford University Press, New York.

Burr DB, Piotrowski G, Martin RB, Cook PN (1982) Femoral mechanics in the lesser bushbaby (*Galago senegalensis*): structural adaptations to leaping in primates. Anat Rec 202:419–429.

Burr DB, Ruff CB, Johnson C (1989) Structural adaptations of the femur and humerus to arboreal and terrestrial environments in three species of macaque. Am J Phys Anthropol 79:357–367.

Burr DB, Milgrom C, Fyhrie D, Forwood M, Nyska M, Finestone A, Hoshaw S, Saiag E, Simkin A (1996) In vivo measurement of human tibial strains during vigorous activity. Bone 18:405–410.

Carlson KJ (2002) Shape and material properties of African pongid femora and humeri: their relationship to observed positional behaviors. PhD dissertation, Indiana University.

Carlson KJ (2005) Investigating the form-function interface in African apes – relationships between principal moments of area and positional behaviors in femoral and humeral diaphyses. Am J Phys Anthropol 127:312–334.

Carlson KJ, Demes B, Franz T (2005) Mediolateral forces associated with quadrupedal gaits of lemurids. J Zool (Lond) 266:261–273.

Carlson KJ, Doran-Sheehy DM, Hunt KD, Nishida T, Yamanaka A, Boesch C (2006) Locomotor behavior and long bone morphology in individual free-ranging chimpanzees. J Hum Evol 50:394–404.

Carlson KJ, Grine FE, Pearson OM (2007) Robusticity and sexual dimorphism in the postcranium of modern hunter-gatherers from Australia. Am J Phys Anthropol 134:9–23.

Carlson KJ, Judex S (2007) Increased non-linear locomotion alters diaphyseal bone shape. J Exp Biol 210:3117–3125.

Carlson KJ, Sumner DR, Morbeck ME, Nishida T, Yamanaka A, Boesch C (2008) The role of non-behavioral factors in adjusting long bone diaphyseal structure in free-ranging chimpanzees. Int J Primatol 29:1401–1420.

Carter ML, Pontzer H, Wrangham RW, Kerbis Peterhans J (2008) Skeletal pathology in *Pan troglodytes schweinfurthii* in Kibale National Park, Uganda. Am J Phys Anthropol 135:389–403.

Chapman CA, Chapman LJ, Wrangham R, Isabirye-Basuta G, Ben-David K (1997) Spatial and temporal variability in the structure of a tropical forest Afr J Ecol 35:287–302.

Currey JD (2002) Bones: Structure and Mechanics. Princeton University Press, Princeton, NJ.

Demes B, Jungers WL (1989) Functional differentiation of long bones in Lorises. Folia Primatol 52:58–69.

Demes B, Jungers WL, Selpien K (1991) Body size, locomotion, and long bone cross-sectional geometry in Indriid primates. Am J Phys Anthropol 86:537–547.

Demes B, Jungers WL (1993) Long bone cross-sectional dimensions, locomotor adaptations and body size in prosimian primates. J Hum Evol 25:57–74.

Demes B, Stern JT, Hausman MR, Larson SG, McLeod KJ, Rubin CT (1998) Patterns of strain in the Macaque ulna during functional activity. Am J Phys Anthropol 106:87–100.

Demes B, Qin Y, Stern JT Jr, Larson SG, Rubin CT (2001) Patterns of strain in the Macaque tibia during functional activity. Am J Phys Anthropol 116:257–265.

Demes B, Carlson KJ, Franz TM (2006) Cutting corners: the dynamics of turning behaviors in two primate species. J Exp Biol 209:927–937.

Demes B, Carlson KJ (2009) Locomotor variation and bending regimes of capuchin limb bones. Am J Phys Anthropol 139:558–571.

Devlin MJ, Lieberman DE (2007) Variation in estradiol level affects cortical bone growth in response to mechanical loading in sheep. J Exp Biol 210:602–613.

Dial R, Bloodworth B, Lee A, Boyne P, Heys J (2004) The distribution of free space and its relation to canopy composition at six forest sites. Forest Sci 50:312–325.

Gentry AH, Emmons LH (1987) Geographical variation in fertility, phenology, and composition of the understory of Neotropical forests. Biotropica 19:216–227.

Goodall J (1986) The Chimpanzees of Gombe. The Belknap Press of Harvard University Press, Cambridge, MA.

Green DJ, Gordon AD, Richmond BG (2007) Limb-size proportions in *Australopithecus afarensis* and *Australopithecus africanus*. J Hum Evol 52:187–200.

Griffin NL (2008) Bone architecture of the hominin second proximal pedal phalanx: a preliminary investigation. J Hum Evol 54:162–168.

Gunji H, Shimizu D, Hosaka M, Huffman A, Kawanaka K, Matsumoto-Oda A, Nishida T (1998) Notes on some chimpanzee skeletons from the Mahale Mountains National Park, Tanzania. In: Nishida T (ed), Resource Use Patterns and Social Structure Among Chimpanzees. Nisshindo Printer, Kyoto, pp 113–130.

Guy F, Brunet M, Schmittbuhl M, Viriot L (2003) New approaches in hominoid taxonomy: morphometrics. Am J Phys Anthropol 121:198–218.

Haeusler M, McHenry HM (2007) Evolutionary reversals of limb proportions in early hominids? Evidence from KNM-ER 3735 (*Homo habilis*). J Hum Evol 53:383–405.

Harms KE, Powers JS, Montgomery RA (2004) Variation in small sapling density, understory cover, and resource availability in four Neotropical forests. Biotropica 36:40–51.

Holt B (2003) Mobility in Upper Paleolithic and Mesolithic Europe: evidence from the lower limb. Am J Phys Anthropol 122:200–215.

Hunt KD (1992) Positional behavior of *Pan troglodytes* in the Mahale Mountains and Gombe Stream National Parks, Tanzania. Am J Phys Anthropol 87:83–105.

Hunt KD, McGrew WC (2002) Chimpanzees in the dry habitats of Assirik, Senegal and Semliki Wildlife Reserve, Uganda. In: Boesch C, Hohmann G, Marchant LF (eds), Behavioural Diversity in Chimpanzees and Bonobos. Cambridge University Press, Cambridge, pp 35–51.

Jungers WL, Burr DB, Cole MS (1998) Body size and scaling of long bone geometry, bone strength, and positional behavior in cercopithecoid primates. In: Strasser E, Fleagle J, Rosenberger A, McHenry HM (eds), Primate Locomotion – Recent Advances. Plenum Press, New York, pp 309–330.

Lieberman DE, Polk JD, Demes B (2004) Predicting long bone loading from cross-sectional geometry. Am J Phys Anthropol 123:156–171.

Lockwood CA, Kimbel WH, Lynch JM (2004) Morphometrics and hominoid phylogeny: support for a chimpanzee-human clade and differentiation among great ape subspecies. PNAS 101:4356–4360.

Madar SI, Rose MD, Kelley J, MacLatchy L, Pilbeam D (2002) New *Sivapithecus* postcranial specimens from the Siwaliks of Pakistan. J Hum Evol 42:705–752.

Marchi D, Sparacello VS, Holt BM, Formicola V (2006) Biomechanical approach to the reconstruction of activity patterns in Neolithic Western Liguria, Italy. Am J Phys Anthropol 131:447–455.

Marchi D (2007) Relative strength of the tibia and fibula and locomotor behavior in hominoids. J Hum Evol 53:647–655.

Martin RB, Burr DB, Sharkey NA (1998) Skeletal Tissue Mechanics. Springer-Verlag, New York.

Montgomery RA, Chazdon RL (2001) Forest structure, canopy architecture, and light transmittance in tropical wet forests. Ecology 82:2707–2718.

Morbeck ME, Zihlman AL, Sumner DR, Galloway A (1991) Poliomyelitis and skeletal asymmetry in Gombe chimpanzees. Primates 32:77–91.

Morbeck ME (1999) Life history of Gombe chimpanzees: the inside view from the skeleton. In: Strum SC, Lindburg DG, Hamburg D (eds), The New Physical Anthropology: Science, Humanism, and Critical Reflection. Prentice-Hall, Upper Saddle River, NJ, pp 18–31.

Morbeck ME, Galloway A, Sumner DR (2002) Getting old at Gombe: skeletal aging in wild-ranging chimpanzees. In: Erwin J, Hof PR (eds), Aging in Nonhuman Primates Interdisciplinary Topics in Gerontology, Vol. 31. Karger, Basel/New York, pp 48–62.

Nagurka ML, Hayes WC (1980) An interactive graphics package for calculating cross-sectional properties of complex shapes. J Biomech 13:59–64.

Nishida T (1990) The Chimpanzees of the Mahale Mountains – Sexual and Life History strategies. University of Tokyo Press, Tokyo.

Ohman JC (1993) Cross-sectional geometric properties from biplanar radiographs and computed tomography: functional application to the humerus and femur in hominoids. PhD dissertation, Kent State University.

Pearson OM, Lieberman DE (2004) The aging of Wolff's "Law": ontogeny and responses to mechanical loading in cortical bone. Yrbk Phys Anthropol 47:63–99.

Polk JD, Demes B, Jungers WL, Biknevicius AR, Heinrich RE, Runestad JA (2000) A comparison of primate, carnivoran and rodent limb bone cross-sectional properties: are primates really unique? J Hum Evol 39:297–325.

Pusey AE, Oehlert GW, Williams JM, Goodall J (2005) Influences of ecological and social factors on body mass of wild chimpanzees. Int J Primatol 26:3–31.

Robling AG, Warden SJ, Shultz KL, Beamer WG, Turner CH (2007) Genetic effects on bone mechanotransduction in congenic mice harboring bone size and strength quantitative trait loci. J Bone Miner Res 22:984–991.

Ruff CB (1987) Structural allometry of the femur and tibia in Hominoidea and Macaca. Folia Primatol 48:9–49.

Ruff CB (1989) New approaches to structural evolution of limb bones in primates. Folia Primatol 53:142–159.

Ruff CB, Runestad JA (1992) Primate limb bone structural adaptations. Annu Rev Anthropol 21:407–433.

Ruff CB (1999) Skeletal structure and behavioral patterns of prehistoric Great Basin populations. In: Hemphill BE, Larsen CS (eds), Prehistoric Lifeways in the Great Basin Wetlands: Bioarchaeological Reconstruction and Interpretation. University of Utah Press, Salt Lake City, pp 290–320.

Ruff CB (2002) Long bone articular and diaphyseal structure in Old World monkeys and apes I: Locomotor effects. Am J Phys Anthropol 119:305–342.

Ruff CB, Holt B, Trinkaus E (2006) Who's afraid of the big bad Wolff?: "Wolff's Law" and bone functional adaptation. Am J Phys Anthropol 129:484–498.

Ruff CB (2008) Femoral/humeral strength in early African *Homo erectus*. J Hum Evol 54:383–390.

Schaffler MB, Burr DB, Jungers WL, Ruff CB (1985) Structural and mechanical indicators of limb specialization in primates. Folia Primatol 45:61–75.

Smith RJ, Jungers WL (1997) Body mass in comparative primatology. J Hum Evol 32:523–559.

Stock J, Pfeiffer S (2001) Linking structural variability in long bone diaphyses to habitual behaviors: foragers from southern African Later Stone Age and Andaman Islands. Am J Phys Anthropol 115:337–348.

Sumner DR, Morbeck ME, Lobick JJ (1989) Apparent age-related bone loss among adult female Gombe chimpanzees. Am J Phys Anthropol 79:225–234.

Sumner DR, Andriacchi TP (1996) Adaptation to differential loading: comparison of growth-related changes in cross-sectional properties of the human femur and humerus. Bone 19:121–126.

Swartz SM, Bertram JEA, Biewener AA (1989) Telemetered in vivo strain analysis of locomotor mechanics of brachiating gibbons. Nature 342:270–272.

Terranova CJ (1995a) Functional morphology of leaping behaviors in galagids: associations between loading, limb use and diaphyseal geometry. In: Atterman L, Izard MK, Doyle GA (eds), Creatures of the Dark: the Nocturnal Prosimians. Plenum Press, New York, pp 473–493.

Terranova CJ (1995b) Leaping behaviors and the functional morphology of strepsirhine primate long bones. Folia Primatol 65:181–201.

Turner CH, Hsieh YF, Muller R, Bouxsein ML, Baylink DJ, Rosen CJ, Grynpas MD, Donahue LR, Beamer WG (2000) Genetic regulation of cortical and trabecular bone strength and microstructure in inbred strains of mice. J Bone Miner Res 15:1126–1131.

Wergedal JE, Sheng MH, Ackert-Bicknell CL, Beamer WG, Baylink DJ (2005) Genetic variation in femur extrinsic strength in 29 different inbred strains of mice is dependent on variations in femur cross-sectional geometry and bone density. Bone 36:111–122.

Whitten A, Goodall J, McGrew WC, Nishida T, Reynolds V, Sugiyama Y, Tutin CEG, Wrangham RW, Boesch C (1999) Cultures in chimpanzees. Nature 399:682–685.

Xiong DH, Shen H, Xiao P, Guo YF, Long JR, Zhao LJ, Liu YZ, Deng HY, Li JL, Recker RR, Deng HW (2006) Genome-wide scan identified QTLs underlying femoral neck cross-sectional geometry that are novel studied risk factors of osteoporosis. J Bone Miner Res 21:424–437.

Yamanaka A, Gunji H, Ishida H (2005) Curvature, length, and cross-sectional geometry of the femur and humerus in anthropoid primates. Am J Phys Anthropol 127:46–57.

Zihlman AL, Stahl D, Boesch C (2008) Morphological variation in adult chimpanzees (*Pan troglodytes verus*) of the Taï National Park, Côte d'Ivoire. Am J Phys Anthropol 135:34–41.

Chapter 10
Field Study Methods for Primate Locomotor Ecology and Biomechanics

Mary L. Blanchard and Robin H. Crompton

Abstract Primates has perhaps the greatest diversity of locomotor behavior of any mammalian Order. This may reflect its predominantly arboreal nature: Primates have to navigate a complexly three-dimensional environment comprised of irregularly spaced, discontinuous, often unstable supports. This complexity offers both challenges and opportunities for those wishing to understand primate locomotor adaptation, and requires understanding of both the biomechanical and the ecological interactions between a primate's locomotor capabilities and the environment(s) to which it is exposed during an individual's life history and that of its clade. We distinguish between phenomena that require study under controlled, usually laboratory, conditions; those that can equally or better be studied in the natural environment; and those that can legitimately be studied only in the natural environment. We suggest methods for field studies and discuss how new technologies are blurring the distinction between laboratory and field and permitting a true "field biomechanics."

Keywords Adaptation • Biomechanics • Ecology • Locomotion • Ontogeny • Postcranial

Abbreviations

a	vertical height change
b	horizontal distance traveled
BMR	basal metabolic rate
c	mechanically effective bout length
COM	center of mass
FMR	field metabolic rate
fps	frames per second
GPS	global positioning system

M.L. Blanchard (✉)
Institute of Aging and Chronic Disease, The University of Liverpool, UK
e-mail: mary.blanchard@Liverpool.ac.uk

K. D'Août and E.E. Vereecke (eds.), *Primate Locomotion: Linking Field and Laboratory Research*, Developments in Primatology: Progress and Prospects,
DOI 10.1007/978-1-4419-1420-0_10, © Springer Science+Business Media, LLC 2011

Introduction

Primates is a quintessentially arboreal order. Thus the special physical qualities of all arboreal milieu (three-dimensional complexity of support orientation, diameter and spacing; support instability and compliance; and support discontinuity) combine with the low food quality of the most ubiquitous food resource, leaves, and the temporal and spatial separation of fruit and flying insects to present Primates with particular locomotor and ecological challenges. These challenges are amplified with increasing body size, but at the other end of the size range primates are more often faced with the challenges of predator avoidance.

It is not surprising then that the order Primates displays probably the greatest diversity of locomotor adaptation among all Mammalia, exhibiting more or less all of the Order's locomotor modes with the exception of burrowing, sustained gliding, and powered flight. Nor is it surprising that Primates should excel in the modes of locomotion that best equip them to handle support discontinuity and now, it increasingly appears, support compliance, such as leaping, bridging, arm-swinging, and indeed, bipedalism.

Although studies of the adaptive basis of locomotor morphology date back at least as far as Mollison's (1911) classification of primates as "runners," "climbers," "leapers," "brachiators," and "bipeds," the behavioral basis of this study was anecdotal observations of travelers. Systematic study of the locomotor behavior of primates in their natural habitat began more recently; among the first was Ripley's (1967) study of locomotion of the Hanuman langur, *Presbytis entellus*, in Ceylon (now Sri Lanka). But though this examined many of the parameters necessary for an understanding of locomotion in its natural context, describing the full locomotor repertoire, substrate use, and even seasonal and ontogenetic variation, observations were not made on a quantitative basis. In the same year, Napier and Walker (1967) established a new category of locomotor behavior, "vertical clinging and leaping," based primarily on Napier's anatomical observations on living and fossil prosimians and Walker's field observations of locomotion in Malagasy prosimians, but again, the latter were qualitative rather than quantitative. It was not until Fleagle's field studies of locomotion of Malaysian foret primates (e.g. 1977) that fully quantitative field studies of locomotion began. And it remains true that quantitative field studies of primate locomotor behavior form only a very small proportion of the published studies of the locomotor system.

Our group has been involved in quantitative field studies of primate locomotion since 1978. Study species include eight prosimians, all classified as vertical clingers and leapers by Napier and Walker (1967): *Galago moholi* (southern lesser bushbaby), *Otolemur crassicaudatus* (thick-tailed greater bushbaby), *Tarsius bancanus* (western tarsier), *Lepilemur edwardsi* (Milne-Edwards' sportive lemur), *Avahi occidentalis* (western woolly lemur), *Hapalemur g. griseus* (eastern lesser bamboo lemur), *Propithecus diadema* (diademed simpona), and *Indri indri* (indri), and one hominoid: the quadrumanous climber *Pongo abelii* (Sumatran orang-utan). The prosimians alone span a 50-fold range in body mass, and diets ranging from obligate animalivory to folivory. The habitats in which we have worked include subtropical bush savannah and riparian woodland in South Africa, lowland evergreen rain forest in Borneo and Sumatra, and Malagasy Western Dry Forest and Eastern Montane Rain Forest. Throughout these 30 years we have used essentially the same observational data schedule of Crompton (1980), elaborated to suit local conditions and modified for greater compatibility with the standard data schedule of Hunt et al. (1996). This chapter reviews the variables that our experience has shown are best collected in the field, and discusses why they are important for an understanding of locomotion of free-ranging primates. Our hope is that this might prove useful both to fieldworkers whose main interest is in primate locomotion and to those with a different central focus who would like to include locomotor variables in their studies.

First of all, what is locomotion? We can define it behaviorally as *all aspects of biology related to movement, or prevention of movement, of an animal's body mass relative to its environment*. We need not specify whether an animal is actually moving or staying immobile relative to its environment. Thus, one kind of locomotion is concerned with locomotion in its narrow sense, i.e., the processes that cause the body to displace itself or its segments relative to the environment, and the other is concerned with keeping the body immobile, preventing external forces from displacing the body relative to its environment. The latter is "posture" and there is no clear division between the two. It is only in a very small number of postures that any animal can remain immobile relative to its environment without the expenditure of energy by muscular contraction. In the real world, postural behavior is constantly being disturbed by shifts in position of the body or its segments, where these act to maintain the center of gravity of the animal in the same position, for example, as a branch sways in the wind. Locomotor behavior can best be understood as the way that an animal uses the type of supports, such as the branches or the ground, available in its habitat, in its daily tasks of avoiding predators and locating food or mates.

Primates use locomotion in a wide variety of behavioral contexts: they use it to locate food, to avoid being eaten, to locate mates, and, being mammals, they use it in play. Therefore, a complete field study of locomotion should involve a study of dietary factors: the type of food animals eat, how and where this food is located in the environment and how long it takes to eat it; it involves studies of ontogeny and social behavior, and of activity patterns. Of particular interest to our group is the interaction between locomotor potential, foraging strategy and the distribution of resources and threats, combined with the physical characteristics (orientation, diameter, compliance, stratification, and continuity or discontinuity) of the supports available to access or avoid them: the factors that influence the locomotor choices made by a primate to address the daily needs of survival and reproduction.

Therefore many studies of primate behavior or ecology automatically collect information pertinent to our understanding of primate locomotor adaptation. The purpose of this chapter is to identify the kinds of data that can most usefully be collected in the field, and to suggest how best this should be done for maximum value. By the same token, we can identify the kinds of information that are best collected in the laboratory, although we believe that technological advances are rapidly blurring the boundaries between fieldwork and laboratory studies.

Bock and von Wahlert (1965) described the elements of the form-function complex by removing any reference to the environment from the definition of "function of an anatomical feature." They regarded this reference as the "biological role" of a feature. In the present context, we might usefully refer to the "performance" rather than the "function" of a feature of the locomotor system. This enables us more readily to distinguish the elements of locomotor adaptation that are perhaps best studied under controlled conditions from those which more or less exclusively require field study The most traditional aspect of laboratory studies is that of skeletal morphology, although even here, other contributors (see Carlson et al., Chapter 9) show how this is starting to move into the field, so as to examine population trends in bony morphology. We may then consider soft tissue characters, which nearly always require invasive study, such as mass, physiological cross-sectional area, moment arm, and fiber population of muscles. Further, there are those that may or must be addressed in the living animal, such as muscle pennation and muscle-tendon unit shortening, physiological measurements of metabolic energy cost, and the kinematics and kinetics of locomotion. At this end of the spectrum accelerometers can be used to gather kinetics in the field, as has been done for flying lemurs (Byrnes et al. 2008), although some questions will require the fine resolution of a force plate and hence at least partially controlled conditions. High-speed still/movie cameras or 200 fps (frames per second) high-definition camcorders (such as, at time of writing, the Casio Exilim Pro EX-F1, with up to 1200 fps in burst mode and good low-light performance; or the Sony HDR-SR10E respectively, at around £500–600) now give little excuse for fieldworkers not to collect samples of gaits and other performances for analysis wherever a laptop can be used. By measuring, or estimating, a few dimensions such as branch thickness, improving software capabilities allow correction for angular distortion (e.g., Stevens et al. 2006) and even some 3D reconstruction (for techniques, see, e.g., Hartley and Zisserman 2003), although this is unlikely ever to reach the accuracy of multicamera motion capture or DLT-based marker-free analysis of multicamera video, e.g., in-house code such as gap (code available in Sellers 1992) and commercial packages such as Vicon Motus, Kwon3D, etc.

But even studies of locomotor behavior may require controlled conditions wherein we seek to determine the "locomotor totipotentiality" (Prost 1965)—more mellifluously, "locomotor plasticity" or simply, capability of an animal, where we need to gather the statistics of locomotor behavior expressed under environmental (e.g., substrate availability or food distribution) conditions that do not currently exist in the wild.

Equally, there are limits to the kinds of information that can be derived in the laboratory or indeed any controlled environment. No natural environment is likely to elicit the complete range of behaviors of which a subject species is capable, since most species, except those in the most stable of habitats, are likely to have experienced and adapted to a range of environments, with gain and loss of food species and predators, and changes in support availability by which to access or avoid them. But the behaviors that can be elicited from a species in captivity, with very limited "home range," support choice, and artificial distribution of resources and threats, are not likely to be informative about species ecology.

Observational Field Methods

Although each of our field studies has adapted the methodology of Crompton (1980)—itself heavily influenced by that of Fleagle (1976, 1977)— to suit the site, species and questions of particular interest, our schedule of locomotor observations has remained largely the same. Table 10.1 is taken from Warren (1994) as adapted by Blanchard (2007).

The typical discontinuity of arboreal milieu favors basing locomotor observations on a *bout* or event, rather than recording for example the duration of a period of activity. Our sampling method is *ad libitum*, i.e., we observe what we see, when we see it as often as we see (and can record!) it, from as many animals as we can observe simultaneously. This decision was made on a purely pragmatic basis. Under forest conditions, more rigorous sampling methods (e.g., Altmann 1974), such as focal-animal observations taken every 5 min, reduce the number of observations of locomotion to statistically meaningless numbers. Even where 10,000 to 20,000 observations of locomotor behavior *sensu lato* (i.e., including postural behaviour) may not be enough to allow the interactions of habitat and locomotor variables to be analyzed in a statistically satisfactory manner; given the need for subsampling, the inevitable absence of parameters from some records, and the fact that in large-bodied species postural observations are likely to greatly outnumber locomotion.

A bout is much easier to define for some forms of locomotion and posture than others. Acyclical modes, such as a single, nonricochetal leap, have an obvious

Table 10.1 Observation schedule for locomotor behavioral data

1.	Date
2.	Species and individual identification where applicable
3.	Time
4.	Locomotor or postural mode
5.	Initial support diameter
6.	Initial support orientation
7.	Terminal support diameter
8.	Terminal support orientation
9.	Initial height (m)
10.	Terminal height (m)
11.	Horizontal distance traveled (m)
12.	Activity
13.	Continuous from previous? (yes/no)
14.	Notes

beginning and end, and any observable delay can be used to separate them. Climbing and quadrupedalism can grade into one another, however. Also, it should not be forgotten that in some primates, using locomotion that flows through a spectrum of labile modes can itself be adaptive, as we suspect is the case for the orang-utan, which combines great size with frequent use of small, unstable supports. Thus, division of a continuum of behavior into separate bouts can be very much a matter of judgment. This is the reason why we include item 13 in our data collection schedule, to signal continuity so that it can later be taken into consideration. However, an arbitrary end to a bout may need to be set for postural behavior: a single posture may be held for hours. In this case we adopt the same cut-off as the time interval chosen for observing ranging behavior: 5 minutes.

Categories 1–3: Date, Species, and Time

It is fairly obvious why one should wish to record (2) *species* in any multispecies comparison, but the inclusion of (1) *date* and (3) *time* deserve some discussion. RHC has been using this schedule for 30 years, and previously *date* has been used just as a marker for seasonal differences and *time* for general discussion of activity patterns over the night. But in 2008 they proved essential to a collaborative analysis of the influence of moon-phase on travel distance in animalivores versus frugivores and herbivores (Curtis et al. 2008), using the whole of our nocturnal database. It is far better to collect extra data than not, unless it interferes with core data collection.

Category 4: Locomotor and Postural Mode

In most cases, it will prove essential to use/create some behavioral categories specifically to encompass the expressed behavior of the subject species. However, it is good practice not only to minimize the number of new categories used to those that are truly essential, but also to compare categories as far as possible with those in the published literature. The most extensive compilation is that by Hunt et al. (1996), who provide codes for each mode and submode which may be useful shorthand referents. Hunt and colleagues (1996) base their categorization of locomotor modes on identifying the weight-bearing limb(s), and Thorpe and Crompton (2005) used their approach to analyze support use in the Sumatran orang-utan.

Categories 5–8: Support Diameter and Orientation

Support diameter affects locomotion in three ways. First, it determines for each species whether a support can be grasped by the feet and hands, or must be encircled by the arms, with the hind limbs used pressed against the support. Second, it affects, through its compliance, the stability it offers for a primate of given body weight—and indeed the free length that can occur before the support bends under its own weight—and third, again through its compliance, the energy costs of its use. Compliance will result in energy loss, unless the cycle time of a locomotor mode is long enough so as to allow most of the energy to be returned. If a branch is set vibrating by a primate walking over it or leaping from it, and continues vibrating after the last contact is lost, then energy remains stored in the branch and is not recovered. Selection of a stiffer take-off support would reduce this energy loss. But if a primate lands on a compliant branch after a leap, the energy lost to the landing branch can be advantageous, as the animal does not have to do as much negative work to decelerate, hence reducing musculoskeletal shock. Of course, it is possible that some primates may manage to use the recoil of the landing support to power the next acceleration, although too compliant a landing support will risk instability and a fall. Thus, we might expect a primate leaper to take off from a stiff support (unless it can recover energy from recoil) and land on a compliant support. But, branch diameter is not the only factor affecting support stiffness. Because it is the rigidity of a branch along the line of application of a force which determines its effective compliance, the *orientation* of supports affects their properties in relation to the direction of applied load.

Orientation obviously interacts with gravity to affect the stability of an animal using the support, and therefore the amount of muscular activity that will be necessary to maintain a posture upon it, but it will also directly affect the postures that may be adopted upon it without loss of stability. Further, the orientation of the support will affect the position of the center of gravity of the animal, and the effectiveness of the limbs in accelerating the center of mass (COM) in the desired direction. This will be greatest if the COM is located along the resultant of the forces exerted against the support, which for a leaper will determine trajectory; and this applies also to deceleration by the limbs at the end of a leap.

Interactions Between Locomotion and Substrate: The Case of Leaping Prosimians

We can illustrate some of these interactions with reference to our own work on prosimian leapers. They are of particular interest to studies of locomotion because prosimian primates can attain the biggest leap distances and height gains of any mammal, and yet they span a 50-fold range of body sizes, so that they have to deal with scaling effects affecting not only locomotion, but also thermoregulation and diet.

Leaping locomotion is acyclic and hence expensive: it may be seven times more expensive than level running (Demes et al. 1995, 1999). Theoretically, energy requirements might be decreased if kinetic energy was stored as elastic energy between subsequent leaps, as might be possible in ricochetal leaping of indriids, or returned from external energy stored in compliant branches (Warren and Crompton 1997, 1998).

Neither possibility has yet been tested experimentally or under field conditions. Return of externally stored energy is probably less likely to occur: energy stored in a compliant branch during foot contact cannot be returned unless recoil occurs before the animal loses contact with the substrate, a situation that so far has been demonstrated only for branch-swinging in orang-utans (Thorpe et al. 2007a, b).

As Demes and Günther (1989) point out, leaping presents different problems to small and large leapers. Small leapers have a large cross section of muscle (and hence, muscle force) relative to body mass, but short limbs with which to accelerate their COM at take-off and (by resistance to limb flexion) decelerate it on landing and hence avoid high peak landing forces, although landing forces are lower than those at take-off (Demes et al. 1999). They tend to accelerate with high forces over short time periods. The inverse applies to large-bodied leapers, but their problems are complicated by the dietary correlates of large body mass. Folivory is more common at large body size and yet folivores, especially arboreal folivores (McNab 1978), tend to have relatively little muscle compared to body mass.

Demes et al. (1995) examined leaping in wild *Indri indri* and *Propithecus edwardsi* (Milne-Edward's simpona) and found a preference for small-diameter supports for both take-off and landing. However, in large leapers problems of limited muscle and force generation capacity combine with limited bone cross-section and hence strength (Demes and Günther 1989). Because support compliance implies loss of energy to the support (increasing take-off costs but reducing peak force on landing), it might be expected that large-bodied primates will use small, compliant supports for landing, to absorb energy, and will use larger, less compliant supports for take-off, to minimize energy loss. Demes et al. (1995) also found that smaller-bodied *Hapalemur g. griseus* had relatively higher take-off and landing forces than the larger *Propithecus verreauxi* (Verreaux's sifaka), indicating that take-off and landing forces decrease with increased body mass (Demes et al. 1999). Similarly, younger animals generate higher relative forces, which could be a factor of their smaller body or lack of neuromuscular experience (Demes et al. 1999).

There has been an assumption that all leaping primates will adopt high-trajectory ballistic paths that minimize both energy costs and loads applied to the body. However, experimental and field studies have shown that this is not necessarily the case, and ecological factors, such as support availability and flight time, may be a primate's main consideration (Crompton et al. 1993; Crompton and Sellers 2007). Further, in a previous study of smaller-bodied leapers, neither *Otolemur crassicaudatus* (thick-tailed greater bushbaby) nor *Tarsius bancanus* (western tarsier) selected larger diameter take-off supports with increasing jump length (Crompton et al. 1993). Leaps greater than 1.6 m in *Galago moholi* (southern lesser bushbaby) were, however, made from larger initial supports than shorter jumps, and *Tarsius bancanus* did use larger terminal supports in longer jumps. Crompton et al. (1993) hypothesized that this was due to the requirements of a stable landing support for this small-bodied species, which lands feet first. For bushbabies and tarsiers, longer jumps in general resulted in a height gain, whereas shorter jumps resulted in a loss of height (Crompton et al. 1993). Crompton et al. (1993) suggested that leaping may be an energetically more efficient way of changing height than climbing.

Demes et al. (1999) claimed that specialist leapers generate less force than generalist species, probably due to longer hind limbs, and found that their specialists (*Indri indri* and *Propithecus* spp.) preferred small vertical substrates, while generalists (*Lemur catta*, ring-tailed lemur) chose horizontal supports.

Crompton et al. (1993) also distinguished mechanically specialized and unspecialized prosimian leapers, with respect to take-off trajectory and choice of take-off and landing supports, and their work suggests that Demes and colleagues' (1999) findings may follow from the choice of trajectory. An apparent tendency for flatter trajectories in unspecialized leapers (Crompton et al. 1993) may relate to body design: animals that are strongly hind limb dominated, and can flex their hind limbs deeply, can more easily locate their COM directly in line with the energetically optimum 45° takeoff directory (Crompton et al. 1993). Incidentally, the link between vertical clinging and leaping observed by Napier and Walker (1967) may arise from a related phenomenon: orthograde body posture makes it easier to accelerate the body COM into a high, energetically optimal trajectory and may direct more of the take-off force along the stiffer axis of the take-off support (Crompton and Sellers 2007). Inversely, horizontal body posture may make it easier to accelerate the body along a low trajectory, and further reduce the loss of energy to a branch in these rapid leaps, as horizontal body posture is likely to occur on a low-angled support, so that the long axis of the support, and hence its stiffest direction, is close to the line of force exerted by the leaper.

Although support diameter and orientation are closely linked in their effects on primate locomotor mechanics, and data on both need to be gathered, the appropriate categories for field data collection are very different. A simple four-way division into horizontal, sloping/oblique, angled, and vertical will suffice for primates of more or less any size, but the diameter categories or intervals chosen will vary entirely according to the size and the locomotor morphology and habits of the species under study. A small species such as a marmoset can access only 30 cm plus diameter tree trunks because it has secondary claws, and for a primate of equal size that lacks such claws or keels on its nails, like those of the needle-clawed bushbaby *Euoticus*, supports of such size will not be viable unless they are characterized by very high surface roughness. Not all observations will fall neatly into a four-way division of orientations and a simple diameter scale: for example, use of the ground is difficult to categorize in a totally satisfactory way, and use of multiple supports (as in the case of use of a tree fork for sitting, or grasping multiple fine twigs in the toes) also needs to be signaled in the data in such a way as not to lead to difficulties upon statistical analysis.

In both cases, though, much or even most useful information about support use will be lost if only the support of first sighting is assessed in this way. Bout-based observation methods are well suited to assessment of initial and terminal supports, and it is difficult to see how a more continuous assessment could be made observationally.

Categories 9 and 10: Initial and Terminal Height

In part, the same reasons lie behind our assessment of heights at the beginning and end of a bout, rather than attempting to assess height change directly, but absolute heights are important information in their own right, inasmuch as they inform about the forest stratum that is being used. Height gain, and hence work done against gravity, is a particular problem for large-bodied animals, because masses increase as the cube of linear dimensions, and metabolic costs of climbing upwards are relatively low for small primates compared to the cost of horizontal movement: the long limbs of large animals reduce their relative costs of horizontal travel compared to their high costs for height change.

Studies of locomotor behavior in the field, as with all field studies, involve particular distinct challenges not faced in controlled or laboratory environments, and height change is a case in point. Is the absolute height of the subject animal more important, or the height difference between the initial and terminal supports of the observed bout? If one is more interested primarily in stratum, probably the first, but if one is interested in the subject's energy expenditure, probably the second. Traditional behavioral studies record height of the subject from the ground, based on the observer's position. In flat, horizontal terrain, height change can be assessed readily on the basis of the initial and terminal height (Fig. 10.1a). But on hillsides, where the observer is following the subject either up or downhill, the animal will not be at the same height from the ground as it appears from the observer's position. Compare two cases: first, one in which a subject is leaping at canopy level across a valley (Fig. 10.1b). In tropical forest, trees in valleys tend to be taller than trees on crests, so that the canopy remains level. Thus, the subject is neither losing nor gaining height. But as it crosses the valley, its height from the ground will appear, from the observer's perspective, to decrease and increase. In the second case, a subject individual is leaping from tree-trunk to tree-trunk, below the canopy, up or down the slope (Fig. 10.1c), while remaining equidistant from the ground. In this case it will be losing (or gaining) height compared to the observer with each leap. On the other hand if the terminal height of a bout is recorded relative to the first height, not to the ground, and a sequence of bouts is recorded for animals down or up slope, the result could be unbalanced height gain or height loss. However one decides to tackle this problem, as with all data collection, the most important thing is to remain consistent throughout the study.

Category 11: "Horizontal" Distance Traveled

This measure assesses the horizontal distance traveled in a locomotor bout or displacement. Together with data on ranging behavior, it enables us to assess the contribution to total travel contributed by each kind of locomotion, usually assessed following Fleagle (1976) as the measure meters of horizontal travel per kilometer

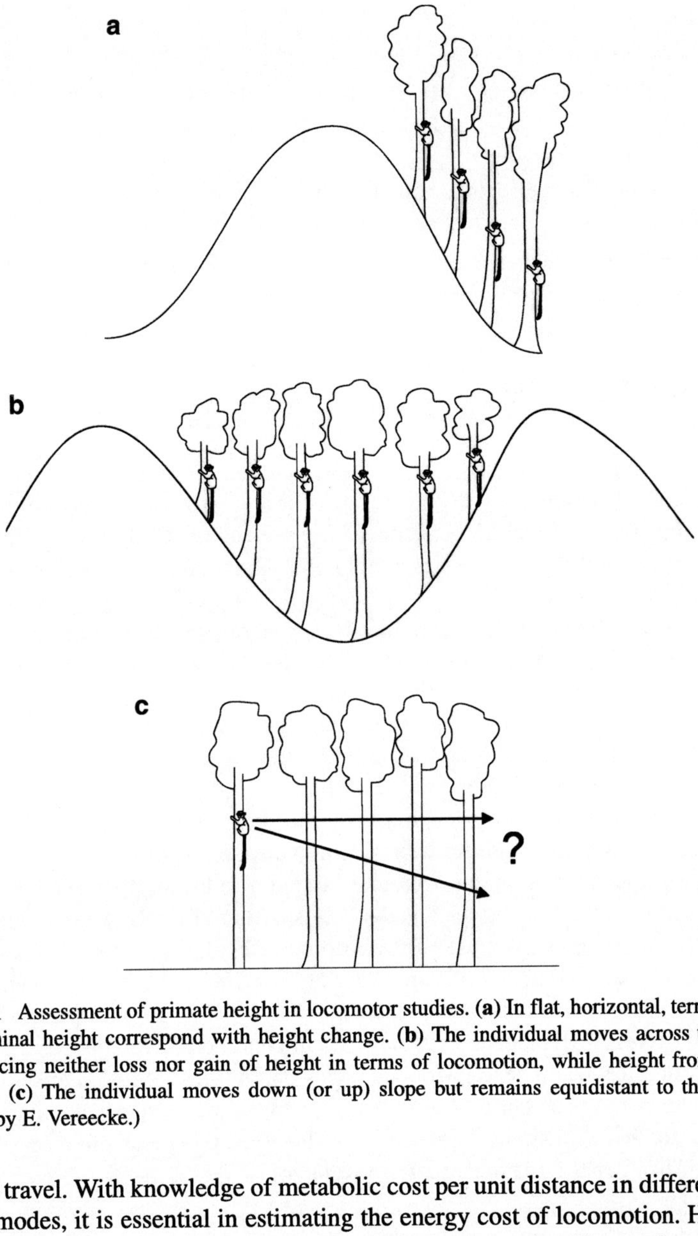

Fig. 10.1 Assessment of primate height in locomotor studies. (**a**) In flat, horizontal, terrain initial and terminal height correspond with height change. (**b**) The individual moves across the valley experiencing neither loss nor gain of height in terms of locomotion, while height from ground changes. (**c**) The individual moves down (or up) slope but remains equidistant to the ground. (Figure by E. Vereecke.)

of total travel. With knowledge of metabolic cost per unit distance in different locomotor modes, it is essential in estimating the energy cost of locomotion. However, although it is easy to assess distance traveled when an animal is moving horizontally or nearly horizontally, as the height difference between the start and end point increases it becomes more important to assess the mechanically effective distance. The mechanically effective distance can be easily calculated using Pythagoras' theorem using the horizontal distance and absolute vertical height change for each locomotor bout:

$$c = \sqrt{(a^2 + b^2)} \tag{10.1}$$

where c is the mechanically effective bout length, a the vertical height change (in metres), and b the horizontal distance traveled (in meters) for any given locomotor bout.

This mechanically effective bout length is of further value in biomechanical analysis of field data, e.g., in determining variables such as the slope of the leap, optimum take-off angle, minimum kinetic energy costs, take-off velocity, flight time and horizontal velocity (e.g., Crompton and Sellers 2007).

Category 12: Activity

It has long been shown that activity influences the choice of locomotor mode (Fleagle and Mittermeier 1980), as different strategies are required for accessing food as opposed to traveling between feed or rest sites. Fleagle and Mittermeier (1980) suggest that during travel, the end point of any bout is indeterminate, while in feeding it is more specific, often involving access to relatively "small" terminal branches. In South American platyrrhines, as in African strepsirrhines, less leaping and more climbing and suspension/cantilevering have been observed during bouts of feeding compared to travel (Fleagle and Mittermeier 1980; Crompton 1984; Gebo and Chapman 1995). This agrees with findings in Malaysian and African strepsirrhines and Central American platyrrhines, where larger (horizontal) supports were used during locomotion than during feeding (Fleagle 1977; Boinski 1989; Gebo and Chapman 1995). During feeding and foraging leaping often decreases, and climbing and quadrupedalism increase (Dagosto 1995).

This begs the question of how to define "activity" in locomotor terms, as locomotion is itself travel: a link between activities. Blanchard (2007) recorded locomotor activity as the assumed purpose of the locomotor bout, i.e., what the bout led to. However, this can lead to complications during foraging when there is a sequence of locomotor bouts to access food in a different area of the feeding tree; should this be classified as travel? Blanchard's (2007) study did not include foraging as a category, but future studies would benefit from its inclusion. Further confusion arises over the classification of "rests" during bouts of travel. In a travel sequence an animal will often rest for a few moments between leaps, but does this mean the purpose of the leap is to "rest" or is it just a pause in movement?

Category 13: Continues from Previous?

The category "Continues from Previous?" allows the observer to record whether the observed bout is standalone, or continuous with the previously recorded bout. This is important in cases such as ricochetal leaping or running leaps, where there may be the possibility of energy conservation between bouts. As we have already observed, continuous-bout recording can prove problematic for statistical tests that assume that cases are independent, and some subsampling regimen will then be necessary. Alternatively, it is possible to isolate the effect by comparing analyses carried out on unrelated bouts with that using the entire data set.

Support Availability

Although not part of our schedule of locomotor observations, the *availability* of supports of different diameters and orientations needs to be assessed over the study area in order to test hypothesis of support selection by the subject species. This is not a particularly straightforward task. Cannon and Leighton (1994) have presented techniques, and Warren (1994, 1997) employed a modified version of these and quantified support characteristics in meter cubes of space above a transect, with the help of long marked poles, and used the statistic Jacob's D to infer selection. However, data collection by this method is near impossible without climbing trees where there is a high canopy, and the inference of selection from availability/use statistics is problematic (e.g., Britt 1996). Simpler 2D photographic assessments along transects or in different forest zones have been employed by Crompton (1980, 1984) and by Blanchard (2007), and analysis of canopy continuity is possible using aerial photographs (Crompton 1980) or by use of fish-eye images analyzed with the aid of standard image-analysis functions contained within commercial packages such as Adobe Photoshop (K.A.I. Nekaris, pers. comm.) and various specialized statistical tools for ecology.

Ranging

Of further importance when examining the energetic cost of locomotion is daily ranging behavior. Ranging is a relatively easy variable to measure, especially with the use of handheld GPS (global positioning systems). Where terrain and forest cover allows, GPS can be used during follows, alternatively, if acquiring a position using the GPS is slower than the travel speed of the animal, the route can be tagged and positions subsequently surveyed in or acquired using GPS. It must, however, be realized that point-to-point ranging distances, and even more totals of bout lengths, will be a substantial underestimate, or at best a minimum estimate, of total travel distances. Much more accurate data collection is possible using GPS radio-collars, which are now of a size where they can be fitted to larger-bodied primates. Some types transmit data to satellite tracking services, but as mobile (cell) phone coverage is spreading rapidly, the type of GPS collar that sends location data as text messages may be more practical and less expensive. In time, GPS units are likely to become available that are suitable for use on all but the very smallest primates.

Field Techniques for Estimation of Mechanical Energy Costs of Locomotion and Estimates of Metabolic Energy Budgets

Wherever radiocollars are employed, subject individuals need to be captured, and morphometric data (mass, segment lengths, etc.) are usually collected. Since force equals mass times acceleration, dynamic analysis can be used to calculate forces, and hence kinetic energy costs (e.g., Winter 1990 for methodology), using kinematics of locomotor performance recorded with high-speed video or still cameras and field collection of data on segment mass distribution and inertial properties of segments (body build). Kinetic energy costs can in turn be used with daily travel distance, height change, and the contribution made by the locomotor mode to each kilometre of travel, to estimate total kinetic energy and metabolic costs (e.g., Warren and Crompton 1998). Acquisition of segment mass properties need not involve cadaveric measurements. If we make the reasonable assumption that, since our bodies are mostly water, density will be fairly uniform across the body, mass distribution can be estimated geometrically from segment lengths and three circumferences, or in the case of elliptically shaped segments such as the trunk, anteroposterior and mediolateral diameters. Method and code for this approach, as well as cadaveric approaches, have been presented by Crompton et al. (1996).

But just as daily ranging distances derived from point-to-point ranging data represent the minimum distance traveled, estimates of energy costs based on observed frequencies and bout lengths of locomotor performances will represent only a small proportion of real values. And just as a technological solution to the ranging problem is now becoming available by using GPS radiocollars, use of 3D accelerometers now offers a far more complete measure of forces exerted during the daily cycle. Sellers and Crompton (2004) tested a prototype harness-mounted telemetered system, but newer technologies using logging accelerometers glued to the subject's hair greatly reduce noise. The unit can be recovered after the glue/hair attachment fails, allowing data to be downloaded; for re-location prior to recovery a secondary standard radio-transmitter with a longer battery life is desirable. Byrnes et al. gathered valuable data on take-off and landing forces were gathered in this way for colugos (flying lemurs), but calibration of accelerometry signals against kinematics (see Sellers and Crompton 2004) should allow a full record of locomotor activity throughout the activity cycle.

Diet and Metabolism

From a locomotor perspective, the main value of understanding the dietary behavior of an animal is gaining an insight into its energy budget. Small mammals have large surface-to-mass ratios and hence high energetic requirements for thermoregulation. They therefore require foods that are high in energy, such as insects and fruit (Mutschler 2002), and therefore folivory is not usually seen in mammals weighing less than 1.5 kg (Kay 1984). Although larger mammals have lower thermoregulatory costs, their costs for locomotion are absolutely larger. These requirements can, however, be met by bulkier, lower quality foods such as leaves, as their gut volumes are relatively large. Large-bodied folivores may, however, adopt an expensive, leaping locomotion, and equally body size below Kay's (1984) threshold is successfully combined with an expensive leaping locomotion and folivory in several small-bodied folivorous leapers such as *Avahi laniger* and *Lepilemur edwardsi* (Warren and Crompton 1997, 1998). It is clear that we have much to learn about energy budgets in such animals.

Measurement of Metabolic Rates

The basal metabolic rate (BMR) is considered the minimum energetic cost of maintaining cells and organs ready for activity, defined as the "*rate of energy metabolism of a non-growing organism at rest under post-absorptive conditions [i.e. no digestion] in a thermoneutral environment*" (Ricklefs et al. 1996). Self-contained portable equipment for measuring metabolic rate via respirometry in the field (e.g., Drack et al. 1999) is now more readily available, and leaky systems, such as the TurboFox, permit gas analysis even in the field, even in nest-holes of *Cheirogaleus* and *Lepilemur*. Energy metabolism measured on a free-living animal is that of an *active* organism and is termed the field metabolic rate (FMR; Ricklefs et al. 1996). The relationship between BMR and FMR can be considered as the relationship between readiness and action (Ricklefs et al. 1996). Although Koteja (1991) concluded that BMR is not a reliable index of energy expenditure in free-ranging animals, Ricklefs et al. (1996) did find a relationship between BMR and FMR in mammals. Nevertheless, measurements of BMR conducted on inactive wild animals, although valuable and necessary, should be treated with caution, as they do not closely reflect the total metabolic rate of an active animal.

Conclusion

This brief review of field methods for the study of locomotion shows there is an increasing blurring of the boundary between laboratory and field. It is no longer the case that forces and energy costs must be measured in the laboratory. Nevertheless, when it is necessary to simplify, modify, or control the environment, to elicit unusual behaviors or obtain large sample sizes for individual performance types —as, e.g., when investigating the relationship between speed and limb kinematics in quadrupedalism—then field research still remains less appropriate. Similarly, it is difficult to justify, ethically or scientifically, invasive procedures in a field setting unless the question posed can be answered in no other way. However, if we wish to understand what Bock and von Wahlert (1965) termed the biological role of an adaptation, field research remains the only way to go, and one that is becoming progressively more powerful and quantitative.

Acknowledgments We thank Kris and Evie for organizing the symposium at the 22nd Congress of the International Primatological Society, and our co-participants for making it a friendly, collegial, and enjoyable event. Our field research has been funded by World Wildlife Fund Hong Kong, The L.S.B. Leakey Foundation, the Royal Society, the Natural Environment Research Council, and The Leverhulme Trust.

References

Altmann J (1974) Observational study of behavior: sampling methods. Behaviour 49:227–267.

Blanchard ML (2007) Locomotor behaviour and ecology of three sympatric lemur species in Mantadia National Park, Madagascar. Unpublished PhD dissertation, The University of Liverpool.

Bock WJ, von Wahlert G (1965) Adaptation and the form-function complex. Evolution 19:269–299.

Boinski S (1989) The positional behavior and substrate use of squirrel monkeys: ecological implications. J Hum Evol 18:659–677.

Britt A (1996) Environmental influences on the behavioural ecology of the black- and-white ruffed lemur (*Varecia variegata variegata*, Kerr 1792). Unpublished PhD dissertation, The University of Liverpool.

Byrnes G, Lim NT-L, Spence AJ (2008) Take-off and landing kinetics of a free-ranging gliding mammal, the Malayan colugo (*Galeopterus variegatus*). Proc R Soc B 275:1007–1013.

Cannon CH, Leighton M (1994) Comparative locomotor ecology of gibbons and macaques: selection of canopy elements for crossing gaps. Am J Phys Anthropol 93:505–524.

Crompton RH (1980) *Galago* locomotion. Unpublished PhD dissertation, Harvard University.

Crompton RH (1984) Foraging, habitat structure and locomotion in two species of *Galago*. In: Rodman PS, Cant JG (eds), Adaptations for Foraging in Nonhuman Primates. Columbia University Press, New York, pp 73–111.

Crompton RH, Sellers WI, Gunther MM (1993) Energetic efficiency and ecology as selective factors in the saltatory adaptation of prosimian primates. Proc R Soc Lond Biol 254:41–45.

Crompton RH, Li Y, Günther MM, Alexander RM (1996) Segment inertial properties of primates: new techniques for laboratory and field studies of locomotion. Am J Phys Anthropol 99:547–570.

Crompton RH, Sellers WI (2007) A consideration of leaping locomotion as a means of predator avoidance in prosimian primates. In: Gursky S, Nekaris KAI (eds), Primate Anti-Predator Strategies. Springer, New York, pp 125–143.

Curtis D, Waters R, Crompton R (2008) The effect of moonlight on activity in nocturnal primates in relation to their dietary ecological niches. Abstracts of the XXII Congress of the International Primatological Society, Edinburgh.

Dagosto M (1995) Seasonal variation in positional behavior of Malagasy lemurs. Int J Primatol 16:807–833.

Demes B, Günther MM (1989) Biomechanics and allometric scaling in primate locomotion and morphology. Folia Primatol 53:125–141.

Demes B, Jungers WL, Gross TS, Fleagle JG (1995) Kinetics of leaping primates: influence of substrate orientation and compliance. Am J Phys Anthropol 96:419–429.

Demes B, Fleagle JG, Jungers WL (1999) Takeoff and landing forces of leaping strepsirhine primates. J Hum Evol 37:279–292.

Drack S, Ortmann S, Buehrmann N, Schmid J, Warren RD, Heldmaier G, Ganzhorn JU (1999) Field metabolic rate and the cost of ranging of the red-tailed sportive lemur (*Lepilemur ruficaudatus*). In: Rakotosamimanana B (ed), New Directions in Lemur Studies. Kluwer Academic/Plenum, New York, pp 83–91.

Fleagle JG (1976) Locomotion posture and anatomy of Malaysian forest primates. Unpublished PhD dissertation, Harvard University.

Fleagle JG (1977) Locomotor behaviour and muscular anatomy of sympatric Malaysian leaf monkeys (*Presbytis obscura* and *P. melalophus*). Am J Phys Anthropol 46:297–308.

Fleagle JG, Mittermeier RA (1980) Locomotor behavior, body size, and comparative ecology of seven Surinam monkeys. Am J Phys Anthropol 52:301–314.

Gebo DL, Chapman CA (1995) Positional behavior in five sympatric old world monkeys. Am J Phys Anthropol 97:49–76.

Hartley R, Zisserman A (2003) Multiple View Geometry in Computer Vision, 2nd ed. Cambridge University Press, Cambridge.

Hunt KD, Gebo DL, Rose MD, Walker SE, Youlatos D, Cant JG (1996) Standardized descriptions of primate locomotor and postural modes. Primates 37:363–387.

Kay RF (1984) On the use of anatomical features to infer foraging behavior in extinct primates. In: Rodman PS, Cant JG (eds), Adaptations for Foraging in Nonhuman Primates. Columbia University Press, New York, pp 21–53.

Koteja P (1991) On the relation between basal and field metabolic rates in birds and mammals. Funct Ecol 5:56–64.

McNab BK (1978) Energetics of arboreal folivores: physiological problems and ecological consequences of feeding on an ubiquitous food supply. In: Montgomery GG (ed), Ecology of Arboreal Folivores. Smithsonian Institute Press, Washington, pp 153–162.

Mollison T (1911) Die Korperproportionen der Primaten. Morph Jahrb 42:79–304.

Mutschler T (2002) Alaotran gentle lemur: some aspects of its behavioral ecology. Evol Anthropol 11:101–104.

Napier JR, Walker AC (1967) Vertical clinging and leaping – a newly recognized category of locomotor behaviour of primates. Folia Primatol 6:204–219.

Prost JH (1965) A definitional system for the classification of primate locomotion. Am Anthropol 67:1198–1214.

Ricklefs RE, Konarzewski M, Daan S (1996) The relationship between basal metabolic rate and daily expenditure in birds and mammals. Am Nat 147:1047–1071.

Ripley S (1967) The leaping of langurs; a problem in the study of locomotor adaptation. Am J Phys Anthropol 26:149–170.

Sellers WI (1992) A study of leaping in prosimian primates. Unpublished PhD dissertation, The University of Liverpool.

Sellers WI, Crompton RH (2004) Automatic monitoring of primate locomotor behaviour using accelerometers. Folia Primatol 75:279–293.

Stevens NJ, Schmitt D, Cole T, Chan K-T (2006) Technical note: out-of-plane angular correction based on a trigonometric function for use in kinematic studies. Am J Phys Anthropol 129:399–402.

Thorpe SKS, Crompton RH (2005) The locomotor ecology of wild orang-utans (*Pongo pygmaeus abelii*) in the Gunung Leuser Ecosystem, Sumatra, Indonesia: a multivariate analysis using log-linear modelling. Am J Phys Anthropol 127:58–78.

Thorpe SKS, Crompton RH, Alexander RM (2007a) Orangutans utilise compliant branches to lower the energetic cost of locomotion. Biol Lett 3:253–256.

Thorpe SKS, Holder RL, Crompton RH (2007b) Origins of human bipedalism as an adaptation for locomotion on flexible branches. Science 317:1328–1331.

Warren RD (1994) Lazy leapers: a study of the locomotor ecology of two species of saltatory nocturnal lemurs in sympatry at Ampijoroa, Madagascar. Unpublished PhD dissertation, The University of Liverpool.

Warren RD (1997) Habitat use and support preference of two free-ranging saltatory lemurs (*Lepilemur edwardsi* and *Avahi occidentalis*). J Zool 241:325–341.

Warren RD, Crompton RH (1997) Locomotor ecology *of Lepilemur edwardsi* and *Avahi occidentalis*. Am J Phys Anthropol 104:471–486.

Warren RD, Crompton RH (1998) Diet, body size and energy costs of locomotion in saltatory primates. Folia Primatol 69:86–100.

Winter DA (1990) Biomechanics and Motor Control of Human Movement, 2nd ed. John Wiley & Sons, New York.

Chapter 11
Gibbon Locomotion Research in the Field: Problems, Possibilities, and Benefits for Conservation

Susan M. Cheyne

Abstract The benefit of understanding primate locomotion in detail has potential application for understanding how primates use their habitat, e.g., the forest canopy layers for arboreal primates. Energy intake (feeding ecology and diet), behavior, travel distances, and energy expenditure all combine to influence a primates's choice of locomotion mode. Increasingly, an arboreal primate's territory is being represented as a 3D image encompassing time, horizontal distance, and vertical distance. Understanding locomotion has conservation benefits relating to how primates adapt to disturbance, i.e., if the dominant locomotion has to change significantly to cope with habitat changes. Also of great importance is relating energetic expenditure on travel to diet, energetic intake, food availability, and travel distance and how these may change seasonally. Using wild gibbons as a case study, I will provide insights into how laboratory techniques can be brought effectively into long-term field studies and the benefits to conservation that can be achieved.

Keywords Energetic • Field studies • Forest canopy • Logging • Natural habitat • Travel

Introduction

During the course of evolution, primates have adapted to deal with the challenges imposed by travelling in their natural habitat. These challenges include speed, acceleration and maneuvrability, endurance, economy of energy, and stability (each contributing to the overall fitness). Within these challenges, the design of the limbs

---

S.M. Cheyne (✉)
Wildlife Conservation Research Unit, Department of Zoology, University of Oxford, UK
e-mail: susan.cheyne@zoo.ox.ac.uk

K. D'Août and E.E. Vereecke (eds.), *Primate Locomotion: Linking Field and Laboratory Research*, Developments in Primatology: Progress and Prospects,
DOI 10.1007/978-1-4419-1420-0_11,

of each primate has constraints, e.g., size of the primate or the habitat in which it lives. Locomotor types have evolved to confer fitness on the primates and to maximize the primate's ability to move in its habitat while minimizing energy loss (summarized in Alexander 2003). However, across primate ranges, the habitat to which they are adapted is being destroyed (see Chapman and Peres 2001 for a full review), with the rate of change of forest to plantation as high as 85.7% in Malaysia.

A combination of logging (legal and illegal), forest conversion for plantations and fire creates forest fragments of varying size. These human activities also impact on contiguous forest, reducing pristine rain forest to degraded secondary forest with altered habitat characteristics due to the loss of the largest, tallest trees to the loggers. Selective logging has been seen as the long-term "compromise" for both humans and primates, but areas that are selectively logged are often overexploited by the timber industries. Johns and Skorupa (1987) recommend that selective logging practices should normally involve the felling of not more than 10% (which causes damage of up to 50%) of the trees, thus enabling the remaining trees to regenerate, and the effects of logging to be mitigated. Nevertheless, there are always exceptions to the rule and some logging operations have been recorded destroying anywhere from 5% to 70% of the forest, thus exceeding the average predicted level of destruction of 45–50% (Johns and Skorupa 1987). Some habitat-disturbance studies have shown that the survival of certain primates communities can be tolerated alongside logging practices (Johns 1987; Arnhem et al. 2008). But the removal of tall trees and the breaking up of the continuous canopy may have serious implications for arboreal species in terms of travel and locomotion.

Locomotion requires energy, and energy comes from food available in the habitat. Primates move through the forest using "learned brain maps" (Carpenter 1972) or "mind-maps" (Menzel 1973; Garber 1989, 2004; Janson 1998; MacDonald and Agnes 1999; Gibeault and MacDonald 2000; Erharta and Overdorff 2008). This "map" is hypothesized to be a record of fruiting tree locations and travel routes the primate has taken and will direct the individual on which pathway to take, i.e., a spatial reference of a known set of routes that will take it to the nearest source of food in any given season. Primates seek to minimize energy expenditure from travel to food sources by selecting the most direct routes (Menzel 1973; Janson 1998; MacDonald and Agnes 1999; Gibeault and MacDonald 2000). If these direct routes have been rendered unusable by logging, primates are forced to travel by longer routes to reach the patchily dispersed food sources. In addition, these changes to the forest structure could have implications for the locomotion patterns of the primates. If the gaps in the forest are too wide to cross by normal locomotion, arboreal primates may be forced to come to the ground to traverse the distance, making them vulnerable to predators (Thorpe et al. 2007; Thompson et al. in press). An understanding of how changes in the habitat are affecting primate locomotion and energy expenditure is important for conservation and management of the ecosystem (Cannon and Leighton 1994; Arnhem et al. 2008).

The Focus of Current Field Studies

Much of the literature from field-based research into locomotion is based on gait and locomotion types, e.g., quadrupedal clambering, tree-sway, brachiation, bipedal walking, and on looking into the number of limbs in contact with a support at any one time (Thorpe et al. 2007; Thompson et al. in press). These variables are reasonably easy to quantify and observe in habituated primates without the need for expensive equipment. Data such as these tell us a great deal about how the primates move, how high they travel, and how they cope with variations in canopy continuity and forest structure, though are not particularly helpful for more detailed biomechanical studies.

Many studies have been conducted on energy expenditure in primates (Taylor et al. 1970; Cant 1985, 1986; Crompton et al. 1993; Rao and van Schaik 1997). Fleagle's (1974) study on the dynamics of a brachiating siamang and the energy transfer involved in brachiation highlighted the importance of the gibbon's body posture in maximizing kinetic energy. Several studies provide information on comparisons into structural preference for many primate species including; siamang and lar gibbons (Fleagle 1976), three lemur species (Dagosto and Yamashita 1998), red colobus monkeys (Clutton-Brock 1973), Japanese macaques (Chatani 2003), cercopithecid monkeys (McGraw 1996), and two species of spider monkey (Mittermeier 1978). Thorpe et al. (2007) report that orang-utans (*Pongo abelii*) use substrate compliance to decrease the energetic costs of locomotion. They do this by tree-swaying (bridging gaps between trees by shifting their weight), a locomotor behavior that gibbons rarely exhibit (pers. obs.). This behavior was discovered to be less than half as energetically costly as jumping or coming to the ground to cross the gap. Thorpe et al. (2007) is the first study of its kind to assess the energetic differences between gap-crossing techniques in orang-utans. Although gibbons do not characteristically display this mode of locomotion, it highlights the different energetic costs involved in different types of locomotion.

Introducing the Gibbons: The True Brachiators

Possibly the most striking part of gibbon behavior is the way they move through the canopy. They are highly specialised brachiators with distinct morphological characteristics, e.g., hand and wrist morphology, elongated forelimb proportions, and specific body-muscle structure, enabling them to travel through the canopy at substantial speeds (Usherwood and Bertram 2003; Usherwood et al. 2003). The gibbon hand is long and slender, with the four fingers acting as specialized hooks during brachiation (van Horn 1972). The thumb is kept out of the way during brachiation by folding it into a deep cleft (Straus 1942). The thumbs of *Colobus* and *Ateles* have either been lost completely or reduced somewhat over the generations, but in *Hylobates* they have evolved to a precision grip for picking small fruit. This thumb is one of the longest primate thumbs relative to body size (Schultz 1930; Fig. 11.1a–b).

Fig. 11.1 Photos of wild gibbons showing the difference in length between thumb and other fingers. (Photos by Susan M. Cheyne.)

True brachiation is unique to the small apes (Hylobatidae). Some New World monkeys (*Ateles, Lagothrix, Alouatta*) have adapted a form of brachiation, using their prehensile tails as a fifth limb during locomotion (Andrews and Groves 1976; Mittermeier 1978). Orang-utans (*Pongo pygmaeus*), chimpanzees (*Pan troglodytes*), and bonobos (*Pan paniscus*) do adopt a similar suspended feeding posture, occasionally combining their travel with forelimb suspensory bouts, but most of their suspended postures are adopted while feeding, although normally two limbs are used for support instead of one (Hollihn 1984), termed "modified brachiation" by Napier and Napier (1967). Gibbons will use two limbs instead of one if hanging in suspensory position for a long period of time; this enables them to maintain balance and lead directly into locomotion (Carpenter 1972).

Gibbons have the advantage of long forelimbs, an adaptation to suspensory locomotion and posture. It has been widely documented that the small apes spend more time in bimanual suspensory postures than any other primate (Chivers 1973; Fleagle 1976; Mittermeier and Fleagle 1976; Mittermeier 1978). They will hang from substrates using both or only one of the forelimbs as their principal or only means of support. Resting, feeding, traveling, and even copulation will take place in this suspended position. The increased arm length allows gibbons to have a wider reach, both when extending to grasp other substrates and for foraging. The long arm-reach allows gibbons to exploit food on the terminal branches, which are unreachable by the average small-bodied primate and inaccessible to the orang-utans (Hollihn 1984). To avoid distortion of the trunk during brachiation, the gibbon rib-cage is extremely stiff, with the lower spine being short and fairly inflexible (Schultz 1933). Gibbons, asleep or awake, will instinctively grip overhead supports to steady themselves and be geared up at any point to commence movement (Carpenter 1972). Brachiation through dense forest canopies requires extensive maneuvrability and injury or amputation can result in severe restrictions to locomotion, leading to modifications, such as one-armed brachiation (Gibbons and Lockwood 1982; Sayer et al. 2007, pers. obs.).

Evolutionary Advantages of Brachiation

Brachiation can be defined as the bimanual sequence involving rhythmic movement along overhead structures over a distance of several meters without the aid of the hind limbs or the use of other forms of locomotion (Napier and Napier 1967; Baldwin and Teleki 1976). There is major rotation of the wrist, elbow, and shoulder joints (Carpenter 1972), adequate muscle strength, and a sufficient blood supply to the extended forelimbs being essential (Preuschoft and Demes 1984, 1985). A brachiating gibbon advances forward in two ways: by means of exchanging handholds, termed *continuous contact brachiation* (Fig. 11.2a), but also by achieving momentum that propels the body through the air, called *ricochetal brachiation* (Fig. 11.2b; Hollihn 1984). With ricochetal brachiation the swing length will intensify and ultimately progress to a "glide" period when the gibbon has no contact with the supports (Carpenter 1972).

Bimanual suspensory behavior has influenced the style and order of many different types of gibbon behavior. It has set the style in which they carry their infants (most other nonhuman primates will carry infants on their backs, but female gibbons always carry infants ventrally) and engage in play activities, copulation, and fighting (Carpenter 1972; Cheyne 2004, 2010).

Foraging activities are hypothesized to play a key role in the development of positional behaviors (Grand 1972; Chivers 1974; Andrews and Groves 1976). While traveling, gibbons will brachiate more than when foraging (climbing being the dominant locomotor mode while foraging), but bimanual suspension is the most common feeding position (Fleagle 1976). This correlates with the fact that primates tend to spend much of their time feeding rather than traveling, thus spending more time in bimanual suspension rather than bimanual locomotion (Hollihn 1984). The advantages of adopting a suspensory posture for feeding are clear; food at the terminal branches is easier to reach (Andrews and Groves 1976; Mittermeier and Fleagle 1976). This is a trait almost exclusively useful to gibbons; smaller primates are incapable of such suspensory behavior and large primates are simply too large to exploit food sources in the terminal branches (Grand 1972).

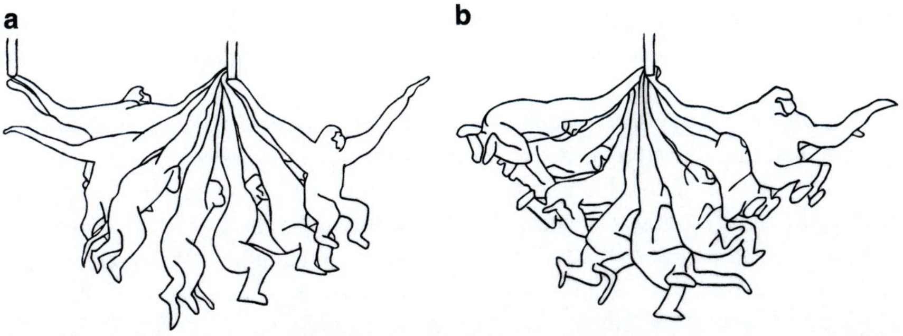

Fig. 11.2 Tracings of a gibbon using (**a**) continuous contact brachiation and (**b**) ricochetal brachiation to move between handholds. (Based on Bertram et al. 1999.)

Another theory for the selective advantage of bimanualism is predator avoidance. Moving bimanually is a quieter and less eye-catching mode of locomotion than quadrupedal jumping/clambering (Grand 1972). There is also much more cover provided in the leafy foliage of the terminal branches, which may be another strategy for avoiding predators (Preuschoft and Demes 1984).

Perceptual Motor Skills

Gibbons are well known for their ability to move at great speeds through the canopy (1.8 m/s; Nigg et al. 2000). What is less known are injuries resulting from falls. Schultz (1969) found that one third of all gibbon specimens collected had skeletal fractures, a testament to the risks associated with their locomotion. Many factors support the perceptual motor skill theory which describes the ability or capacity for learning involving the interaction of perception and voluntary movement (Alexander 2003). The most fundamental factors are the speed of uninterrupted travel, the accuracy of the countless corrective acts, and the limited time required for making choices between alternative travel routes and substrate options.

Why Study Gibbon Locomotion in the Field?

The standardized laboratory setup often does not accurately reflect the conditions in the wild. The initial theory of gibbon locomotion as a pendulum swing (Chang et al. 1997) was later refined by Bertram et al. (Bertram et al. 1999; Bertram and Chang 2001; Bertram 2004). These authors incorporated a force transducer into one of the handholds in the gibbon's enclosure (Chang et al. 2000). The gibbon behaved less like a simple pendulum and more like one with an incorporated spring mechanism. Fleagle (1974) found that gibbons can increase their energy when brachiating by bending and then extending their legs during the contact phase, similar to human children on playground swings. This flexion and extension of the legs provides additional momentum for the gibbon, especially going into a richochetal brachiation phase, so that they have enough energy to reach the launch angle (Bertram et al. 1999).

The above theory assumes equally spaced supports of equal height above the ground, and that the supports are rigid. In reality, supports are of unequal distance apart, may not be of equal height above the ground, will be of varying width, and will not be rigid. Nonrigid substrates are left vibrating by the gibbon's passage, the energy of which is lost to the gibbon. The work the gibbon must do to replace this lost energy makes the metabolic (thus energetic) costs of brachiation in the wild much higher than if the gibbons were using rigid supports (Alexander 2003).

From the above example, it is clear that though laboratory-based studies allow scientists to formulate models of locomotion and metabolic expenditure, these models can only be a guideline and are not always representative of the situation for the primates in the wild.

Equally, field-based studies often do not account for the differences in substrate type that the primates are using, or for the fact that arboreal primates are moving in two planes with each travel sequence involving not only a change in horizontal distance but often a change in vertical distance (height within the tree; Fig. 11.3).

Habitat disturbance presents a major problem for primates, and particularly for primates that travel exclusively in the canopy. Efficient travel through the canopy requires the primate to take the most direct route available and minimise any divergence from its travel path, be this horizontally or vertically by traveling through the most continuous canopy layer (Thompson et al. in press). Natural, e.g., tree-falls, and unnatural canopy gaps, e.g., logging, logging canals, clearings for hunting flying foxes, and fire damage, are posing a problem as the canopy is becoming highly uneven, thus producing less direct travel paths than were originally available. Gaps in the forest present a significant challenge associated with increased energy expenditure for arboreal primates. Engstrom (2000) found that an increased presence of gaps was related to a reduction of orang-utan density in disturbed forest. She proposed that this was due to a higher predation risk, as the nests were built at lower heights in disturbed forest than in primary forest. Further, an increase in gaps also increased the energy expenditure of orang-utans, as they were compelled to travel further to reach food sources (Rao and van Schaik 1997; Felton et al. 2003). It is reasonable to assume that the above consequences would also affect gibbons. Although gibbons are extremely versatile when it comes to moving on the

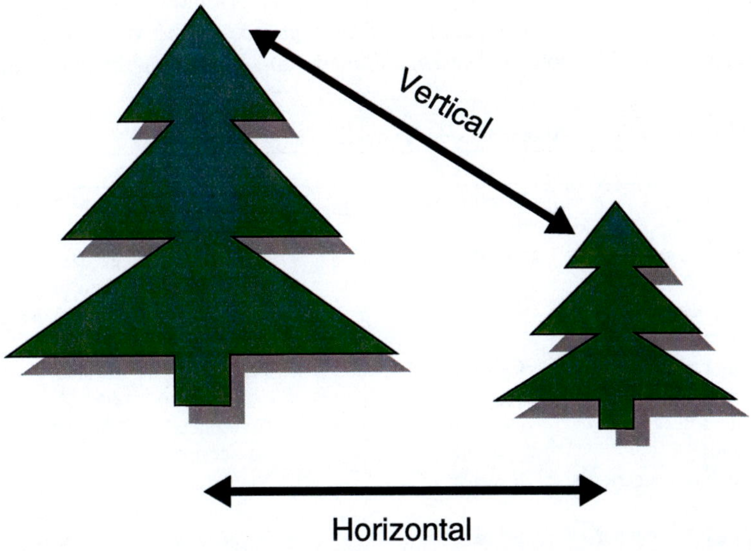

Fig. 11.3 Diagram showing changes in position of a traveling gibbon relevant to vertical and horizontal distance

ground (their elongated forelimbs having not limited their terrestrial abilities; Vereecke et al. 2006), this presents additional potential problems in the form of disease transmission, susceptibility to hunters, and increased predation risks.

Thus, studying locomotion in the field allows us to evaluate the laboratory-based models, understand more about the complex environment in which the gibbons live and how changes to this environment may have detrimental affects to gibbon locomotion and ability to traverse the forest canopy.

Methods: Problems and Solutions for Field-Based Locomotor Studies

Methods for studying locomotion in primates can be split broadly into several categories, many of which are invasive and, thus, unsuitable for use on wild primates (Table 11.1). Given this list it would seem that many of the techniques are inapplicable to the field, leaving only two possibilities. I will now look at these in more detail and explain how they can be incorporated into field studies.

Video Cameras

Many behavioral field studies already use handheld video cameras to record behaviors and to identify individual animals (pers. obs.). Calibrating such videos for biomechanical studies is possible with habituated primates. Calibration of the height of the animal on substrates and the amplitude of the vibrations of these substrates is possible from measurements made on support lengths and diameters

Table 11.1 List of common techniques for studying locomotion

Method	Invasive/non-invasive	Common method
Video recording	Noninvasive	Two video cameras to capture full range of movement
Stationary locomotion	Invasive	Using wind tunnels, tread-mills etc to control primate's movement
Measuring energy consumption	Invasive	Fitting mask or hood to measure O_2 consumption
Observing flow	Invasive	Primates are in artificial wild tunnels or rivers
Measuring forces and pressure	Non-invasive	Transducers or strain gauges on substrates or
Recording muscle action	Invasive	Electromyography involved fitting electrodes to the primate
Recording movement at a distance	Invasive	Radio collars/data loggers/ transmitters fitted to primate

Adapted from Alexander (2003).

at ground level (Thorpe et al. 2007; Thompson et al. in press). Video sequences also do not need to be done with high-speed cameras, a recording speed of 25 frames per second is adequate for most purposes (Thorpe et al. 2007; Thompson et al. in press).

Filming arboreal primates does have some problems: the video is often shot up into the canopy, meaning much of the footage is of the animal's posterior, not of its limbs on an even plane. In addition, there is the issue of filming against a bright background (the sky), which can render the subject difficult to distinguish; and it is seldom possible to obtain good footage of the primates' locomotion without the view being obstructed by branches. To be able to get good footage, especially with arboreal primates, the subjects should be habituated to human presence. Ground-dwelling primates may be easier to film if not habituated, but it is possible that obtaining usable footage of unhabituated arboreal primates would be difficult, though towers could be employed (see later). Building platforms within a gibbon group's territory is also a possibility to ensure the camera is at the same level as the gibbons, which is ideal but not strictly necessary. Again, with habituated animals, the presence of a human in the canopy will be relatively unobtrusive. In addition, remote sensor camera traps can be programmed to shoot video footage. If suitable areas of forest can be identified, remote cameras can also be positioned in the canopy, e.g., Cuddeback Expert Scouting Cameras (Non-Typical, Park Falls, WI) in use in Sebangau National Park (pers. obs.). For a quantitative analysis, the footage should best be from tripod-mount cameras, and the field of view should include a calibration object or clearly visible habitat features that can be used for calibration by measurement before or after the actual recordings.

Measuring Forces and Pressure

Long-term studies can gain valuable information which can be used by biomechanical studies, e.g., tensile strength, amplitude of vibrations under different weight conditions, stiffness and strain on natural substrates, and even under different seasonal conditions, e.g., wet and dry. Experiments with transducers or strain gauges can be carried out at different heights and distances from the main tree trunk based on observational data from the primates, without the need for invasive procedures on the primates. These data also provide an indication of what size of substrate different age (weight) classes can effectively travel along.

Again, if the primates are habituated, this method is easier. Using GPS units, tags and maps, researchers can identify areas where the gibbons have traveled, and which substrate was used. Experiments can then be carried out on the substrates once the gibbons have left. Samples of the substrates can be compared between species of tree, and behavioral observations can yield information on the physical properties of substrates and if the gibbons have a preference for smaller, more supple substrates or more rigid ones. Data such as these are very useful for redesigning laboratory-based experiments to more closely match the wild situation.

Behavioral Data

Field researchers can collect valuable behavioral data that can be used by biomechanics scientists. Data on travel height, gait, substrate type, and tree species can all be useful for biomechanics. The study of travel choice (Cannon and Leighton 1994; Thompson et al. in press) can help understand how primates select travel routes based on the available habitat. Thus, habitat characteristics data are important.

Recent Findings from the Field

Data from a study by Cannon and Leighton (1994) are not consistent with the results of Thompson et al. (in press) which found that for both canopy height and forest type, the gibbons used the most available category. Both of these recent results indicate that gibbons are actively selecting bigger, taller trees with a more uniform canopy than is predominantly available. One likely explanation for this is that Cannon and Leighton (1994) describe their study site as a "mosaic of pristine rainforest formations or habitats" whereas the more recent study was carried out in an ex-logging concession consisting of regenerating secondary peat-swamp forest.

Despite the low abundance of big trees or high canopy in the available habitat, the peat-swamp gibbons are actively avoiding traveling in areas where the canopy is under 10 m. They neither selected nor avoided midway canopy heights of 11–20 m, probably as they use these as platforms to reach higher heights and prefer canopy heights of 21 m and above. Although broken canopy is the dominant forest type, gibbons were indifferent to this during travel.

Uneven canopy and canopy gaps pose a crucial problem for arboreal primates, as they present either a very large break in the canopy or a succession of smaller breaks (uneven canopy). Efficient, cost-effective travel through the canopy, in terms of reducing distance (and time) of direct travel between two points, is heavily constrained by the presence of gaps (Cannon and Leighton 1994). It is generally assumed that gibbons will select the more continuous forest types over the less continuous ones, and higher canopies over lower ones. During travel, gibbons tend to follow established routes through the trees, referred to as "arboreal highways" (Chivers 1974). These routes minimize their chance of encountering gaps and also provides support for the theory that they appear to be selecting actively certain structures for travel. But, gibbons are limited in the distances they can cross with each locomotor mode, with the maximum distance seen crossed by brachiation being 12 m and 6 m by leaping Thompson et al. (in press).

Despite these interesting results on selection of travel routes, we have no data on whether the gibbons are also selecting the least energetically costly route, based on diet at the time, energy expenditure based on travel mode, and distance traveled. All these data require biomechanical data.

Conservation Implications

Gibbons are a highly adaptable species and appear capable of tolerating various levels of disturbance (Chivers 1974, 1977; Thompson et al. in press). But we have no data on how their energetics affect this, or on how their locomotion may vary in relation to food availability. Long-term studies are starting to address energetics more by looking at energy intake and diet (Cheyne 2008) but this has yet to be compared to biomechanics.

In conclusion, there are many ways that the field-based study of primate biomechanics can benefit conservation and science. A gibbon's energetic needs not only comes from the energy needed to perform the various locomotor patterns, but also depends on food availability, seasonality, and energy intake in the wild. Biomechanics has a use in understanding adaptation of locomotion to disturbance and changes in the canopy/main habitat. Finally, the application of understanding primate locomotor needs in the wild can be translated to zoos and rehabilitation centers to help with welfare and for preparing primates for a return to the wild: rigid substrates are not good.

Biomechanics specialists and field researchers need to improve cross-speciality collaborations to address some of the issues raised in this chapter, and throughout this volume. Solving the problems of laboratory-style data collection in the field does present difficulties but, with more dialogue, new and innovative ways can be developed to answer some of the many questions regarding primate locomotion.

Acknowledgments S.M. Cheyne conducted this research as part of the Orang-utan Tropical Peatland Project in collaboration with Dr. Suwido H. Limin and the Center for International Cooperation in Sustainable Management of Tropical Peatland (CIMTROP), University of Palangka Raya. Thank you to Claire Thompson, who provided many interesting discussions about gibbon locomotion and all the Indonesian behavior team for help collecting the data. Funding for SMC was provided by the George Washington University, USA and Primate Conservation Inc.

References

Alexander RMcN (2003) Principles of Animal Locomotion. Princeton University Press, Princeton, NJ.

Andrews P, Groves CP (1976) Gibbons and brachiation. In: Rumbaugh DM (ed), Gibbon and Siamang, Vol. 4: Suspensory Behaviour, Locomotion and Other Behaviours of Captive Gibbons; Cognition. Karger, Basel, pp 167–218.

Arnhem E, Dupain J, Vercauteren Drubbel R, Devos C, Vercauteren M (2008) Selective logging, habitat quality and home range use by sympatric gorillas and chimpanzees: a case study from an active logging concession in Southeast Cameroon. Folia Primatol 79(1):1–14.

Baldwin LA, Teleki G (1976) Patterns of gibbon behaviour on Hall's Island, Bermuda. In: Rumbaugh DM (ed), Gibbon and Siamang, Vol. 4: Suspensory Behaviour, Locomotion and Other Behaviours of Captive Gibbons; Cognition. Karger, Basel, pp 21–105.

Bertram JEA, Ruina A, Cannon CE, Chang YH, Coleman MJ (1999) A point-mass model of gibbon locomotion. J Exp Biol 202:2609–2617.

Bertram JEA, Chang YH (2001) Mechanical energy oscillations of two brachiation gaits: measurement and simulation. Am J Phys Anthropol 115:319–326.

Bertram JEA (2004) New perspectives on brachiation mechanics. Am J Phys Anthropol 47:100–117.

Cannon CH, Leighton M (1994) Comparative locomotor ecology of gibbons and macaques: selection of canopy elements for crossing gaps. Am J Phys Anthropol 93:505–524.
Cant JGH (1985) Locomotor and postural behavior of orangutan (*Pongo pygmaeus*) in Borneo and Sumatra. Am J Phys Anthropol 66:153.
Cant JGH (1986) Locomotion and feeding postures of spider and howling monkeys: field study and evolutionary interpretation. Folia Primatol 46:1–14.
Carpenter CR (1972) Suspensory behaviour of gibbons (*Hylobates lar*): a photo-essay. In: Rumbaugh DM (ed), Gibbon and Siamang, Vol. 4: Suspensory Behaviour, Locomotion and Other Behaviours of Captive Gibbons; Cognition. Karger, Basel, pp 1–18.
Chang YH, Bertram JEA, Ruina A (1997) A dynamic force and movement analysis for brachiation. J Exp Biol 200:3013–3020.
Chang YH, Bertram JEA, Lee DV (2000) External forces and torques generated by the brachiating white-handed gibbon (*Hylobates lar*). Am J Phys Anthropol 113:201–216.
Chapman CA, Peres CA (2001) Primate conservation in the new millennium: the role of scientists. Evol Anthrop 10(1):16–33.
Chatani K (2003) Positional behaviour of free-ranging Japanese macaques (*Macaca fuscata*). Primates 44(1):13–23.
Cheyne SM (2004) Assessing rehabilitation and reintroduction of captive-raised gibbons in Indonesia. PhD dissertation, University of Cambridge, UK.
Cheyne SM (2008) Feeding ecology, food choice and diet characteristics of gibbons in a disturbed peat-swamp forest, Indonesia. XXII Congress of the International Primatology Society, Edinburgh, UK.
Cheyne SM (2010) Behavioural ecology and socio-biology of gibbons (*Hylobates albibarbis*) in a degraded peat-swamp forest., In Indonesian Primates. eds J. Supriatna, S.L. Gursky, pp. 121–156. Springer, New York.
Chivers DJ (1973) An introduction to the socio-ecology of Malaysian forest primates. In: Michael RP, Crook JH (eds), Comparative Ecology and Behaviour in Primates. Academic Press, London, pp 101–146.
Chivers DJ (1974) The Siamang in Malaya: A Field Study of a Primate in Tropical Rainforest. Karger, Basel.
Chivers DJ (1977) The ecology of gibbons: some preliminary considerations based on observations in the Malay Peninsula. In: Prasad MRN, Anand Kumar TC (eds), The Use of Non-Human Primates in Biomedical Research. Indian Science Research Academy, New Delhi, pp 85–105.
Clutton-Brock TH (1973) Feeding levels and feeding sites of red colobus in the Gombe National Park. Folia Primatol 19:368–379.
Crompton RH, Sellers WI, Gunther MM (1993) Energetic efficiency and ecology as selective factors in the saltatory adaptation of prosimian primates. Proc Roy Soc Lond B 254:41–45.
Dagosto M, Yamashita N (1998) Effect of habitat structure on postural behaviour and support use in three species of lemur. Primates 39(4):459–472.
Engstrom L (2000) Effects of selective hand-logging on orang-utan habitat quality and population density in Gunung Palung National Park, Borneo, Indonesia; with focus on forest structure and nest resources. MSc dissertation, Committee for Tropical Ecology, Uppsala University.
Erharta EM, Overdorff DJ (2008) Spatial memory during foraging in prosimian primates: *Propithecus edwardsi* and *Eulemur fulvus rufus*. Folia Primatol 79:185–196.
Felton AM, Engstrom LM, Felton A, Knott CD (2003) Orangutan population density, forest structure and fruit availability in hand-logged and unlogged peat swamp forests in West Kalimantan, Indonesia. Biol Conserv 114:91–101.
Fleagle J (1974) Dynamics of a brachiating siamang (*Hylobates (Symphalangus) syndactylus*). Nature 248(445):259–260.
Fleagle JG (1976) Locomotion and posture of the Malayan siamang and implications for hominid evolution. Folia Primatol 26:245–269.
Garber PA (1989) Role of spatial memory in primate foraging patterns: *Saguinus mystax* and *Saguinus fuscicollis*. Am J Primatol 19:203–216.
Garber PA (2004) New perspectives in primate cognitive ecology. Am J Primatol 62(3): 133–137.

Gibbons EF, Lockwood R (1982) One–armed brachiation in gibbons (*Hylobates lar*). Am J Primatol 3:167–177.
Gibeault S, MacDonald SE (2000) Spatial memory and foraging competition in captive western lowland gorillas (*Gorilla gorilla gorilla*). Primates 41(2):147–160.
Grand TI (1972) A mechanical interpretation of terminal branch feeding. J Mammal 53:198–201.
Hollihn U (1984) Bimanual Suspensory Behaviour: Morphology, Selective Advantages and Phylogeny. Edinburgh University Press, Edinburgh.
Janson CH (1998) Experimental evidence for spatial memory in foraging wild capuchin monkeys, *Cebus apella*. Anim Behav 55:1229–1243.
Johns AD (1987) The use of primary and selectively logged rainforest by Malaysian hornbills (*Bucerotidae*) and implications for their conversion. Biol Conserv 40:179–190.
Johns AD, Skorupa J (1987) Responses of rain-forest primates to habitat disturbance: a review. Int J Primatol 8(2):157–191.
MacDonald SE, Agnes MM (1999) Orangutan (*Pongo pygmaeus abelii*) spatial memory and behavior in a foraging task. J Comp Psychol 113(2):213–217.
McGraw WS (1996) Cercopithecid locomotion, support use and support availability in the Tai forest, Ivory Coast. Am J Phys Anthropol 100:507–522.
Menzel EM (1973) Chimpanzee spatial memory organisation. Science 182(4115):943–945.
Mittermeier RA, Fleagle JG (1976) The locomotor and postural repertoires of *Ateles geoffroyi* and *Colobus guereza*, and a re-evaluation of the locomotor category semi-brachiation. Am J Phys Anthropol 45:235–256.
Mittermeier RA (1978) Locomotion and posture in *Ateles geoffroyi* and *Ateles paniscus*. Folia Primatol 30:161–193.
Napier JR, Napier PH (1967) A Handbook of Living Primates. Academic Press, London.
Nigg BM, MacIntosh BR, Mester J (2000) Biomechanics and Biology of Movement. Human Kinetics, Champaign, IL.
Preuschoft H, Demes B (1984) Biomechanics of brachiation. In: Preuschoft H, Chivers DJ, Brockleman WY, Creel N (eds), The Lesser Apes. Edinburgh University Press, Edinburgh, pp 96–118.
Preuschoft H, Demes B (1985) Influence of size and proportions on the biomechanics of brachiation. In: Jungers WL (ed), Size and Scaling in Primate Biology. Plenum Press, New York, pp 383–399.
Rao M, van Schaik CP (1997) The behavioural ecology of Sumatran orangutans in logged and unlogged forest. Trop Biodiv 4:173–185.
Sayer EC, Whitham JC, Margulis SW (2007) Who needs a forelimb anyway? Locomotor, postural and manipulative behavior in a one-armed gibbon. Zoo Biol 26:215–222.
Schultz AH (1930) Skeleton of the trunk and limbs of higher primates. Hum Biol 2:303–428.
Schultz AH (1933) Observations on the growth, classification and evolutionary specialisation of gibbons and siamangs. Hum Biol 5(2–3):212–255.
Schultz AH (1969) The Life of Primates. Universe Books, New York.
Straus WL (1942) Rudimentary digits in primates. Q Rev Biol 17:228–243.
Taylor CR, Schmidt-Nielsen K, Raab L (1970) Scaling of energetic cost of running to body size in mammals. Am J Phys Anthropol 291:1104–1107.
Thompson CJH, Cheyne SM, Chivers DJ (in press) Locomotor ecology and adaptation of gibbons (*Hylobates albibarbis*) in a degraded secondary forest. Folia Primatol
Thorpe SKS, Crompton RH, Alexander RMcN (2007) Orangutans use compliant branches to lower the energetic cost of locomotion. Biol Lett 3(3):253–256.
Usherwood JR, Bertram JEA (2003) Understanding brachiation: insight from a collisional perspective. J Exp Biol 206:1631–1642.
Usherwood JR, Larson SG, Bertram JEA (2003) Mechanisms of force and power production in unsteady ricochetal brachiation. Am J Phys Anthropol 120:364–372.
Van Horn RN (1972) Structural adaptations to climbing in the gibbon hand. Am Anthrop 74(3):326–334.
Vereecke EE, D'Août K, Aerts P (2006) Locomotor versatility in the white-handed gibbon (*Hylobates lar*): a spatiotemporal analysis of the bipedal, tripedal and quadrupedal gaits. J Hum Evol 50:552–567.

Chapter 12
Posture, Ischial Tuberosities, and Tree Zone Use in West African Cercopithecids

W. Scott McGraw and Paul W. Sciulli

Abstract Ischial callosities are present in all cercopithecoid primates and hylobatids while callosity-like structures are occasionally found in gorillas, chimpanzees and orangutans. The size and shape of calloused sitting pads and the underlying tuberosities vary among cercopithecoids and multiple explanations for this variation have been proposed. We examined the functional context of ischial callosity use in three colobine and four cercopithecine species ranging in the Ivory Coast's Taï forest. Innominates of 96 individuals were collected opportunistically and tuberosity size was determined by digitizing the outline of each bone and calculating the circumscribed area. The square root of this dimension was scaled with acetabular diameter to create a species index which was compared to several behavioral variables. Data from the field and laboratory indicate that the largest tuberosities are found in monkeys that feed most often from seated postures in the small branch milieu of tree crowns. Our results provide strong support for Vilensky's (1978) hypothesis that ischial callosities are adaptations for exploiting the terminal branch niche.

Keywords Feeding • Innominate • Positional Behavior • Support Use

Abbreviations

FS	forelimb support
PCA	Principal Component Analysis
QS	quadrupedal stand

> "Among Cercopithecidae and Hylobatidae, long periods of sitting, particularly during night-resting (Washburn 1957), are facilitated by the presence of specialized pads surmounting the ischium: the ischial callosities. These pads are lacking in Cebidae which usually adopt a horizontal sleeping posture. Ischial callosities show considerable variation between genera (Napier and Napier 1967:36). An adaptation of the ischia, which exhibit varying degrees of

W.S. McGraw (✉)
Department of Anthropology, The Ohio State University, USA
e-mail: mcgraw.43@osu.edu

K. D'Août and E.E. Vereecke (eds.), *Primate Locomotion: Linking Field and Laboratory Research*, Developments in Primatology: Progress and Prospects,
DOI 10.1007/978-1-4419-1420-0_12, © Springer Science+Business Media, LLC 2011

expansion, occurs in the suprajacent skin, which is modified by increased cornification and loss of hair to produce bare sitting pads (ischial callosities)." (Washburn 1957).

"Callosities vary in size, being small in Cercopithecus and gibbons and large in macaques and baboons, where in males the two are confluent and surrounded by a variable amount of hairless but cornified skin. These differences, however, are indications of taxonomic affinity rather than correlated with habitat or ecology." (Hill 1972:72).

"Ischial tuberosities lacking (in *Proconsul*), implying sitting and sleeping behaviors similar to platyrrhines, rather than to extant catarrhines." (Harrison 2002:317).

Introduction

A principal reason for conducting field studies is to observe the adaptive context of behavior: how, when, where and in what frequencies are behaviors used? Long-term studies of animals in their natural habitats document behavioral tendencies and establish the brackets on variation associated with particular morphologies. Field studies can also provide source material for detailed investigations under controlled conditions, which enable more fine-grained observations and experimental manipulation. In many cases, laboratory-based experimental studies can provide more detailed entries into the biomechanical or neural basis for behavior, and it is by integrating results from field and laboratory studies that our ability to associate specific anatomical complexes with corresponding behaviors in living organisms is enhanced. Ultimately, these associations allow for more reliable interpretations of behavior in the fossil record.

The complementary strategy of combining ecological field data with those derived from experimental and morphological studies is becomingly increasingly routine in primatology (e.g., Vinyard et al. 2008). This is particularly true in investigations of primate positional behavior, where a battery of experimental techniques has been used to explore aspects of functional anatomy associated with locomotor questions prompted by observations made in the field (e.g., Tuttle and Basmajian 1974; Jungers and Stern 1981; Stern and Susman 1981; Demes et al. 1995, 2001; Schmitt 1999; Larson and Stern 2006). As Napier (1967:333) famously remarked, "locomotor adaptations have provided the principal milestones along the evolutionary pathway of primates," and to that end, the diversity of experimental techniques used outside of the field has yielded a rich literature devoted toward establishing reliable associations between positional behavior, corresponding anatomies, and their evolutionary significance within the order Primates. Prominent examples include investigating the evolution of arboreality, vertical clinging and leaping, suspensory behavior, and bipedality (Washburn 1950; Napier 1963; Napier and Walker 1967; Jenkins 1974; Fleagle et al. 1981; Susman et al. 1984; Jones 2008). Understanding these adaptive breakthroughs is critical not only for reconstructing the paleobiology of fossil species, but also, as Napier noted, for defining primate grades themselves. In other words, the study of primate positional behavior has ecological, taxonomic, and evolutionary significance.

To date, the majority of positional studies have investigated locomotor differences between taxa (Garber 2007; Dagosto and Gebo 1998). This is reasonable given the common belief that the forces involved in locomotion are largely responsible for shaping postcranial form. It is also understandable given that no group of mammals displays a greater variety of locomotor adaptations to tree-dwelling than do primates (Fleagle 1978). But there are several dimensions to an animal's positional repertoire, and the nonlocomotor ones have received far less attention than those involving movement of the limbs and trunk. Postural elements may be less conspicuous than locomotor ones and, in most cases, involve smaller forces than those used during movement, however, stationary behaviors may be no less significant for an animal's fitness (Stern and Oxnard 1973; Rose 1974a; Fleagle 1980). Postural behaviors not only dominate a species' positional repertoire and in many cases involve sustained forces, but for most taxa, the crucial act of feeding is much more closely associated with postural tendencies than with any single locomotor mode (Fleagle 1984; McGraw 1998a). Nevertheless, although the anatomical correlates of several postures have been established and these associations extended to the fossil record (e.g., Anemone 1990; Hamrick et al. 1995; Hartwig and Cartelle 1996; Gebo et al. 1997; Crompton et al. 2008), attempts to examine the ecology of posture or to identify morphological correlates of postural tendencies are rare. The dearth of studies is particularly significant among Old World monkeys. This is curious because the cercopithecoid clade is defined (in part) by one anatomical feature—ischial callosities—that is almost certainly related to an element of postural behavior that is readily quantifiable and of potentially great ecological significance.

Ischial callosities are avascular, calloused areas of cornified tissue that overlie expansions of the ischial bones (Rose 1974b). Commonly referred to as sitting pads, true ischial callosities are present in all cercopithecoid primates and hylobatids while callosity-like structures are occasionally found in gorillas, chimpanzees and orangutans (Pocock 1925; Schultz 1936; Miller 1945; Napier and Napier 1967; Rose 1974b; Sigmon and Farslow 1986; Tuttle 1986). The presence of ischial callosities has figured prominently in several recent phylogenetic debates involving fossil catarrhines (Ward et al. 1993; McCrossin and Benefit 1994, 1992; Harrison and Sanders 1999; Benefit and McCrossin 2002; Harrison 2002; Walker and Shipman 2005), and this feature is one of the most recognizable and diagnostic cercopithecoid traits (Strasser and Delson 1987). It is therefore surprising how little is known about the function of callosities beyond their weight-bearing role during sitting. Several authors have noted that the size and shape of ischial calloused pads and their underlying tuberosities vary widely in cercopithecoids (Schultz 1936; Miller 1945; Napier and Napier 1967; Hill 1972; Sigmon and Farslow 1986; Ankel-Simons 2000). However, no quantitative analysis of tuberosity variation has been attempted for a group of cercopithecoids, nor has the ecological context of callosity use been systematically assessed in the field or in captivity. Although it is widely accepted that these pelvic adaptations are related to some aspect of postural behavior, it is not clear what the positional correlate is nor how it might covary with the size and shape of the tuberosities themselves.

Washburn (1957) was the first to offer a specific hypothesis for the origin and function of ischial callosities beyond their obvious use during sitting. He reasoned that because many other animals—including whole primate radiations, e.g., platyrrhines, prosimians—sit frequently yet have not developed expanded tuberosities, there must be something distinctive about the sitting context of Old World monkeys. Based on observations of baboons in Zimbabwe and of several other primate species in captivity, Washburn argued that callosities evolved to facilitate upright sitting postures during sleep in trees, most likely as a response to predation pressure. Trees probably afford greater safety from predators than the ground, and Washburn noted that baboon males tended to sleep on large supports at lower tree levels while smaller group members were generally found in higher branches. Predators approaching from the ground would therefore be likely to encounter large males in defensive positions before the rest of the group could be reached. Rose (1974a) expanded Washburn's reasoning, making specific reference to sleep-sitting postures in the small branch setting. He posited that upright postures in which the callosities form one part of a stable, triangular base can be passively maintained during periods of muscular relaxation while allowing for rapid sleep arousal. Fast arousal would be beneficial during encounters with danger, and by sleep-sitting amid slender branches, monkeys can limit the size of predators approaching through the trees. According to Rose, the need for callosities is diminished in New World monkeys because most platyrrhines sleep lying down and on larger supports than those used by monkeys of the Old World (Rose 1974b).

In addition to their role as a sleeping adaptation, Rose (1974a) discussed callosity use in the small branch setting during daytime activities. He argued the use of the characteristic cercopithecoid "foot-prop" posture, in which one or both of the callosities assume the majority of body weight with feet propped on supports at or above head or shoulder level, provides a stable solution to the problem of negotiating slender supports by large primates during feeding (Fig. 12.1). The need for such postures and accompanying sitting pads is decreased among platyrrhines because 1) the body size to branch ratio is smaller in most platyrrhines, and 2) large New World monkeys employ suspensory adaptations in the small branch setting rather than sitting.

Fig. 12.1 Variations of the foot prop posture in *Colobus polykomos* (**left**) and *Procolobus badius* (**right**) described by Rose (1974a)

Vilensky (1978) followed Rose (1974a, b) in exploring the selective advantage of ischial callosities in the small branch setting. Vilensky surveyed the available literature and concluded that most arboreal cercopithecoids and hylobatids feed from seated positions among the small branches of tree peripheries. Such positions, particularly when adopted for extended periods of times as in colobines, are facilitated by the frictional properties of callosities, which decrease the risk of slipping off a branch. Vilensky hypothesized that callosities served a function similar to that of the prehensile tail of platyrrhines, i.e., as a feeding adaptation amid the terminal branches of tree crowns. The terminal branches are a resource-rich foraging zone, and their exploitation is believed to have played an important role in the origin and radiation of primates (e.g., Cartmill 1972, 1992; Jenkins 1974; Sussman 1991; Bloch and Boyer 2002). Early primates were small-bodied foragers that could effectively negotiate the small branch niche because of their comparatively diminutive body size. Increased body size characterizes the anthropoid radiation, and as the body size to branch ratio became larger through time, continued exploitation of tree peripheries required novel strategies. For medium to large primates, one strategy is a mechanism that enables a monkey to distribute weight below the support: suspension by tail or forelimb. An alternative solution is a mechanism for comfortably distributing weight across one or more supports by concentrating the majority of weight in one area (on one or both callosities), with the limbs providing stability and security on neighboring supports as in the foot-prop posture described by Rose (1974b).

Napier and Napier (1967) posited that callosity size and shape variation was a function of tail length which, in turn, covaried with overall habitat use (arboreality *vs.* terrestriality). This argument was explored in *Macaca* by Wilson (1972), who showed that macaque species with longer tails had relatively smaller callosities. According to Hill (1972), variation in callosity size and shape is best explained via phylogeny, and any functional association is, at best, weak. A thorough testing of these hypotheses requires a broad, quantitative analysis of tuberosity variation, habitat use, and tail length across Cercopithecoidea, a task outside the scope of this chapter. What is possible is an exploration of the relationship between callosity use and exploitation of the small branch setting within a group of cercopithecoids that shares a single habitat but differs in their use of micro-habitat zones and associated positional behaviors.

Vilensky's (1978:368) statement that "ischial callosities and prehensile tails both evolved in response to similar selective pressures of feeding in a small branch setting" provides a hypothesis that can be tested with information on postural frequencies and tree zone use combined with those on the anatomy of the tuberosities themselves. In this chapter, we investigate the relationship between tuberosity size (surface area) and several variables related to positional behavior and habitat use during feeding among cercopithecid monkeys from the Ivory Coast's Taï forest. The Taï forest is home to a large community of cercopithecid species that have been under continuous study since 1991 (McGraw et al. 2007; McGraw and Zuberbuhler 2008). The seven Taï cercopithecids discussed here represent an array of dietary and support preferences, postural profiles, and body sizes. Several aspects of the positional behavior of these species are known (McGraw 1996, 1998a, b, c, 2000, 2003, 2007); however, the use of different postures, including those involving callosities in different tree zones, has not been addressed. The novel behavioral data presented here are complemented with measurements taken on a collection of skeletal elements from the Taï monkeys. If, as Vilensky and Rose predict, callosities have evolved—at least in part—to facilitate feeding in the terminal branches, then it is reasonable to expect that tuberosity size will covary with the extent that monkeys of different size adopt sitting postures in this critical habitat zone. We predict that larger arboreal monkeys will adopt seated positions more frequently in the small branch setting because the greater body size to branch size ratio generally precludes standing postures. Smaller primates should require fewer seated postures in the terminal branches, and will employ more quadrupedal and bipedal (with forelimb assistance) postures. Variation in callosity size will mirror this basic feeding contrast, and we predict that larger primates will have relatively larger callosities as a function of the need for prolonged seated postures in the small branch niche. The need for expanded callosities is decreased in smaller monkeys, which will, consequently, have ischial tuberosities of modest size. Thus, the general question addressed here is: to what extent is variation in tuberosity size associated with body size, postural behavior and the exploitation of the small branch setting?

Methods

We collected data on postural behavioral and habitat use on seven cercopithecid species inhabiting the Ivory Coast's Taï forest (McGraw et al. 2007). The focal species [including average male (♂) and female (♀) body weights taken from Smith and Jungers (1997) and Oates et al. (1990)] are *Procolobus badius* (Western red colobus, ♂ = 8.36 kg, ♀ = 8.21 kg), *Procolobus verus* (olive colobus, ♂ = 4.7 kg, ♀ = 4.2 kg), *Colobus polykomos* (Western black-and-white colobus, ♂ = 9.9 kg, ♀ = 8.3 kg), *Cercopithecus diana* (Diana monkey, ♂ = 5.2 kg, ♀ = 3.9 kg), *Cercopithecus petaurista* (lesser spot-nosed monkey, ♂ = 4.4 kg, ♀ = 2.9 kg), *Cercopithecus campbelli* (Campbell's monkey, ♂ = 4.5 kg, ♀ = 2.7 kg), and *Cercocebus atys* (sooty mangabey, ♂ = 11.7 kg, ♀ = 6.2 kg). The first six species are primarily arboreal, whereas the last—*Cercocebus atys*—spends the majority of its foraging and traveling time on the ground. We used an instantaneous time point scheme at 3-min intervals to record all positional and habitat use data. We maintained an interval of at least 15 min between consecutive samples of single individuals to ensure data independence and avoid temporal autocorrelation. On every time point we recorded the following information: 1) individual identification; 2) maintenance activity involving posture (rest, social, feed); 3) postural behavior [sit, quadrupedal stand (QS), sprawl (prone-lie), lie (including supine, back and lateral lie), stand/forelimb-suspend (FS), and other (including bipedal stand, vertical cling, quadrumanous suspend, extended hind limb suspend, tail/hind limb-suspend, etc. (see McGraw 1998a for sample sizes and figures of these postures)]; 4) support type (based on support diameter): bough, branch, twig, other; and 5) tree zone [Trunk = trunk and portions of boughs immediately adjacent to trunk, i.e., spine of tree; Core = network of supports comprising tree crown's skeleton (distal to tree spine) between trunk and smaller, terminal branches; Periphery = region of terminal branches forming the outer, lateral perimeter of crown; Top = region of terminal branches forming top of crown; and Other = all other zones, e.g., ground, treefall, lianas, etc.; Fig. 12.2]. Tree zones were not scaled to height, however, the purpose of this scheme is to evaluate the mechanical constraints of foraging versus exposure to predators, feeding height preferences, distribution of actual resources, etc.

Over the past 18 years, members of the Taï Monkey Project have collected innominates from approximately 100 individual monkeys that died of natural causes at Taï. These skeletal elements were collected in accordance with US Fish and Wildlife Regulations and transported to the Department of Anthropology at The Ohio State University, where they are currently housed. The size and shape of the ischial tuberosities vary interspecifically (Fig. 12.3) and we assume the size of

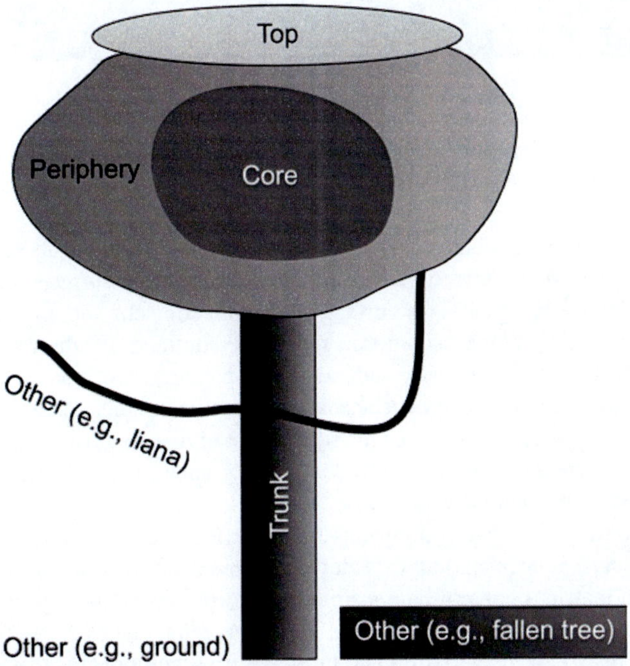

Fig. 12.2 Diagram used to describe exploitation of different tree portions. Trunk = trunk and portions of boughs immediately adjacent to trunk, i.e., spine of tree; Core = network of supports comprising tree crown's skeleton or framework (distal to spine) between trunk and smaller, terminal branches; Periphery = region of terminal branches forming the outer, lateral perimeter of crown; Top = that portion of tree periphery forming top of crown; Other = all other zones including ground, tree falls, lianas, and other places that do not fit in previous zones

Fig. 12.3 Ischial tuberosities of six cercopithecid species from the Ivory Coast's Taï forest. Male on left, female on right in all species pairs. Note the globular shape in the two large colobine species (*Procolobus badius* and *Colobus polykomos*) and the more rectangular configuration in the guenons (*Cercopithecus diana* and *Cercopithecus campbelli*)

the overlying callosity corresponds closely to that of the bony tuberosity (Miller 1945; Rose 1974a). We traced the outline of each tuberosity surface, defined as the portion of the overlying callosity surface that would contact a support during sitting positions. In all cases, tuberosity borders were well defined. We digitized the traced outline of each ischial tuberosity and calculated the circumscribed area, yielding a measure of the total tuberosity surface area that could contact a support during all possible seated postures. We scaled the square root of this dimension with acetabular diameter to create a tuberosity area index for each species. We then compared this index to several behavioral variables.

We performed Principal Components Analysis (PCA) on the species ($n = 7$) times *postures in periphery* ($n = 5$) and species times *tree-zone use* ($n = 5$) data matrices.

Our data matrices contain compositional data in which, for a species, the variables are proportions that sum to 1.0. This constraint places restrictions on the correlation matrix of the variables and produces PCA components that are difficult to interpret. To overcome these limitations and to allow a customary interpretation of the principal components, Aitchison (1983, 1982) recommended the following transformation of the compositional data matrix, which we employed:

$$X_i^* = \ln(m_i) - (1/p)\sum \ln(m_j)\, and\ i = 1,\ 2, \ldots p$$

where m is the observed proportion and p is the number of variables. We then performed PCA on the covariance matrix of the X_i^*. We plotted ischial tuberosity index $\left(\sqrt{ischial\ tuberosity\ /\ acetabular\ diameter}\right)$ for each species against the PCA score for that species derived from the tree-zone use PCA and postures in periphery PCA. We then performed simple linear regression of ischial tuberosity index on principal components score.

Results

Overall postural profiles as well as those postures observed during feeding, resting, and social behavior are presented for each species in Table 12.1. These data have been discussed before (McGraw 2000, 1998a), and the main points are summarized as follows: 1) In all species, sitting is the primary postural activity overall as well as during each individual maintenance activity; 2) colobines sit much more frequently than do cercopithecines; and 3) arboreal cercopithecines employ postures such as quadrupedal stand and stand/forelimb suspend that allow for faster, more efficient movement to the next feeding site. These postures, which may entail weight distribution across multiple (and small) supports, are facilitated by comparatively small body size (see later); and 4) colobines frequently adopt reclining postures whereas cercopithecines do not.

Data on the use of different zones during feeding, resting, and social behavior are presented in Table 12.2. The sooty mangabey (*Cercocebus atys*) is the obvious outlier, spending the majority of its foraging time on the ground. During feeding, the arboreal species differ from each other in significant ways, and each can be characterized by a zone preference. Red colobus (*Procolobus badius*) obtain the majority of their food from the terminal branches of tree tops and tree peripheries. Black-and-white colobus (*Colobus polykomos*) spend half as much time as red colobus feeding in the peripheral zone and rely to a large extent on zones classified as "Other." The use of supports in "Other" zones reflects the tendency of *Colobus polykomos* at Taï to feed on liana leaves (Korstjens 2001). Olive colobus (*Procolobus verus*) divide the majority of their feeding time between tree peripheries and "Other" zones. The "Other" zones preferred by the cryptic olive colobus include dense vine tangles, tree falls, and understory supports that afford cover and safety. Diana monkeys (*Cercopithecus diana*) forage throughout all zones but devote the greatest amount of their feeding time (56.9%) to the tops and peripheries of trees. Both Campbell's monkey (*Cercopithecus campbelli*) and the lesser spot-nosed monkey (*Cercopithecus petaurista*) exploit all zones for feeding and use the Trunk, Periphery, and Other zones in similar frequencies.

Despite species-specific tendencies, all species feed and forage in every zone to some extent, and no species is prevented from exploiting any zone, as defined here, to a significant degree. In fact, despite difference in body weight, the terminal branch zones (Periphery and Top) account for at least 30% of the foraging space used by each species. Moreover, these small-branch zones account for approximately 40% of the foraging zones exploited by the largest arboreal taxa, *Colobus*

Table 12.1 Postural behavior of Tai cercopithecids

	Sit	QS	Sprawl	Lie	Stand/FS	Other	*n*
Procolobus badius							
Overall	87.0	1.4	4.2	6.1	0.55	0.75	2690
Resting	80.4	0	9.1	10.5	0	0	1248
Social	72.5	13.5	0	13.2	0	0.8	244
Feeding	97.3	0.17	0	0	1.2	1.33	1198
Colobus polykomos							
Overall	89.5	0.19	2.9	6.3	0.84	0.27	2495
Resting	82.8	0	6.4	10.8	0	0	1191
Social	80.1	0	0	19.9	0	0	181
Feeding	97.3	0.41	0	0	1.8	0.49	1123
Procolobus verus							
Overall	90.7	1.3	3.8	3.7	0.28	0.22	1081
Resting	86.9	0	7.4	5.7	0	0	557
Social	81.0	9.5	0	7.6	0	1.9	103
Feeding	98.3	0.72	0	0	0.72	0.26	421
Cercopithecus diana							
Overall	59.6	22.1	0.16	0.77	15.2	2.17	1908
Resting	87.3	8.7	0.5	3.5	0	0	403
Social	86.5	10.8	0	0	0	2.7	37
Feeding	51.5	25.9	0	0	19.9	2.7	1468
Cercopithecus campbelli							
Overall	70.9	21.9	0.16	0.77	4.8	1.47	838
Resting	97.2	2.4	0	0	0	0.4	290
Social	92.5	0	0	7.5	0	0	40
Feeding	54.1	34.8	0	0	7.9	3.2	508

Cercopithecus petaurista							
Overall	81.9	12.9	0	<1.0	3.1	1.3	1530
Resting	91.5	7.9	0	0	0	<1.0	662
Social	75	11.4	0	13.6	0	0	88
Feeding	74.6	17.4	0	0	6.2	1.8	1530
Cercocebus atys							
Overall	80.9	15.7	0	0	1.9	1.5	854
Resting	98.4	1.2	0	0	0	0.4	248
Social	89.6	10.3	0	0	0	0.1	87
Feeding	71.1	23.5	0	0	3.1	2.3	519

From McGraw (1998a, 2000).

Table 12.2 Zone use during feeding, resting and social behavior (postural activities)

	Ground	Trunk	Core	Periphery	Top	Other
Procolobus badius						
Feeding	0.27	11.5	26.9	42.1	14.4	4.8
Resting	0	26.5	58.9	12.9	0.97	0.65
Social	0	14.3	70.5	13.3	0	1.9
Colobus polykomos						
Feeding	0	17.5	17.5	21.4	16.5	27.2
Resting	0	16	47.5	18.5	16.7	1.2
Social	0	7.7	53.8	26.9	11.5	0
Procolobus verus						
Feeding	1.9	13.5	7.7	30.5	12.5	33.8
Resting	3.2	36.2	28.1	8.1	0.73	23.7
Social	2.6	24.7	44.2	14.3	1.3	13
Cercopithecus diana						
Feeding	3.6	16.5	14.9	40.6	16.3	8.2
Resting	1.2	37.1	34.1	22.9	1.8	2.9
Social	4.2	8.3	50	29.2	4.2	4.2
Cercopithecus campbelli						
Feeding	12.2	21.3	11.8	24.4	6.4	23.9
Resting	10.8	25.4	31.3	9.7	0.4	22.4
Social	28.9	18.4	26.3	5.3	0	21.1
Cercopithecus petaurista						
Feeding	2.6	28.6	16.5	28.4	7.9	16
Resting	—	—	—	—	—	—
Social	—	—	—	—	—	—
Cercocebus atys						
Feeding	62	17.4	3.6	4.8	5.4	6.9
Resting	59.3	12	9.3	0	0	19.3
Social	83.9	3.2	6.2	0	0	0

polykomos and *Procolobus badius*. In other words, the largest arboreal monkeys forage in the small-branch niche more than do the smaller arboreal monkeys. The small-branch niche is used less frequently during nonforaging activities by all species. The major exception to this generalization is the comparatively high frequency that *Colobus polykomos* was observed in social activities in tree peripheries. Black-and-white colobus monkeys often groom one another while sunning in tree tops early in the morning. Otherwise, the Trunk and Core zones of trees—areas that tend to contain larger, more stable support—are used more often by all species during resting and social activities.

The frequencies that postures were employed by each monkey in each of the different zones are summarized in Table 12.3. Three key points are apparent in these data: 1) there are significant interspecific differences in the use of different postures (and see McGraw 1998a), 2) the use of individual postures is not distributed evenly throughout each zone, and 3) during virtually all maintenance activities observed across species, the most frequently-adopted posture in all zones is sitting. In tree peripheries and tree tops, sitting is used more frequently by colobines than by arboreal cercopithecines; quadrupedal stand and supported stand are employed much less frequently—or not at all—by colobines in these tree zones. The reclining postures frequently adopted by colobine monkeys (McGraw 1998a) tend to occur in tree cores where supports tend to be larger and more stable. Black-and-white colobus also adopt reclining postures in tree peripheries and tree tops, most often during morning hours when they sunbathe. During these sunbathing sessions, black-and-white colobus often position themselves in the forks of branches exposed to the sun. Neither large colobus monkey employs the quadrupedal stand or supported stand in tree peripheries or tree tops with great regularity; these postural behaviors are used much more frequently by the guenon (*Cercopithecus*) species.

Data on tuberosity area, body weight, and acetabular diameter are presented in Table 12.4. Several differences in overall tuberosity shape are apparent (Fig. 12.3): tuberosities of the two large colobines are ovoid or globular in shape, while those of the guenons are more rectangular or linear. Tuberosities of the sooty mangabey are the most distinctive with an inverted, teardrop shape. In addition to these shape differences, the average tuberosity size varies across taxa. Tuberosity area significantly correlates with both the species means for body size ($r = 0.86$, $p < 0.0001$; Fig. 12.4) and acetabular diameter as the proxy for body size ($r = 0.93$, $p < 0.001$; Fig. 12.5). Based on the stronger association between the latter variables, we calculated the tuberosity index as the square root of the tuberosity area scaled by acetabular diameter. Mean values of this index for each species are reported in Table 12.4. These data indicate that ischial tuberosity surface area is positively associated with body size: large monkeys tend to have relatively larger tuberosities. Regression (data from Table 12.4) of ln tuberosity area on ln body size for all species and both sexes (tuberosity area = $8.24 \times$ body mass$^{0.91}$; standard error of exponent = 0.14) and excluding *Cercocebus atys* (tuberosity area = $6.41 \times$ body mass$^{1.10}$; standard error of exponent = 0.11) shows tuberosity area scales with (body mass)$^{1.0}$ indicating the callosities function in load bearing and all monkeys should have equal force on them.

Table 12.3 Postures used in each zone

	Ground	Trunk	Core	Periphery	Top	Other
Procolobus badius						
Sit	100.0	76.4	79.6	93.0	91.9	90.9
Recline	0	18.2	15.73	3.8	0	0
Quadrupedal stand	0	0	4.0	2.3	0	0
Supported stand	0	0	0.29	0.94	0	4.5
Other	0	5.4	0.38	0	8.1	4.6
Colobus polykomos						
Sit	0	91.3	72.2	84.7	87.2	96.6
Recline	0	4.3	27.7	13.6	12.8	0
Quadrupedal stand	0	2.2	0	0	0	0
Supported stand	0	2.2	0	1.7	0	3.4
Other	0	0	0.1	0	0	0
Procolobus verus						
Sit	85.7	94.7	79.8	95.7	97.6	86.6
Recline	4.8	3.4	10.4	0.72	0	5.4
Quadrupedal stand	9.5	0.5	3.3	0.72	0	3.6
Supported stand	0	0	0	0.72	2.4	0
Other	0	1.4	6.5	2.14	0	4.4
Cercopithecus diana						
Sit	50	76.4	62.8	62.0	67.7	73.8
Recline	0	0.64	2.02	0	0	0
Quadrupedal stand	50	13.4	29.1	33.3	23.7	26.2
Supported stand	0	7.0	4.7	3.8	8.6	0
Other	0	2.56	1.38	0.9	0	0
Cercopithecus campbelli						
Sit	74.7	74.3	87.1	50.7	56.7	71.2

Recline	1.1	0	0.6	0	0	0.56
Quadrupedal stand	23.2	17	11.6	36.2	40	22
Supported stand	0	4.7	0.7	12.3	3.3	3.9
Other	1.0	4.0	0	0.8	0	2.34
Cercopithecus petaurista						
Sit	72.6	77.4	81	54.5	58.2	72.2
Recline	0	0	0.4	0.1	0	0
Quadrupedal stand	26.6	13.8	14.7	32.7	34.8	22.6
Supported stand	0	6.1	3.0	12.6	7.0	3.6
Other	0.8	2.7	0.9	0.1	0	1.6
Cercocebus atys						
Sit	74.1	92.2	93.3	87.5	94.4	91.1
Recline	0	0	0	0	0	0
Quadrupedal stand	24.4	2.6	3.3	6.3	5.6	7.1
Supported stand	0.59	1.3	0	6.3	0	1.8
Other	0.91	3.9	3.4	0	0	0

Table 12.4 Tuberosity indices and body masses for Tai cercopithecids

	χ Tuberosity area (*n*)	Acetabular diameter	Index	Body mass (kg)[a]
Procolobus badius				
Male	69.14 (28)	19.5	42.97	8.36
Female	65.4 (17)	18.6	43.24	8.21
Colobus polykomos				
Male	80.24 (5)	21.27	42.01	9.9
Female	77.47 (7)	20.47	42.9	8.3
Procolobus verus				
Male	40.48 (4)	14.97	42.5	4.7
Female	30.4 (2)	14.6	37.8	4.2
Cercocebus atys				
Male	51.85 (9)	19.25	37.2	11.7
Female	31.55 (6)	16.33	34.2	6.2
Cercopithecus diana				
Male	32.13 (4)	17.15	33.05	5.2
Female	25.35 (7)	15.2	32.9	3.9
Cercopithecus petaurista				
Male	24.72 (3)	14.16	34.9	4.4
Female	21.96 (5)	12.83	36.6	2.9
Cercopithecus campbelli				
Male	31.29 (4)	15.9	35.39	4.5
Female	24.19 (3)	12.6	38.97	2.7

[a] From Oates et al (1990); Smith and Jungers (1997).

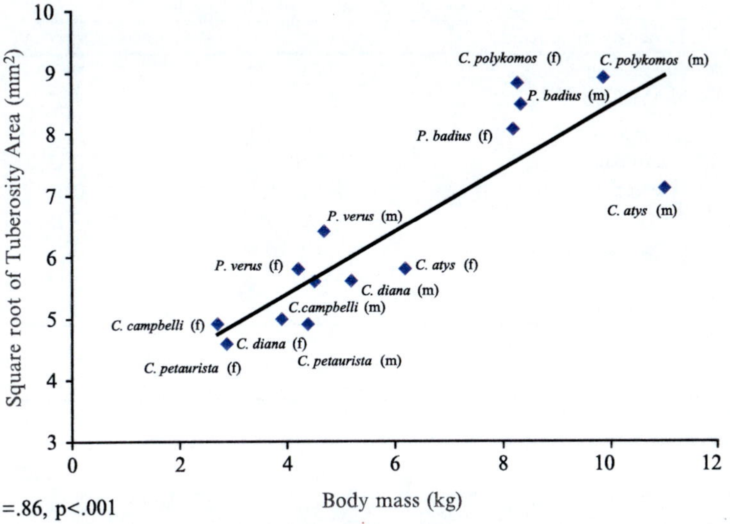

r =.86, p<.001

Fig. 12.4 Relationship between average male and female body mass (kg) and ischial tuberosity surface area (mm²) in the seven Taï cercopithecids

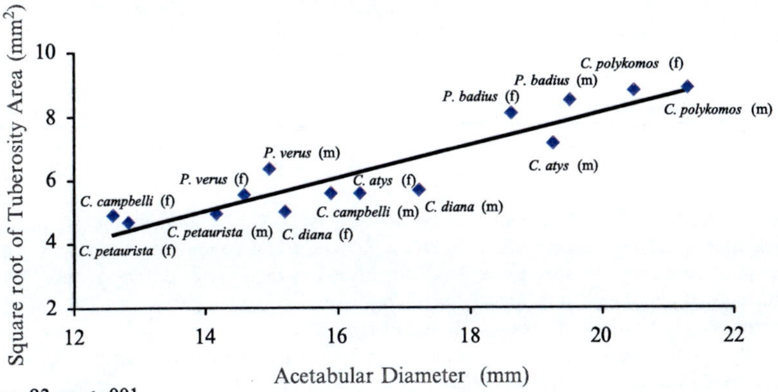

r= .93, p < .001

Fig. 12.5 Relationship between acetabular diameter (mm) and ischial tuberosity area (mm²) in the seven Taï cercopithecids

Results of the Principle Components Analysis of zones used during postural behavior are presented in Table 12.5 and Fig. 12.6. The first two components (Table 12.5) represent 95% of the variation in the transformed data. The first eigenvector is a contrast primarily between zones "Core," "Periphery," "Top," and "Ground," while the second eigenvector is primarily a contrast between "Other" and zones designated "Core," "Periphery" and "Top." Figure 12.6, the plot of the principal components scores on the first and second components, indicates that tree zone use, independent of stratum, separates taxa. The three guenons and olive colobus are generalists, using all zones in similar proportions. Despite their larger size,

Table 12.5 Principal components analysis of tree zone use

Principal component	1	2
Eigenvalue	5.25	0.59
% Variance	85.68	9.55
Eigenvector		
Ground	0.88	–0.13
Trunk	–0.01	0.06
Core	–0.28	–0.36
Periphery	–0.27	–0.29
Top	–0.22	–0.13
Other	–0.11	0.86

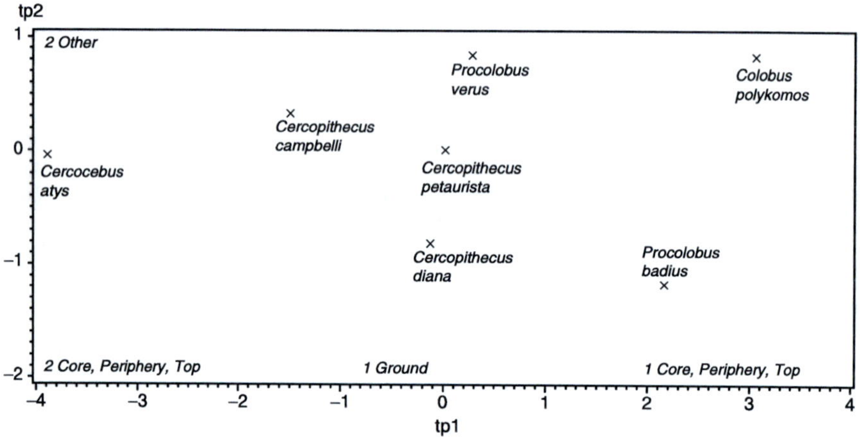

Fig. 12.6 Results of the Principal Components Analysis (PCA) examining the use of different "zones" during all postural activities (*x*-axis = first principal component and *y*-axis = second principal component). The species can be distinguished by their use of different zones. The three guenon species and the olive colobus (*Procolobus verus*) are habitat generalists that use all zones in approximately similar proportions. The two large colobine species —*Colobus polykomos* and *Procolobus badius*—tend to be found most often in tree peripheries, tree tops, and, to a lesser extent, tree cores. Sooty mangabeys (*Cercocebus atys*) are predominantly terrestrial

Procolobus badius and *Colobus polykomos* tend to be found most often in tree tops, tree peripheries, and, to a lesser extent, cores. In addition to using large supports while feeding, such as those routinely used by *Colobus polykomos* during prolonged processing of *Pentaclethera macrophylla* pods (McGraw 1998a), these large colobines feed frequently in the terminal zones of trees. Sooty mangabeys are outliers, feeding largely on the ground.

We next examine the relationship between tuberosity size (tuberosity index) and the use of each tree zone. Regressing the tuberosity index against the PCA scores of the first principal component of tree zone use during feeding reveals a strong relationship among the monkey taxa: species that feed most often in tree tops and tree peripheries (where support sizes are smallest) tend to have the largest tuberosities

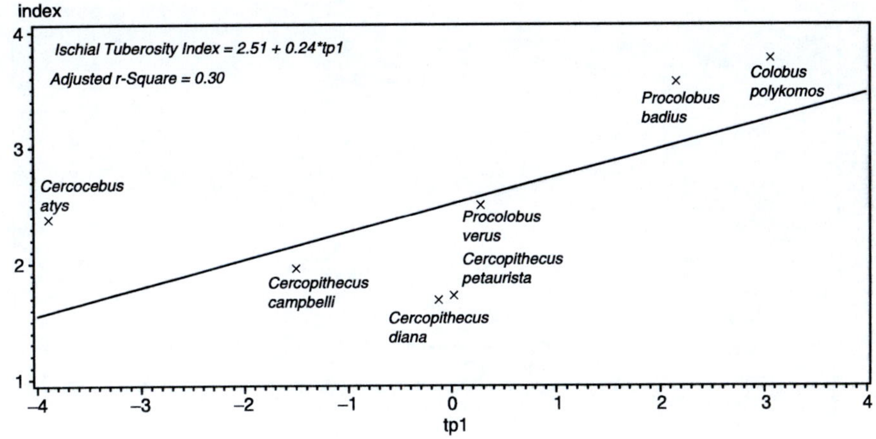

Fig. 12.7 Regression of the ischial tuberosity index (*y*-axis) against scores from PCA of zones used during feeding (*x*-axis = first principal component). There is a significant relationship between these values: the taxa that feed most often in the tree tops and tree peripheries have the largest ischial tuberosities for their body size

Table 12.6 Principal components analysis of postures used in tree peripheries

Principal component	1	2
Eigenvalue	9.54	1.93
% Variance	78.93	16.01
Eigenvector		
Sit	0.15	–0.04
Recline	0.73	–0.25
Quadrupedal stand	–0.63	–0.16
Supported stand	–0.23	–0.41
Other	–0.01	0.86

(Fig. 12.7). The large colobines frequently exploit not only the cores of trees (see Fig. 15 in McGraw 1998a), where average support diameters are expected to be large, but also frequent the small branch niche at rates higher than expected for their body size. How are the large, arboreal monkeys able to safely position themselves for sustained periods of time given the disparity in branch to body size?

Table 12.6 contains the results of the PCA of the transformed postural behaviors used in tree peripheries data. The first two components represent almost 95% of the variation in the transformed data. The first eigenvector is a contrast primarily between recline and sit on one hand and quadrupedal stand and supported stand on the other. The second eigenvector contrasts "other" and supported stand. Figure 12.8 is the plot of the principal component scores on the first and second components for the postural behaviors used in tree periphery analyses. This plot indicates a strong relationship between body size and the adoption of specific postures. The largest arboreal monkeys

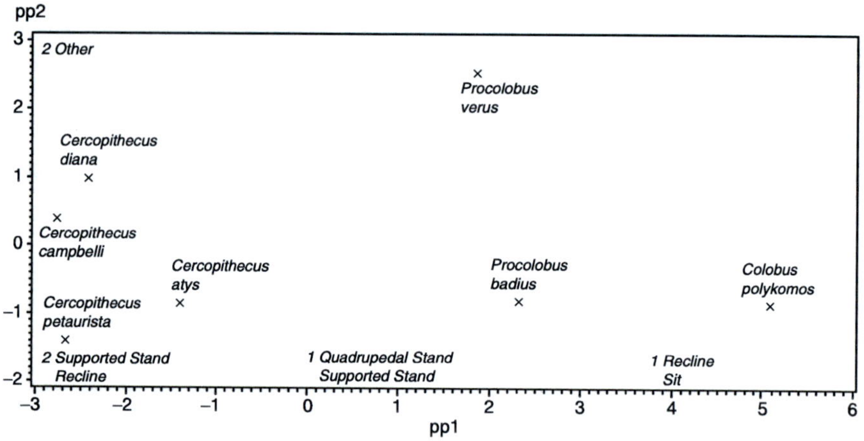

Fig. 12.8 Results of the Principal Components Analysis (PCA) examining the use of different postural behaviors in the tree peripheries (*x*-axis = first principal component and *y*-axis = second principal component). The analysis reveals a strong association between body size and posture. Large arboreal monkeys (*Colobus polykomos* and *Procolobus badius*) most often use seated and reclining postures in the peripheral zone; seated postures are used during feeding while reclining postures are used during morning sunning, especially by *Colobus polykomos*. The smaller arboreal monkeys frequently employ quadrupedal and supported stand postures

tend to sit and recline in tree peripheries: the former posture is used during feeding while the latter is used almost exclusively (though not often) during periods of sunning (rest and social activities). *Colobus polykomos* and *Procolobus badius*, both of whom weigh approximately twice as much as the guenons, rarely feed from quadrupedal postures or adopt the stand-forelimb suspend in any tree zone, particularly in the periphery or tops of crowns. These latter postures are the principal foraging postures of guenons and are likely related to the spatial arrangement of preferred food items and the need to move quickly to the next feeding location.

This relationship is more closely examined by regressing the ischial tuberosity index against scores on the first principal component of the PCA of postures used in the periphery (Fig. 12.9). The data show that those species that most frequently employ sitting postures in the terminal branches of tree crowns—the large colobine species—have the largest tuberosities. In contrast, the monkey species that do not sit often in tree peripheries, the guenons, have relatively small tuberosities. Taken together, these data suggest a functional relationship among large body size, sitting, use of tree peripheries, and expanded ischial tuberosities.

Finally, it is reasonable to ask whether body weight alone is enough to explain variance in callosity size. If members of the Taï monkey assemblage were behaving randomly, i.e., there was no common element in their use of callosities, then body mass is all that would be required to predict both tuberosity size and callosity use. However, size and use are related, and we argue that to understand why the association between tuberosity size and body size exists, it is necessary to look at context:

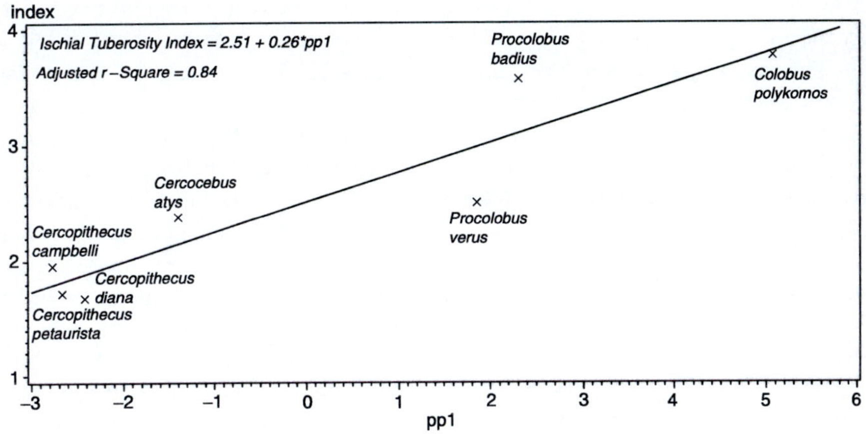

Fig. 12.9 Regression of the ischial tuberosity index (*y*-axis) against scores from the PCA of postures employed in the tree peripheries (*x*-axis = first principal component). There is a strong positive relationship between these variables: the species that sit most frequently in the tree peripheries have the relatively largest ischial tuberosities

when, where, and how is the feature used? For example, we might have expected relatively modest sized tuberosities—a presumed adaptation to arboreality—in sooty mangabeys (*Cercocebus atys*) since these monkeys spend the great majority of their time on the ground. Data on positional behavior and tree zone use indicate that when these monkeys do feed and rest in trees, they sit on their callosities in terminal branches. Our review of the scant literature on tuberosity size and body size in other cercopithecids bolsters our conclusion that the behavior informs the size relationship: several langur species surveyed are larger and more terrestrial than Taï colobines, yet they have smaller callosities and, presumably, tuberosities. We would predict decreased exploitation of the small branch niche in these forms (Ankel-Simons 2000; Hill 1953).

Discussion

The primary goal of this chapter was to explore the context of sitting associated with ischial callosity use in a group of sympatric monkeys that differ in their exploitation of tree zones. Two major hypotheses have been proposed to explain the function of callosities among cercopithecoids: that they provide a comfortable, stable base for sleep-sitting (Rose 1974a, b; Washburn 1957) and that they are principally feeding adaptations that allow primates to access resources in the small branch niche, thus serving a function similar to prehensile tails in atelines and *Cebus* sp. (Rose 1974a, b; Vilensky 1978). These hypotheses, which are not necessarily mutually exclusive, have rarely been tested, and our first challenge was to establish the extent that available data can assess the probability of each scenario.

The manner of sleep in free-ranging primates has been a topic of interest for years. Many studies have addressed the selection of sleeping sites across a habitat landscape, and there is increasing evidence that primates chose sleeping localities according to the protection they may provide against predators (e.g., Buxton 1951; Lumsden 1951; Anderson and McGrew 1984; Day and Elwood 1999; Anderson 2000, 1998, 1984; Schreier and Swedell 2008). Though a good number of studies have examined the characteristics of primate sleeping trees, e.g., certain species are known to select sleeping sites in emergent trees with fewer lower branches in order to make it more difficult for predators approaching from the ground, we know far less about the positioning of primates *within* trees or about the actual postures adopted during nocturnal resting. One of the earliest accounts is from Lumsden (1951), who described the sleeping habits of several primate species in Uganda's Semliki Forest (see also Haddow 1952). He noted:

> "*Cercocebus (Lophocebus) albigena johnstoni* and *Colobus abyssinicus ituricus* sleep among small branches on the periphery of the crown of the tree…*Colobus abyssinicus ituricus* prefers usually small branches outstanding from the main crown either above it or at its sides…*Cercopithecus ascanius schmidti* was reported also to sleep usually among small twigs. On the other hand, *Papio doguera tessellates* prefers large horizontal, or nearly horizontal branches near the main tree forks and therefore at a lower average level than the preceding species.…the monkeys do not take up their actual sleeping positions till about sunset. Generally they occupy positions in the lower part of the crown at first, sometimes moving short distances among the lower, larger limbs. At this time some may rest prone on large branches with their limbs hanging down on each side of it, but usually they sit. At about sunset, however, they move to the top sides of the crown and take up their positions for the night." (Lumsden 1951:29)

We have found few similarly detailed descriptions of sleeping postures and sleep sites for cercopithecoids (e.g., Roonwal and Mohnot 1977; Wolfheim and Rowell 1972; Rahaman and Parthasarathy 1969; Hall 1965), and while Napier and Napier (1985:32) may be correct in their assertion that, "Old World monkeys sleep sitting up," the comparative data needed to verify such a broad claim have yet to be collected. Further, observations on the Taï monkeys suggest that the positional behavior adopted during sleep in one cercopithecoid community is more varied. All species at Taï (including terrestrial mangabeys) have been observed sleeping in sitting positions on a variety of supports and in various tree zones; however, such postures are observed less often in the large colobines than in the three cercopithecus monkeys or the olive colobus. During daytime resting periods, Tai colobines frequently adopt prone and other reclining positions on a variety of support types (Fig. 12.10), some of which place moderate weight on the callosity, e.g., a large bough bears the majority of the body weight while a smaller support acts as a brake to prevent the body from sliding down the primary support (Fig. 12.11), while others involve all or most weight directly on the callosities (Fig. 12.12). We do not have the long-term data needed to describe the night resting patterns of Taï colobines, however, we can say with confidence that reclining—not sitting—postures are the preferred sleep positions of large colobines during the day. On the other hand,

Fig. 12.10 Red colobus (**top**) and black-and-white colobus (**bottom**) in prone postures typically adopted during daytime resting periods. Note that no weight is being placed on the callosities

Cercopithecus at Taï have rarely been observed in reclining postures of any sort during the day or during the few observations of guenon sleep habits we made at night. If callosity size is associated with sleeping behavior, we might expect those taxa that rest/sleep using upright and seated postures most frequently to have tuberosities that reflect greater use of this posture. This does not appear to be the case. The fact that guenons sleep upright during the day (and perhaps at night) and have relatively small tuberosities, whereas large colobines that frequently rest without placing weight on their callosities (at least during the day) have relatively large tuberosities, leads to at least two possible scenarios: 1) sleep postures adopted by large colobines at night are distinct from those most frequently used during day resting, and their expanded tuberosities reflect the sleep-sitting adaptations described by Washburn (1957) and Rose (1974a, b), or 2) there is little association between sleeping postures and callosity size. We lack the information required to test these alterna-

Fig. 12.11 Prone postures employed during daytime resting in red colobus (**top**) and olive colobus (**bottom**) in which moderate weight is placed on at least one callosity. In these positions, a branch growing off from the primary, weight-bearing support (arrow), acts as a "brake" to prevent the body from sliding

tives, however, the notion that all cercopithecoids sleep upright during the night (Anderson 2000; Napier 1985, 1967, Hill 1972) is, we strongly suspect, an overstatement.

Several authorities argued that New World monkeys differed from Old World monkeys in that platyrrhines sleep using reclining postures (Napier 1967; Hill 1972; Napier and Napier 1985). A review of the night-resting habitats of several platyrrhine species indicates that reclining postures do not characterize the sleep behavior of all New World monkeys (e.g., Eisenberg and Kuehn 1966; Rose 1974a; Mendel 1976; Mittermeier and Fleagle 1976; Ramirez 1988; Zhang 1995; Anderson 2000). The fact that several platyrrhine species are known to sleep from sitting

Fig. 12.12 Daytime resting postures in olive colobus (**top**) and black-and-white colobus (**bottom**) in which significant weight is placed directly on the ischial callosities. These postures, routinely adopted by African colobines, are distinct from the sleep-sitting postured described by Washburn (1957) and Rose (1974a)

postures weakens the notion that ischial callosities evolved as a mechanism to facilitate upright postures during sleep (Vilensky 1978). Nevertheless, the absence of comparative data precludes us from directly testing the hypothesis that ischial callosities are an adaptation for cercopithecoid sleep-sitting in terminal branches of trees, and that the need for callosities during sleep is diminished in New World monkeys because they do not sleep from sitting positions. This is a significant and surprising gap in our knowledge of an otherwise well studied group of mammals, and one that could be easily addressed by combining observations of night-sleeping primates in the field with manipulation of sleeping site options, i.e., the support environment, under captive conditions. Fortunately, a large body of data exists on postures employed during feeding and foraging that can be used to assess the functional similarity between callosities and prehensile tails as discussed by Vilensky (1978) and Rose (1974b).

The feeding and locomotor advantages of a prehensile tail have been well established (Carpenter and Durham 1969; Grand 1972; Mendel 1976; Mittermeier and Fleagle 1976; Cant 1986; Bergeson 1998). In addition to discussions of their suspensory functions, several studies have explored nonsuspensory prehensile tail use, particularly during terminal branch feeding bouts. These observations provide support for the arguments of Vilensky (1978) and Rose (1974b) that grasping tails play important roles during feeding, even when not being used in suspensory manners. Lawler and Stamps (2002) reported that prehensile tail use in mantled howlers is most frequent during feeding and that the majority of feeding in their focal group occurred in tree peripheries, where the dominant posture was sitting. In more than 70% of the feeding bouts involving seated postures, the prehensile tail was still employed in a grasping manner, i.e., attached to a support as an additional anchor. The challenges of maintaining balance and stability in the small branch setting should be compounded by increasing body size, and one would expect more frequent prehensile tail use —even during seated postures— among larger monkeys during feeding bouts in terminal branches. Observations on black howlers (*Alouatta caraya*) support this prediction. Bicca-Marques and Calegaro-Marques (1993) discussed the age-related changes in foraging behavior among black howlers and showed that as individuals age and grow larger, black howlers decrease their use of suspensory postures while increasing their use of seated postures during feeding. Other studies have confirmed that a significant amount of feeding by howlers is carried out from seated postures while the prehensile tail attached (Bergeson 1998; Gebo 1992; Schon-Ybarra 1984). These studies highlight a strong association between terminal branch feeding, sitting, body size, and prehensile tail use.

The advantages of a suspensory appendage are great, however, what these studies highlight —and what Vilensky (1978) predicted— is that grasping tails are often used in nonsuspensory contexts, particularly during feeding in the small branch niche. Moreover, it appears that the demands for increased security increase as body size increases and that at least several platyrrhines have adapted to these constraints by increasing their use of seated postures and prehensile tail use, as body size increases. We can extend this reasoning to monkeys of the Old World. An alternative to a prehensile tail is a feature that provides for more comfortable weight distribution across small supports and that helps prevent slipping. If you cannot anchor yourself with additional limbs (tail), selection should favor features that help prevent animals from skidding off supports. Based on data presented here, we argue that ischial callosities are primarily related to foraging constraints, i.e., the use of sitting postures during feeding amid small supports in tree crown peripheries. Arboreal primates that have the smallest tuberosities are small-bodied, habitat generalists that employ postures, e.g., supported stand, quadrupedal stand, that allow efficient movement to the next feeding site. Guenons are agile primates, obviously capable of sitting and feeding amid the slender supports of tree crowns, however, they do not often do so. Compared to colobus monkeys, Taï *Cercopithecus* monkeys feed from sitting positions at much lower overall frequencies. The species that feed in tree peripheries most often are the large colobines (*Procolobus badius* and *Colobus polykomos*), and the posture adopted most frequently by these species during feeding is sitting (Fig. 12.13). For large-bodied primates that rely on the small-branch setting and lack any suspensory capabilities, seated postures are required provided the individual's weight can be effectively distributed across one or more supports. A feature such as ischial callosities that decreases opportunities for slipping out of tree crowns while facilitating more comfortable and extended feeding

Fig. 12.13 A red colobus monkey feeding on small to medium-sized branches. The monkey's ischial callosities bear the majority of body weight while feet grasp neighboring, slender supports

bouts should have adaptive value. We argue that the ischial callosities are such a feature and that the postural behavior of the Taï monkeys provides strong support for Vilensky's (1978) hypothesis that, "ischial callosities in Old World primates and prehensile tails in New World primates are different anatomical adaptations that developed to solve similar problems faced in both hemispheres of *peripheral branch feeding with increased body size*."

Conclusions

Based on anatomical and behavioral data presented here, ischial callosities and the underlying bony tuberosities evolved as adaptations for feeding in the terminal branches of tree peripheries. At Taï, the largest tuberosities are found in monkeys that feed most often from seated postures in the small branch milieu of tree crowns. The literature contains broad generalizations about differences in sleeping postures between New and Old World monkeys. Some of these discussions are relevant to the evolution of ischial callosities, and though there is some evidence to the contrary, we have been unable to locate the comparative data needed to test the notion that the ischial callosities of cercopithecoids are primarily related to sleep-sitting in the terminal branches and that the absence of these pelvic adaptations in platyrrhines is due to their reliance on reclining sleep postures. We are therefore reluctant to generalize about comparative sleeping behaviors and their association with ischial callosities, however, we hope the absence of data will serve as a call for investigations of sleep behavior, specifically those that focus on postures, support use, and positions within trees. It has been more than 50 years since Washburn (1957) made a similar call for studies of primate sleep postures using what was then novel technology: night vision goggles. Such a study on a broad range of both New and Old World monkeys, combining sophisticated optic technologies with observations of animals in their natural habitats, is long overdue and would be highly informative.

Acknowledgments We thank Kristin Smith, Catherine Cooke, Elisabeth Hellmer, and the dedicated assistants of the Taï Monkey Project for help with various phases of the project. William Jungers and Clark Larsen provided helpful suggestions on the manuscript and financial support was provided by the Yerkes National Primate Center, Conservation International, and the Ohio State University. Sincere thanks to Kristiaan D'Août and Evie Vereecke for the invitation to the Edinburgh (IPS) symposium and for their careful editing skills. Kristiaan, in particular, provided particularly insightful comments for which we are grateful.

References

Aitchinson J (1982) The statistical analysis of compositional data. J Roy Stat Soc B (Method) 44:139–177.

Aitchinson J (1983) Principal component analysis of compositional data. Biometrika 70:57–65.

Anderson JR (1984) Ethology and ecology of sleep in monkeys and apes. Adv Stud Behav 14: 165–230.

Anderson JR, WC McGrew (1984) Guinea baboons (*Papio papio*) at a sleeping site. Am J Primatol 6:1–14.

Anderson JR (1998) Sleep, sleeping sites, and sleep-related activities: awakening to their significance. Am J Primatol 46:63–75.

Anderson JR (2000) Sleep-related behavioral adaptations in free-ranging anthropoid primates. Sleep Med Rev 4:355–373.

Anemone RL (1990) The VCL hypothesis revisited: patterns of femoral morphology among quadrupedal and saltatorial prosimian primates. Am J Phys Anthropol 83:373–393.

Ankel-Simons F (2000) An Introduction to Primate Anatomy. Academic Press, New York.

Benefit BR, McCrossin ML (2002) The Victoriapethicidae, Cercopithecoidea. In: Hartwig WC (ed), The Primate Fossil Record. Cambridge University Press, Cambridge, pp 241–253.

Bergeson DJ (1998) Patterns of suspensory feeding in *Alouatta palliata, Ateles geoffroyi*, and *Cebus capucinus*. In: Strasser E, Fleagle J, Rosenberger A, McHenry H (eds), Primate Locomotion: Recent Advances. Plenum Press, New York, pp 45–60.

Bicca-Marques JC, Calegaro-Marques C (1993) Feeding postures in the black howler monkey, *Alouatta caraya*. Folia Primatol 60:169–172.

Bloch JI, Boyer DM (2002) Grasping primate origins. Science 298:1606–1610.

Buxton AP (1951) Further observations of the night-resting habits of monkeys in a small area on the edge of the Semliki Forest, Uganda. J Anim Ecol 20:31–32.

Cant JGH (1986) Locomotion and feeding postures of spider and howling monkeys: field study and evolutionary interpretation. Folia Primatol 46:1–14.

Carpenter CR, Durham NM (1969) A preliminary description of suspensory behavior in nonhuman primates. Proceedings of the 2nd International Congress on Primatology, Vol. 2. Karger, Basel, pp 147–154.

Cartmill M (1972) Arboreal adaptations and the origin of the order Primates. In: Tuttle RH (ed), Functional and Evolutionary Biology of Primates. Aldine-Atherton, Chicago, pp 97–122.

Cartmill M (1992) New views on primate origins. Evol Anthropol 1:105–111.
Crompton RH, Vereecke EE, Thorpe SKS (2008) Locomotion and posture from the common hominoid ancestor to fully modern hominins, with special reference to the last common panin/hominin ancestor. J Anat 212:501–543.
Dagosto M, Gebo DL (1998) Methodological issues in studying primate positional behavior. In: Strasser E, Fleagle J, Rosenberger A, McHenry H (eds), Primate Locomotion: Recent Advances. Plenum Press, New York, pp 5–29.
Day RT, Elwood RW (1999) Sleeping site selection by the golden-handed tamarin *Saguinus midas midas*: the role of predation risk, proximity to feeding sites, and territorial defence. Ethology 105:1035–1051.
Demes B, Jungers WL, Gross TS, Fleagle JG (1995) Kinetics of leaping primates: influence of substrate orientation and compliance. Am J Phys Anthropol 96:419–429.
Demes B, Qin Y, Stern JT, Larson SG, Rubin CT (2001) Pattern of strain in the macaque tibia during functional activity. Am J Phys Anthropol 116:257–265.
Eisenberg JF, Kuehn RE (1966) The behavior of *Ateles geoffroyi* and related species. Smith Misc Coll 151:1–63.
Fleagle JG (1978) Locomotion, posture, and habitat utilization in two sympatric, Malaysian leaf-monkeys (*Presbytis obscura* and *Presbytis melalophos*). In: Montgomery GG (ed), The Ecology of Arboreal Folivores. Smithsonian Institution Press, Washington DC, pp 243–252.
Fleagle JG (1980) Locomotion and posture. In: Chivers DJ (ed), Malayan Forest Primates. Plenum Press, New York, pp 191–207.
Fleagle JG, Stern JT, Jungers WL, Susman RL, Vangor AK, Wells JP (1981) Climbing: a biomechanical link with brachiation and with bipedalism. In: Day MH (ed), Vertebrate Locomotion. Symp Zool Soc Lond 48:359–375.
Fleagle JG (1984) Primate locomotion and diet. In: Chivers DJ, Wood BA, Bilsborough A (eds), Food Acquisition and Processing in Primates. Plenum Press, New York, pp 105–118.
Garber P (2007) Primate locomotor behavior and ecology. In: Campbell CJ, Fuentes A, Mackinnon KC, Panger M, Bearder SK (eds), Primates in Perspective. Oxford University Press, Oxford, pp 543–560.
Gebo DL (1992) Locomotor and postural behavior in *Alouatta seniculus* and *Cebus capucinus*. Am J Primatol 26:277–290.
Gebo DL, Maclatchy L, Kityo R, Deino A, Kingdon J, Pilbeam D (1997) A hominoid genus from the early Miocene of Uganda. Science 276:401–404.
Grand TI (1972) A mechanical interpretation of terminal branch feeding. J Mammal 53:198–201.
Haddow AJ (1952) Field and laboratory studies on an African monkey, *Cercopithecus ascanius schmidti* Matschie. Proc Zool Soc Lond 122:297–394.
Hall KRL (1965) Behavior and ecology of the wild patas monkey, *Erythrocebus patas*, in Uganda. J Zool 148:15–87.
Hamrick MW, Meldrum DJ, Simons EL (1995) Anthropoid phalanges from the Oligocene of Egypt. J Hum Evol 28:121–145.
Harrison T, Sanders WJ (1999) Scaling of lumbar vertebrae in anthropoid primates; its implications for positional behavior and phylogenetic affinities of *Proconsul*. Am J Phys Anthropol S28:146.
Harrison T (2002) Late Oligocene to middle Miocene catarrhines from Afro-Arabia. In: Hartwig WC (ed), The Primate Fossil Record. Cambridge University Press, Cambridge, pp 241–253.
Hartwig WC, Cartelle C (1996) A complete skeleton of the giant South American primate *Protopithecus*. Nature 381:307–311.
Hill WCO (1953) Primates: Comparative Anatomy and Taxonomy, Vol. 1. Edinburgh University Press, Edinburgh.
Hill WCO (1972) Evolutionary Biology of the Primates. Academic Press, New York.
Jenkins FA (1974) Tree shrew locomotion and the origins of primate arborealism. In: Jenkins FA (ed), Primate Locomotion. Academic Press, New York, pp 85–115.
Jones AL (2008) The evolution of brachiation in Ateline primates, ancestral character states and history. Am J Phys Anthropol 137:123–144.

Jungers WL, Stern JT (1981) Preliminary electromyographical analysis of brachiation in gibbon and spider monkey. Int J Primatol 2:19–33.
Korstjens AH (2001) The Mob, the Secret Sorority, and the Phantoms: An analysis of the socio-ecological strategies of the three colobines of Taï. PhD Dissertation, University of Utrecht.
Larson SG, Stern JT (2006) Maintenance of above-branch balance during primate arboreal quadrupedalism: coordinated use of forearm rotators and tail motion. Am J Phys Anthropol 129:71–81.
Lawler R, Stamps C (2002) The relationship between tail use and positional behavior in *Alouatta palliata*. Primates 43(2):147–152.
Lumsden WHR (1951) The night-resting habits in a small area on the edge of the Semliki Forest, Uganda. A study in relation to the epidemiology of Sylvan Yellow Fever. J Anim Ecol 20:11–30.
McCrossin ML, Benefit BR (1992) Comparative assessment of the ischial morphology of *Victoriapithecus macinnesi*. Am J Phys Anthropol 87:277–290.
McCrossin ML, Benefit BR (1994) Maboko Island the evolutionary history of Old World monkeys and apes. In: Corruccini RS, Ciochon RL (eds), Integrative Paths to the Past: Paleoanthropological Advances in Honor of F. Clark Howell. Prentice-Hall, Englewood Cliifs, NJ, pp 95–122.
McGraw WS (1996) Cercopithecid locomotion, support use and support availability in the Tai Forest, Ivory Coast. Am J Phys Anthropol 100:507–522.
McGraw WS (1998a) Posture and support use of Old World Monkeys (Cercopithecidae): the influence of foraging strategies, activity patterns and the spatial distribution of preferred food items. Am J Primatol 46:229–250.
McGraw WS (1998b) Comparative locomotion and habitat use of six monkeys in the Tai Forest, Ivory Coast. Am J Phys Anthropol 105:493–510.
McGraw WS (1998c) Locomotion, support use, maintenance activities and habitat structure: the case of the Tai cercopithecids. In: Strasser E, Fleagle J, Rosenberger A, McHenry H (eds), Recent Advances in Primate Locomotion. Plenum Press, New York, pp 79–94.
McGraw WS (2000) Positional behavior of *Cercopithecus petaurista*. Int J Primatol 21:157–182.
McGraw WS (2003) Diversity of guenon positional behavior. In: Glenn M, Cords M (eds), The Guenons: Diversity and Adaptation in African Monkeys. Kluwer Academic Press, New York, pp 113–131.
McGraw WS (2007) Positional behavior and habitat use of Tai monkeys. In: (McGraw WS, Zuberbuhler K, Noe R (eds), The Monkeys of the Tai Forest: An African Primate Community. Cambridge University Press, Cambridge, pp 223–255.
McGraw WS, Zuberbuhler K, Noe R (2007) Monkeys of the Tai Forest: An African Primate Community. Cambridge University Press, Cambridge.
McGraw WS, Zuberbuhler K (2008) Socioecology, predation, and cognition in a community of West African monkeys. Evol Anthropol 17:254–266.
Mendel F (1976) Postural and locomotor behavior of *Alouatta palliatta* on various substrates. Folia Primatol 26:36–53.
Miller RA (1945) The ischial callosities of primates. Am J Anat 76:67–91.
Mittermeier RA, Fleagle JG (1976) Locomotor and postural repertoires of *Ateles geoffroyi* and *Colobus guereza*, and a reevaluation if the locomotor category semibrachiation. Am J Phys Anthropol 45:235–256.
Napier JR (1963) Brachiation and brachiators. Symp Zool Soc Lond 10:183–195.
Napier JR (1967) Evolutionary aspects of primate locomotion. Am J Phys Anthropol 27:333–341.
Napier JR, Napier P (1967) A Handbook of Living Primates. Academic Press, New York.
Napier JR, Walker AC (1967) Vertical clinging and leaping – a newly recognized category of locomotor behavior of primates. Folia Primatol 6:204–219.
Napier JR, Napier P (1985) The Natural History of the Primates. The MIT Press, Cambridge.
Oates JF, Whitesides GH, Davies AG, Waterman PG, Green SM, Dasilva GL, Mole S (1990) Determinants of variation in tropical forest primate biomass: new evidence from West Africa. Ecology 71:328–343.

Pocock RI (1925) The external characters of the catarrhine monkeys and apes. Proc Zool Soc Lond 1479–1579.
Rahaman H, Parthasarathy MD (1969) Studies on the social behavior of bonnet monkeys. Primates 10:149–162.
Ramirez M (1988) The woolly monkeys, genus *Lagothrix*. In: Mittermeier RA, Rylands AB, Coimbra-Filho AF, da Fonseca GAB (eds), Ecology and Behavior of Neotropical Primates, Vol. 2. World Wildlife Fund, Washington DC, pp 539–575.
Rasmussen DT (1990) Primate origins: lessons from a neotropical marsupial. Am J Primatol 22:263–277.
Roonwal ML, Mohnot SM (1977) Primates of South Asia. Harvard University Press, Cambridge, MA.
Rose MD (1974a) Postural adaptations in New and Old World monkeys. In: Jenkins FA (ed), Primate Locomotion. Academic Press, New York, pp 201–222.
Rose MD (1974b) Ischial tuberosities and ischial callosities. Am J Phys Anthropol 40:375–384.
Schmitt DO (1999) Compliant walking in primates. J Zool 248:149–160.
Schon-Ybarra MA (1984) Locomotion and postures of red howlers in a deciduous forest-savanna interface. Am J Phys Anthropol 63:65–76.
Schreier A, Swedell L (2008) Use of palm trees as a sleeping site for hamadryas baboons (*Papio hamadryas hamadryas*) in Ethiopia. Am J Primatol 70:107–113.
Schultz AH (1936) Characters common to higher primates and characters specific for man. Q Rev Biol 11:259–283.
Sigmon BA, Farslow DL (1986) The primate hindlimb. In: Swindler DR, Erwin J (eds), Comparative Primate Biology, Vol. 1: Systematics, Evolution, and Anatomy. Alan R Liss, New York, pp 671–718.
Smith R, Jungers WL (1997) Body mass in comparative primatology. J Hum Evol 32:523–559.
Stern JT, Oxnard CE (1973) Primate locomotion: Some links with evolution and morphology. Primatologica 4:1–93.
Stern JT, Susman RL (1981) Electromyography of the gluteal muscles in *Hylobates, Pongo*, and *Pan*: implications for the evolution of hominid bipedality. Am J Phys Anthropol 60:279–317.
Strasser E, Delson E (1987) Cladistic analysis of cercopithecid relationships. J Hum Evol 16:81–99.
Susman RL, JT Stern, Jungers WL (1984) Arboreality and bipedality in the Hadar hominids. Folia Primatol 43:113–156.
Sussman RW (1991) Primate origins and the evolution of angiosperms. Am J Primatol 23:209–233.
Tuttle RH, Basmajian JV (1974) Electromyography of brachial muscles in *Pan gorilla* and hominoid evolution. Am J Phys Anthropol 41:71–90.
Tuttle RH (1986) Apes of the World. Noyes Publications, Park Ridge, NJ.
Vilensky JA (1978) The function of ischial callosities. Primates 19(2):363–369.
Vinyard CJ, Yamashita N, Tan C (2008) Linking laboratory and field approaches in studying the evolutionary physiology of biting in bamboo lemurs. Int J Primatol 29:1421–1439. DOI:10.1007/s10764–007–9178–9.
Walker A, Shipman P (2005) The Ape in the Tree: An Intellectual and Natural History of Proconsul. Belknap Press, Cambridge, MA.
Ward CV, Walker A, Teaford MF, Odhiambo I (1993) Partial skeleton of *Proconsul nyanzae* from Mfangano Island, Kenya. Am J Phys Anthropol 90:77–112.
Washburn SL (1950) The analysis of primate evolution with particular reference to the origin of man. Cold Spring Harbor Symp Quant Biol 15:67–78.
Washburn SL (1957) Ischial callosities as sleeping adaptations. Am J Phys Anthropol 15:269–276.
Wilson DR (1972) Tail reduction in *Macaca*. In: Tuttle R (ed), The Functional and Evolutionary Biology of Primates. Aldine, Chicago, pp 241–261.
Wolfheim J, Rowell T (1972) Communication among captive talapoin monkeys (*Miopithecus talapoin*). Folia Primatol 18:224–255.
Zhang SY (1995) Sleeping habits of brown capuchin monkeys (*Cebus apella*) in French Guiana. Am J Primatol 36:327–335.

Chapter 13
Forelimb Suspensory Gait Characteristics of Wild *Lagothrix poeppigii and Ateles belzebuth*: Developing Video-based Methodologies in Free-ranging Primates

Denise Guillot

Abstract The understanding of interspecific locomotor performance differences has been greatly expanded by video-based, kinematic analyses conducted in captivity. However, these techniques have rarely been applied to the study of free-ranging subjects as they negotiate the challenges of their natural habitat. In this study, the unrestricted movements of wild *Lagothrix poeppigii* and *Ateles belzebuth* were filmed from canopy level platforms at heights where these species habitually travel and feed. I used these video data to evaluate the capacity for *Ateles* and *Lagothrix* to perform forelimb suspensory behaviors in the wild. I analyzed a combination of temporal gait parameters and several other descriptive parameters thought to reflect categorical differences in suspensory ability and compared the results to kinematic data previously collected in captivity in an effort to address two questions: 1) Are the results between captive and wild subjects similar and, if not, 2) how does the performance of forelimb suspension differ in the wild? In general, temporal gait parameters derived from videos of wild subjects are comparable in direction, though not magnitude, to significant results previously reported for captive subjects. In both environments, suspensory behaviors of *Ateles* tend to involve longer, less frequent strides. In contrast to captive data, results from wild subjects indicate that *Ateles* exhibits a significantly longer no-hands phase relative to stride duration. Data also suggest that stride asymmetry and diversity in handhold patterns are particularly important to wild subjects, likely reflecting an important behavioral response to the heterogeneous nature of their arboreal environment. Reported interspecific differences in performance variables likely reflect differing capacities to respond efficiently to the challenges of pliant and unreliable supports using forelimb suspension.

Keywords Atelines • Forelimb suspension • Gait parameters

D. Guillot (✉)
University of Michigan, USA
e-mail: dguillot@umich.edu

K. D'Août and E.E. Vereecke (eds.), *Primate Locomotion: Linking Field and Laboratory Research*, Developments in Primatology: Progress and Prospects,
DOI 10.1007/978-1-4419-1420-0_13, © Springer Science+Business Media, LLC 2011

Abbreviations

fps	frames per second
LO	lift-off (hand or tail)
PPRS	Proyecto Primates Research Site
TBS	Tiputini Biodiversity Station
TD	touchdown (hand or tail)

Introduction

Numerous studies have examined locomotor behavior in reptilian and mammalian models by quantifying performance capacity (e.g., Larson and Stern 1987; Garland et al. 1990; Hirasaki et al. 1993; Losos et al. 1994; Irschick and Garland 2001; Vanhooydonck and Van Damme 2001; Van Damme et al. 2002; Irschick et al. 2005; Isler 2005; Lawler 2006; Vereecke et al. 2006; Kohlsdorf and Navas 2007). As it is used here, locomotor performance may be best defined as the phenotypic outcome of intrinsic neurological, physiological, and morphological features of the total, integrated locomotor system (Arnold 1983; Aerts et al. 2000a, b). This approach evaluates individual ability to negotiate environmental challenges effectively through a variety of measures including gait parameters, maximal speed, endurance, leaping distance, etc., and emphasizes that selection is essentially blind to the underlying features of the locomotor system. Significantly, measures such as temporal gait characteristics not only illustrate whether or not there is a correlation between structure and function, but they also indicate the "*magnitude* of effects" of intrinsic features such as skeletal structure, muscle fiber orientation, or physiology on extrinsic features such as leaping ability or endurance (Arnold 1983). The perspective provided by these studies views performance as an intermediate variable between fitness and the underlying traits that contribute to adaptive complexes (Arnold 1983; Aerts et al. 2000b) and, thus, comparative studies that evaluate performance capacity provide a framework for exploring the evolution of positional behavior and functional anatomy.

Intraspecific studies suggest that several features of gait, such as mechanistic responses to changes in speed, are predictable and uniform and likely reflect "species-specific spatiotemporal gait characteristics" (Aerts et al. 2000a; see also Cartmill et al. 2002). However, reliably determining species-specific differences in locomotor performance and understanding their functional relevance is not a straightforward endeavor. Locomotor capacity appears to vary in response to genetic, ontogenetic, and environmental influences (Vanhooydonck and Van Damme 2001). To add to the confusion, the tightness of fit between form and ecological function "appears to be highly taxon-dependent" (Aerts et al. 2000b). Clear relationships between "ecomorphs" (here referring to species level variation in morphological phenotype reflecting adaptation to local microhabitats) and performance capacity are not always evident. For example, lacertid lizards occupying distinct microhabitats (arboreal vs. ground-dwelling) revealed little interspecific difference in their capacity to climb or sprint on flat surfaces under captive conditions despite the tendency of these species to vary in the frequency that they performed these behaviors in the wild (Vanhooydonck and Van Damme 2001). Moreover, although sand dwelling versus rock dwelling species of lizards (subfamily Tropidurinae) vary dramatically in their performance of leaping, they appear to lack obvious morphological differences typically correlated with leaping ability (e.g., differences in limb lengths, limb proportions, body size, etc.) (Kohlsdorf and Navas 2007).

The explicit use of terms like locomotor "capacity" or "performance" is uncommon in studies of primate positional behavior; however, the idea has previously been introduced. Cant (1992) used the term "effectiveness" in his study of arboreal primates as a broad concept reflecting differences in the ability to solve particular environmental problems. For example, interspecific variation exists in the ability to perform specific tasks, such as gap crossings initiated and terminated from similar supports, but the idea may also be extended to include behavioral differences that minimize energy or time expenditures (Cant 1992). The study presented here uses the term "performance capacity" in an effort to encourage continuity between ecologists and primatologists interested in the evolution of locomotion in a variety of animal models.

As a first step toward sorting out the complex array of variables contributing to differences in locomotor capacity, it is essential to clarify that the notion of "species-specific" positional behavior must be viewed as a *range of potential responses* to environmental factors (Strier 2009). Quantifying that range is fundamental to discussions of locomotor performance capacity and necessarily involves analyses at a number of different levels. The most prevalent form of existing data on positional behavior includes frequency data collected in the wild on modes used, canopy height occupied, support size and orientation, etc. (e.g., Cant 1987; Hunt 1991; Cannon and Leighton 1994; Defler 1999; Cant et al. 2001, 2003; Thorpe and Crompton 2006). In addition, differences in locomotor performance have been further evaluated by examining kinematic and kinetic data collected almost exclusively in a captive setting. These analyses examine characteristics of locomotion including joint angles, contact time, duty factor (percentage of stride duration that a single limb is in contact with the substrate), stride length, stride frequency, velocity, etc. Such studies extend the analysis of locomotor performance from the "*how often?*" questions addressed by frequency data and reveal similarities and differences in "*how?*" a given behavior is actually performed by different species. Both frequency data and kinematic data are relevant to understanding the factors associated with morphological adaptive complexes.

Until recently, kinematic and kinetic analyses have been limited largely to captive subjects. A significant disadvantage of these captive studies is the small number of available subjects. Frequently, data are collected only on one or a few individuals of a given species (e.g., Turnquist et al. 1999). Moreover, experimental environments are rather simplified and do not mimic the complex environment that has likely shaped anatomy and behavior in the wild. Data collected on few subjects in relatively impoverished environments make it extremely difficult to fully appreciate intra- or interspecific variation in performance capacity. Despite the constraints of the experimental set-ups, these studies have provided valuable information on the mechanics of specific behaviors, offering a point of departure for further analyses in the field.

Although video recordings have been used in previous field-based studies to define locomotor modes or aid in qualitative descriptions (Avis 1962; Cant et al. 2001, 2003; Byron and Covert 2004; Thorpe and Crompton 2006), relatively few investigators/primatologists have attempted to quantify performance by applying video-based kinematic analysis techniques to wild subjects (cf. Demes et al. 1996; Isler and Thorpe 2003; De Silva 2007). Simply collecting un-obscured views of focal animals is extremely difficult in complex forest environments, particularly for arboreal primates that habitually occupy the upper levels of the canopy. Collecting data in the wild, such as joint angles or stride lengths, is complicated further by several factors including: 1) the use of video images that are typically un-calibrated, 2) subjects that are free to move in any direction, 3) the challenge of consistently filming anatomical markers typically used in video analyses, and 4) the use of subjects that have not been captured and for which there is no available data on body segment lengths. As a result, exact angles (cf. Stevens et al. 2006), travel distances and kinetic data cannot be readily quantified under these conditions.

Despite these limitations, it remains possible to collect a wealth of data from video sequences of wild subjects, including stride duration, footfall pattern (e.g., diagonal or lateral sequence), contact time, duty factor, and gait symmetry. Video sequences that involve lateral views of subjects moving in a plane roughly parallel to the camera position also allow for the evaluation of range of motion of the limbs during specific behaviors (e.g., Demes et al. 1996; Stevens et al. 2006; De Silva 2007). For example, lateral views derived from videos of wild, free-ranging chimpanzees have been used to estimate the degree of foot dorsiflexion during vertical climbing (De Silva 2007), and categorical data have been used to describe the position of the knee relative to the vertebral column (knee positioned ventral, mid-axillary, or dorsal to the torso) during vertically clinging and leaping behavior in free-ranging lemurs (Demes et al. 1996). In addition, data collected under natural conditions are likely to include a larger number of subjects, allowing for data sets that reveal intraspecific variation in performance capacity. Video images of unrestrained movements occurring on arboreal pathways that are habitually used by the subjects also capture contextual data, such as feeding activities versus traveling activities. These kinds of data have the potential to contribute substantially to understanding the ecological functions tied to differences in performance capacity. Finally, examining performance measures in the wild can serve to validate and extend results produced by lab-based analyses of locomotion. Comparisons between both conditions reveal intriguing differences in the performance of locomotor behaviors by captive and wild individuals. For example, wild individuals tend to rely on a larger number of simultaneously supporting limbs and to exhibit longer stride durations, shorter stride lengths, and greater gait variability compared to captive individuals (Isler and Thorpe 2003). In light of the fact that locomotor performance may be affected by the mechanical properties of substrates and the spatial complexity of the environment (Dickinson et al. 2000), kinematic data collected under natural conditions may be essential to revealing functional distinctions between species.

The combination of frequency data collected in the wild and kinematic and kinetic data collected in captivity has contributed greatly to the understanding of the ecological function of positional behavior. The addition of video-based, kinematic analyses of wild, free-ranging subjects is an essential next step serving to expand our understanding of the role of behavioral differences in the very environment that selected for and maintains these adaptations.

Study Objectives

This study compares the forelimb suspensory locomotor capacities of wild *Lagothrix poeppigii* (Poeppig's woolly monkey) and *Ateles belzebuth* (white-bellied spider monkey; ateline molecular phylogeny, behavior, and ecology are summarized in Di Fiore and Campbell 2007). Kinematic studies of captive subjects combined with frequency data collected in the wild indicate that these two genera vary not only in how often they perform suspensory locomotor behaviors (Cant et al. 2001, 2003), but also in performance style (Turnquist et al. 1999; Schmitt et al. 2005). For example, comparative kinematic analyses of *Lagothrix lugens* and *Ateles fusciceps* indicate that *Lagothrix* exhibits a more upright body orientation coupled with a shorter pendulum length during brachiation (Turnquist et al. 1999). Further analyses suggest that the tail is used by *Ateles* to stabilize the body against lateral sway (Schmitt et al. 2005). These results, based on just one captive adult of each species, suggest that *Lagothrix* and *Ateles* may have different strategies for sustaining forelimb suspensory locomotion that are likely to impact efficiency and energetic costs during suspensory activities.

As an initial objective, this study aims to further expand the use of video data collected in the wild. The use of video in field-based studies is particularly valuable when quantifying variables that occur too quickly to be measured in real-time by an observer. This is particularly true for observers on the ground attempting to collect data on fast moving, arboreal primates. In this study, the suspensory capabilities of *Lagothrix* and *Ateles* are evaluated by quantifying temporal gait parameters (stride duration, frequency, duty factor, etc.), the use of flexed or extended forelimbs and versatility in suspensory landing behaviors and in hand and tail touchdown patterns. These measures correspond to intrinsic factors, e.g., morphological, physiological, neurological, etc., influencing the capacity for forelimb suspensory locomotion in these two species. In some instances, comparisons are also made to *Alouatta seniculus*. This deliberate quadruped provides a useful contrast that highlights the distinct forelimb suspensory abilities of *Lagothrix* and *Ateles*.

The second objective is to determine whether or not lab performance duplicates ecological performance. Where possible, results reported here for wild *Lagothrix poeppigii* and *Ateles belzebuth* are compared to published temporal and spatial gait parameters collected on captive *Lagothrix lugens* and *Ateles fusciceps* (Turnquist et al. 1999). Performance variables may differ between captive and wild subjects for two important reasons. First, differences in the captive and natural environments may impact each species equally such that we see changes in the magnitude of a performance variable but no change in the pattern of observed differences between the study species. For example, *Lagothrix* may always exhibit higher stride frequencies, on average, than *Ateles*, regardless of experimental conditions. Second, the focal species may not be equally capable of performing a particular behavior, and these differences in underlying performance capacities may emerge only when subjects are responding to the complexity of their natural habitat. These results may occur if the environment in the laboratory, or even in enriched zoo settings, lacks sufficient complexity to reveal limitations to locomotor behavior or to motivate the subjects to perform as they would in the wild. The capacity to maneuver, for example, is likely to contribute to the efficiency of locomotion in the wild, but differences in maneuverability are rarely examined in comparative studies conducted in captivity (cf. Demes et al. 2006). It is possible that the very nature of experimental setups, relying, e.g., on rigid, straight supports that remain consistent between trials conceals functionally important differences in performance capacity (Higham et al. 2001).

The third objective of this study focuses on the link between performance capacity and ecology (Aerts et al. 2002). This approach is particularly illuminating when comparing closely related, sympatric species, such as *Ateles belzebuth* and *Lagothrix poeppigii*, which overlap considerably in body size, diet, and microhabitat preference (Di Fiore 2004, summarizes general ateline ecology). To this end, observed performance differences are discussed in terms of their adaptive role in minimizing energetic cost or the risks of falling or by enhancing maneuverability during locomotion. For example, previous studies clearly indicate that *Ateles* performs suspensory behaviors more frequently than *Lagothrix* (Cant et al. 2003). It is likely that the increased frequency observed in *Ateles* will be associated with greater versatility in hand and tail touchdown patterns and greater diversity in how suspension is initiated and terminated reflecting greater maneuverability overall when relying on this locomotor mode in the wild. Moreover, comparisons of captive and wild gait parameters will likely reveal ways in which *Ateles* and *Lagothrix* differ in their capacity to sustain suspensory locomotion that are not evident during observations made under captive conditions.

Methods

Study Site and Subjects

I collected video data at two sites in the lowland rain forests of northeastern Ecuador of free ranging, sympatric populations of *Alouatta seniculus, Lagothrix poeppigii*, and *Ateles belzebuth*. Tiputini Biodiversity Station (TBS) consists of 650 ha of pristine, seasonally flooded rain forests. The absence of local human populations or roads makes this an ideal site to study the undisturbed behavior of primates, and I collected the majority of video data at this location (ca. 18 h). Additional video data of *Ateles belzebuth* (ca. 2 h) were collected at Proyecto Primates Research Site (PPRS) located ca. 40 km northwest of TBS. This study site is similar in size and in the primates species represented but it has the added benefit of having particularly well habituated groups of *Ateles*. Only adults and large, independently traveling sub-adults were included in this study.

Data Collection

Video data were collected using a Canon GL2 digital camcorder (20× optical zoom, 30 fps) during multiple field seasons from November 2002 to January 2006. Every effort was made to maximize the filming of individuals as they traveled and fed in the canopy. TBS, where most of the filming occurred, has a number of permanent multilevel canopy structures as well as a mobile platform, facilitating the collection of relatively un-obscured video at heights ranging from 20 to 40 m. In the absence of such structures, trees were climbed using arborist's equipment and techniques or filming was conducted from natural elevations in the terrain to maximize filming opportunities. Table 13.1 summarizes the video data collected.

Table 13.1 Summary of video data

	Alouatta seniculus	*Lagothrix poeppigii*	*Ateles belzebuth*
Number unobscured video bouts	52	53	51
Total bout time (s)	3923	1142	1398
Number travel bouts	33	43	22
Total travel bout time (s)	1006	850	424
Number feeding/foraging/resting bouts	18	8	29
Total feeding/foraging/resting bout time (s)	2917	293	974
Minimum number of individuals observed	7	8	12
Females	4	6	4
Males	2	1	4
Subadults	1	1	4
Study sites	TBS[a]	TBS	TBS and PPRS[b]

Note that the minimum number of individuals refers to the minimum number of individuals that could be distinguished with certainty. It is likely that this number underestimates the total number of individuals filmed. Times are rounded to the nearest whole second.

[a] Tiputini Biodiversity Station.

[b] Proyecto Primates Research Site.

Behavioral data collected in the canopy are an important extension to previous studies of primate positional behavior because many behaviors that occur high in trees may be underrepresented in datasets collected by observers on the ground (e.g., Stanford 2004). In addition, data collected at multiple natural sites add considerable complexity to the arboreal challenges faced by the subjects that have not yet been duplicated in captive settings. Even in the wild, subjects have been observed to approach familiar feeding sites from the same direction using the same supports. Collecting data at a single site, particularly a provisioned location where substrates used tend to be reliable and predictable, would fail to capture the diversity of behavioral responses characterizing a species under natural conditions. The use of multiple, unprovisioned filming locations attempts to avoid skewed data sets that may result from substrate familiarity.

Evaluating Suspensory Locomotion

A combination of descriptive and temporal gait parameters was used to evaluate the performance of forelimb suspension by *Lagothrix poeppigii* and *Ateles belzebuth.* It is important to emphasize that both species frequently exhibit quadrupedal walking and climbing (Cant et al. 2001), and variably integrate forelimb suspensory behaviors into locomotor sequences. At times, forelimb suspension may be limited to single forelimb swings (pendular motion below a supporting forelimb) or single steps (defined as the period between contralateral handholds). It is less common to observe sustained forelimb suspension involving sequential full strides (defined as the period between consecutive handholds of the same hand, i.e., two successive steps), which may or may not involve the torso rotation characteristic of true brachiation. I evaluated the capacity for *Ateles* and *Lagothrix* to perform these forelimb suspensory behaviors (swings, steps and strides) in the wild via a combination of temporal gait parameters, typically collected only in captivity, and several other descriptive parameters thought to reflect categorical differences in suspensory ability.

The data presented and a brief rationale for their collection are described in the following sections.

General Arm Raising Ability

The capacity for effective forelimb suspension is functionally linked to adaptations that allow the arm to be raised overhead. Arm-raising ability was quantified as the frequency of postural and locomotor events in which the elbow closely approximated full extension, and the shoulder was flexed and/or abducted relative to the total time the focal individual was observed (total number of positional behaviors involving arm-raising divided by total bout time). In many cases, where the arm was judged to be fully extended at the elbow and flexed/abducted at the shoulder, the humerus was also laterally rotated. Note that overhead "straight" arm behaviors were not limited to observations of forelimb suspension. The subjects also exhibited general arm raising ability during a variety of postural and locomotor behaviors including bridging and clambering. This capacity was quantified for *Lagothrix* and *Ateles* as well as *Alouatta seniculus* in an effort to relate general arm raising ability to varying degrees of commitment to forelimb suspensory locomotion.

Relative Degree of Elbow-Flexion During Forelimb Swings, Steps, and Strides

The tendency for *Ateles* and *Lagothrix* to perform "flexed-elbow" suspension was evaluated by categorizing each handhold as "flexed" or "extended" for all arm-swings, steps, and strides recorded. Handholds were considered "extended" if the forearm was estimated to be positioned between 135° and 180° relative to the humerus. Smaller elbow joint angles were considered "flexed" (<135°). This variable is relevant to evaluating differing capacities for sustained forelimb suspension. Prior studies have demonstrated that flexion of the elbow at the beginning of pendular forelimb suspension serves to elevate the body, thus maximizing the potential energy of the movement (e.g., Jungers and Stern 1984; Turnquist et al. 1999).

Tail Use During Suspension

I examined the use of the tail in a supporting role during every step of a full stride. Tail use may contribute to the efficiency of forelimb suspensory locomotion by minimizing lateral displacement during forward progression (Schmitt et al. 2005). Differences in the frequency of tail contact may reflect differences in the capacity to sustain efficient forelimb suspension.

Gait Symmetry and Hand and Tail Touchdown/Lift-off Patterns

I considered full strides "symmetrical" when each supporting limb was in contact ca. 50% of the time (Cartmill et al. 2002). I also documented touchdown/lift-off patterns for each full stride. The involvement of three supporting limbs has the potential to produce a variety of touchdown/liftoff patterns. To quantify these patterns, I documented the sequence of each hand and tail touchdown (TD) and lift-off (LO) for all sequences containing a full stride. For example, TD-L, LO-L, TD-R, TD-T, TD-L refers to a full stride where the left hand touches down first, then lifts off the substrate before the right hand makes contact, followed by tail touchdown and left hand touchdown. Note that in this example the first step (between touchdown of the left and right hand) involves a "no-hands" phase and no tail support. Such a stride is likely to be asymmetrical in terms of the proportion of time that each hand contacts the support. Hand and tail touchdown/lift-off patterns and gait asymmetry are likely to vary in captive versus wild settings owing to differences in the predictability and stability of supports in these two environments. Comparisons between wild *Lagothrix* and *Ateles* may also reveal differing capacities to respond to the challenges of a complex environment in a versatile manner while using suspensory modes.

Initiation and Termination of Forelimb Swings, Steps, and Strides

Forelimb swings, steps, and strides are often initiated and terminated from a variety of other locomotor or postural modes, and in many cases these behaviors grade fluidly into one another. However, the capacity to initiate and terminate suspension in a highly variable manner may differ interspecifically. I documented whether suspension was preceded or followed by a postural or locomotor mode for each observed occurrence of suspensory locomotion. I also collected data on the frequency of bipedal landings, as this ability may reflect differences in overall capacity to perform orthograde behaviors.

Forelimb Suspensory Temporal Gait Parameters

I further quantified differences in the performance of full stride forelimb suspension by *Lagothrix* and *Ateles* by comparing several temporal gait parameters, including stride duration, swing and support phase durations, duty factor, and "no-hands" phase (e.g., Hildebrand 1967; Turnquist et al. 1999; Larson et al. 2000; Lemelin et al. 2003; Vereecke et al. 2006; Wallace and Demes 2008).

I treated forelimb swings, steps, and strides as distinct suspensory behaviors, and quantified durations as the average time elapsed for each. Initiation of forelimb swings, steps, and strides was determined to begin at touchdown of the first supporting limb or, in the case of forelimb swings that began from a suspended posture involving multiple supporting limbs, at the first video frame in which the body began to fall along its pendular arch. Touchdown of the first limb to contact the destination support was defined as the termination of the behavior. Duty factor is the period of time that a supporting limb is in contact with the substrate as a percentage of stride duration. "No-hands" phase is reported as the percentage of stride duration in which the body is supported only by the tail or is completely airborne. While temporal gait parameters such as duration, stride frequency, and duty factor are certainly influenced by limb and stride lengths, they are also likely to reflect behavioral strategies and/or underlying adaptations of the total locomotor system that improve the efficiency of locomotion. For example, long periods of limb contact during a stride (reflected by duty factor) provide longer periods during which force can be applied to the substrate. This is relevant to sustaining suspensory locomotion, increasing speed, and improving the stability of these movements.

Both the slow rate of the video (30 fps) and the variable nature of unrestricted locomotion present challenges to the analysis of video collected in the wild. Individual video frames are snapshots in time that may or may not capture actual start and stop times accurately, thus limiting the quality of temporal gait measurements. This temporal discrepancy, known as signal distortion or aliasing error, is exacerbated by slower rates of recording (Polk et al. 2005). Moreover, measurement errors are likely to be very large for brief events, e.g., contact times (Polk et al. 2005). To a degree, these sources of error have a relatively smaller impact on temporal data collected on larger bodied primates moving at relatively slower speeds, as seen in this study, since larger bodied subjects have absolutely larger gait parameters (Polk et al. 2005). Accuracy of temporal gait characteristics derived from videos of wild subjects is constrained further by the frequent changes in direction and speed that accompany locomotion along irregular, pliant, and discontinuous supports. The distinctive manner in which these transitions are accomplished is likely to be informative, reflecting differing capacities to efficiently compensate for and react to the challenges in their environment.

Statistical Analysis

I analyzed frequency data and mean durations using Resampling Stats for Excel (Blank et al. 2001). Data were sampled with replacement to generate 95% confidence intervals based on 1000 reiterations. These randomization techniques were adopted because, unlike more traditional statistical tests, they do not assume a normal distribution of the data and do not require large sample sizes (Blank et al. 2001). All significant differences reported in the results below are based on nonoverlapping 95% confidence intervals with $p \leq 0.05$. It should be emphasized that, in many case, the statistics reported here are intended to highlight general trends because the small sample sizes, the slow video sampling rate, and the lack of control over subject speed do not allow for a rigorous statistical analysis of every comparison presented (Demes et al. 1994; Larson 1998).

Results

Reaching Overhead During Ateline Positional Behavior

Ateles and *Lagothrix* were similar in the proportion of time that they used postural and locomotor behaviors relying on overhead, "straight" arm positions, and both exhibited these behaviors significantly more often than *Alouatta* (Table 13.2, all activities: *Alouatta* 1.5%, *Lagothrix* 9.4%, *Ateles* 10%). *Alouatta* showed no tendency to shift the frequency of overhead, "straight" arm positions used in the

Table 13.2 Frequency (%) of arm-raising behaviors for *Alouatta seniculus* (*A.s.*), *Lagothrix poeppigii* (*L.p.*) and *Ateles belzebuth* (*A.b.*)

				Are pairwise comparisons significantly different?		
Context	*A.s.*	*L.p.*	*A.b.*	*A.s.* vs *L.p.*	*A.s.* vs *A.b.*	*L.p.* vs *A.b.*
All activities	1.5 (1.1, 1.9)	9.4 (7.7, 11.2)	10.0 (8.7, 11.8)	Yes	Yes	No
Travel only	2.0 (1.2, 3.1)	10.9 (8.8, 12.9)	13.4 (10.4, 16.7)	Yes	Yes	No
Feeding/foraging and resting	1.3 (0.9, 1.7)	4.8 (2.4, 7.6)	8.8 (7.1, 10.6)	Yes	Yes	No

Arm-raising is quantified as the number of all overhead, "straight" arm behaviors relative to total bout time. "Straight" arms were extended at the elbow AND flexed or abducted at the shoulder. 95% confidence intervals are reported in parentheses.

contest of feeding, foraging, or resting in comparison to the context of travel (Table 13.2: *Alouatta* feeding/foraging/resting 1.3% vs. travel 2%). In contrast, *Ateles* and *Lagothrix* used their arms overhead more often while traveling than while feeding/foraging or resting (Table 13.2: *Lagothrix* travel 10.9% vs. feeding/foraging/resting 4.8%; *Ateles* travel 13.4% vs. feeding/foraging/resting 8.8%). This difference was significant for *Lagothrix* and nearly so for *Ateles.*

Descriptive Characteristics of Forelimb Suspension in **Lagothrix poeppigii** *and* **Ateles belzebuth**

Table 13.3 provides descriptive parameters that tend to distinguish the suspensory behavior of *Lagothrix* and *Ateles*, although sample sizes were often too small to reveal significant differences. In the wild, forelimb suspensory locomotion is dynamically integrated into sequences involving a variety of other locomotor modes; however, bipedal landings appear to be relatively more common for *Ateles* (Table 13.3).

Suspensory locomotion was most often videotaped as the focal individual was approaching a fruiting tree or traveling short distances between nearby resting sites. As result, these sequences are likely to be much slower than maximal speeds that may be attained during directed travel. Even at these slower speeds, 40% of *Ateles* strides involved a no-hands phase. In contrast, only one stride with a no-hands phase was documented for *Lagothrix* (Table 13.3: 11.1%). *Lagothrix* maintained an

Table 13.3 Summary of forelimb suspensory locomotion for *Lagothrix poeppigii* and *Ateles belzebuth*

	Lagothrix	*Ateles*
Bipedal landings (forelimb swings, steps and strides) (*L.p.* $n = 18$; *A.b.* $n = 48$)	16.7 (0, 33.3)	27.1 (14.6, 39.6)
Strides with a no-hands phase[a] (*L.p.* $n = 9$; *A.b.* $n = 10$)	11.1 (0, 33.3)	40.0 (10.0, 70.0)
Strides with tail-support during every step (*L.p.* $n = 9$; *A.b.* $n = 10$)	100	50.0 (22.0, 80.0)
Non-repeating handhold/tail hold patterns[b] (*L.p.* $n = 4$ distinct patterns in 9 strides; *A.b.* $n = 7$ distinct patterns in 10 strides)	44.4 (11.1, 77.8)	70.0 (40.0, 100)

Frequencies are reported as percentages with 95% confidence intervals in parentheses.
[a] Note that the number of strides for *Ateles* with a no-hands phase differs in this table compared to the number reported in Table 13.6 This is due to the fact that no-hands duration could not be quantified in one of the strides.
[b] This proportion is a measure of the versatility of forelimb suspensory behavior. The hand and tail touchdown and liftoff sequence was documented for all forelimb suspensory bouts involving a full stride. Nonrepeating handhold/tail hold patterns are sequences that were observed only once.

existing tail hold or established a new hold in each step of every documented stride. In contrast, a tail hold was involved in every step in only 50% of strides performed by *Ateles*.

The sequences of successive contacts by the two forelimbs and the prehensile tail during forelimb suspensory locomotion were "un-patterned" and even idiosyncratic (in some cases successive handholds were actually made by the same hand). In *Lagothrix*, five of the nine strides repeated the exact same pattern of hand and tail touchdowns (Table 13.3). For example, the sequence TD-L, TD-T, TD-R, LO-L, LO-T, TD-L indicates a single stride initiated and terminated by the left hand. The tail touches down during the first step of the stride and lifts off during the second step of the stride. An alternative and distinct pattern observed involved the following sequence: TD-L, TD-T, TD-R, LO-L, LO-T, TD-T, TD-L. This stride is also initiated and terminated by the left hand, but an important difference in terms of support is that the tail touches down during both the first and second steps of the stride.

In contrast to *Lagothrix*, *Ateles* performed seven distinct patterns in 10 full strides (Table 13.3: 70% of handhold/tail hold patterns were unique). Although these comparisons between species are not significantly different, *Ateles* appears to perform forelimb suspensory locomotion in a more versatile manner.

Interspecific differences in forelimb suspension were evaluated further by categorizing each handhold as either "flexed" (forelimb bent to ca. 135° or less relative to the humerus) or "extended" (135°–180°). *Ateles* tended to incorporate flexion into the support phase of each handhold more often than *Lagothrix* (Table 13.4). Interestingly, for *Ateles*, flexed elbows seem to play an important role in initiating and terminating bouts of suspension (Table 13.4). *Lagothrix* was significantly more likely to initiate and terminate suspension with a locomotor mode rather than a postural mode (Table 13.5; within row intraspecific comparisons: Initiation: 82.1% locomotor mode vs. 17.9% postural mode; Termination: 76.2% locomotor mode vs. 23.8% postural mode). *Ateles* forelimb suspension was also more frequently preceded by locomotor bouts (Table 13.5: Initiation: 68.9% locomotor mode vs. 31.1% postural behavior); however, postural and locomotor behaviors were equally common when terminating suspension (Table 13.5).

Table 13.4 Extended elbow vs. flexed elbow forelimb suspensory locomotion for *Lagothrix poeppigii* and *Ateles belzebuth*

	Lagothrix	*Ateles*
Flexed elbow suspenion (armswings, steps and full strides combined) (*L.p.* $n = 54$; *A.b.* $n = 58$)	27.8 (16.1, 39.3)	44.8 (32.8, 56.9)
Flexed elbow suspension (1st handholds only) (*L.p.* $n = 22$; *A.b.* $n = 31$)	18.2 (4.5, 31.9)	45.2 (29.0, 61.4)
Flexed elbow suspenion (2nd handhold; strides only) (*L.p.* $n = 9$; *A.b.* $n = 10$)	22.2 (0, 55.6)	20.0 (0, 50.0)
Flexed elbow suspension (final handhold; steps and strides only) (*L.p.* $n = 11$; *A.b.* $n = 14$)	18.2 (0, 45.5)	50.0 (21.4, 78.6)

Frequencies are reported as percentages with 95% confidence intervals in parentheses.

Table 13.5 Frequency (%) of postural and locomotor modes used in the initiation and termination of forelimb suspensory locomotion

	Initiation			Termination		
	Postural	Locomotor	*n*	Postural	Locomotor	*n*
Lagothrix	17.9 (7.1, 32.1)	82.1 (67.8, 96.4)	28	23.8 (9.5, 42.9)	76.2 (57.1, 90.5)	21
Ateles	31.1 (17.8, 44.4)	68.9 (55.6, 82.2)	45	47.7 (34.1, 63.6)	52.3 (38.6, 66.0)	44

Frequencies are reported as percentages for *Lagothrix poeppigii* and *Ateles belzebuth* with 95% confidence intervals in parentheses.

Table 13.6 Temporal gait variables during forelimb suspensory locomotion by *Lagothrix poeppigii* and *Ateles belzebuth*

	Lagothrix	*n*	*Ateles*	*n*	*p*
Armswing duration	0.49 (0.44, 0.54)	8	0.56 (0.49, 0.64)	19	
Step duration	0.53 (0.48, 0.60)	5	0.60 (0.54, 0.66)	5	
Stride duration	1.47 (1.33, 1.64)	8	1.61 (1.43, 1.81)	10	*
Stride frequency (strides/s)	0.68 (0.62, 0.76)	8	0.62 (0.57, 0.72)	10	*
Stride swing phase duration	0.76 (0.63, 0.96)	8	0.81 (0.71, 0.90)	10	*
Stride support phase duration	0.71 (0.64, 0.77)	8	0.80 (0.63, 1.00)	10	*
Duty factor	0.48 (0.43, 0.54)	8	0.50 (0.42, 0.54)	10	**
Stride no-hands phase duration	0.03	1	0.21 (0.07, 0.50)	3	***
No-hands phase/stride duration	0.02	1	0.15 (0.04, 0.33)	3	***

All times are reported in seconds with 95% confidence intervals in parentheses.
* The nonsignificant differences found in these parameters are comparable in direction, though not magnitude, to significant results published by Turnquist et al. (1999) in their comparison of the forelimb suspensory capabilities of captive *Lagothrix lugens* and *Ateles fusciceps*. Note that they found *Lagothrix* had a significantly longer no-hands phase relative to stride duration whereas data reported here suggest that it is *Ateles* that has a significantly longer no-hands phase relative to stride duration. Further, a no-hands phase was rare for both species under wild conditions in contrast to captivity.
** Duty factor did not differ between species in this study or in Turnquist et al. (1999). Once again, values for wild subjects were higher than for captive subjects, and *Ateles* showed slightly higher duty factors in both studies.
*** $p < 0.05$.

Temporal-gait Characteristics of Forelimb Suspension in **Lagothrix poeppigii** *and* **Ateles belzebuth**

Table 13.6 summarizes temporal gait characteristics that tend to distinguish *Lagothrix* style of suspension from *Ateles*, although most comparisons are nonsignificant owing to small sample size. *Ateles* arm-swings, steps, and strides tend to have longer durations while *Lagothrix* strides tend to occur at higher frequencies (Table 13.6). The swing phase tends to be longer, and hence slower, when compared to the support phase for full strides performed by *Lagothrix*. In contrast, there was almost no difference for swing and support phase durations for *Ateles*. A "no-hands" phase is not only rare for *Lagothrix* (Table 13.3), it is also extremely brief, with significantly shorter airborne durations compared to *Ateles* (Table 13.6: *Lagothrix* "no-hands" phase is 2% of stride duration; *Ateles* "no-hands" phase is 15% of stride duration).

Only three out of nine full strides (33.3%) for *Ateles* were symmetrical, and the second step tended to be longer in duration in 83.3% of asymmetrical strides. Only one (12.5%) of all *Lagothrix* strides approached symmetry, and in 85.7 % of nonsymmetrical strides, the second step was longer than the first.

Discussion

Kinematic studies of captive subjects aim to characterize and quantify species-specific differences in locomotor capacity. In many cases, only one or a few individuals are sampled and gaits deemed asymmetrical or sequences involving changes in speed or direction are excluded (Turnquist et al. 1999; Wallace and Demes 2008). This deliberate effort to compare apples to apples allows for meaningful interspecific comparisons of kinematic performance variables such as stride length or frequency or joint angles when performing categorically similar behaviors. However, the exclusion of asymmetrical gaits or sequences that are in some way atypical may also obscure differing locomotor performance capacities that are functionally relevant. Studies of other animal models illustrate this point. For example, Aerts et al. (2000b) found that a well-known climbing species of gecko (*Gecko gecko*) could be induced to perform as well on flat surfaces as they did on their preferred vertical substrates despite observations indicating that these animals clearly avoided using flat surfaces under lab conditions. The authors suggest that they have the capacity to perform well in both experimental conditions but avoid doing so due to the expense (energetic costs, fatigue, risk of falling, etc.) of performing a behavior that they are not anatomically, physiologically, or neurologically well tuned to do (Aerts et al. 2000b). In other words, captive studies may not readily distinguish behaviors that are possible from those that are used habitually in the wild.

The nonsignificant temporal gait parameters reported in this study are comparable in direction, i.e., the pattern and/or rank-order are similar, though not magnitude, i.e., the values or frequencies are different, to the results reported for captive subjects by Turnquist et al. (1999). Wild and captive *Ateles* tend to take less frequent strides that are longer in duration with longer swing phases relative to support phases in comparison to either wild or captive *Lagothrix* (Table 13.6). The focal species seemed to be equally affected by the differences in habitats in that both moved more slowly in the wild. The tendency to move more slowly in the wild is likely to be a behavioral response to the complexities of their natural habitat, but it may also reflect filming conditions. Subjects were typically filmed as they approached a fruiting tree or other commonly visited destination. It is likely that subjects were not traveling at maximal speeds under these circumstances. Despite this caveat, the slower speeds documented in wild subjects may also be ecologically relevant, reflecting the need to pause briefly and evaluate available substrates. Even familiar paths are likely to be less reliable in the wild in contrast to the very predictable environments provided in captivity. Moreover, the gait patterns of both *Ateles* and *Lagothrix* are frequently asymmetrical in the wild, suggesting that the diversity in substrate availability in their natural habitat requires greater versatility in hand placement and contact time durations. These findings are similar to those reported in Isler's (2003) study of captive and wild orang-utans in which cycle durations were longer and gaits more asymmetrical in the wild.

The most notable differences between captive and wild settings emerge when comparing the versatility of handhold/tail hold touchdown patterns and the no-hands phase. Atypical strides are often excluded in captive studies, making it impossible to determine if the two species differ in handhold/tail hold touchdown patterns. Data presented here suggest that diverse handhold/tail hold touchdown patterns are common for both species in the wild (Table 13.3); however, these patterns tended to be more versatile for free-ranging *Ateles* than *Lagothrix*, suggesting that *Ateles* is better able to respond to the complexity of their environment when performing suspensory movements and therefore use suspensory movements more often and in more diverse ways.

In captivity, *Ateles* and *Lagothrix* consistently achieved periods of free-flight and did not differ in the duration of the no-hands phase; however, the proportion of no-hands phase relative to stride duration was significantly longer for *Lagothrix* (Turnquist et al. 1999). A very different pattern was observed in the wild. First, in contrast to captivity, casual field observations suggest that neither species consistently achieves a no-hands phase, and second, *Ateles* exhibited significantly longer no-hands phases that made up a larger proportion of the total stride duration. These differences may be related to differences in the ability to generate powerful muscular contractions from a pliant support that deforms or bends in response to applied forces. Though both species may find force generation to be more challenging in the wild, these environmental factors may have a larger impact on *Lagothrix* because it may be less specialized, neurologically, physiologically, or in terms of muscle fiber type, composition, orientation, or attachment, to suspensory movements. Such ecologically functional differences have been demonstrated in other animal models. For example, faster species of lizards demonstrate different strategies for increasing speed compared to slower species (e.g., Aerts et al. 2000b; Vanhooydonck et al. 2002). The faster species took longer strides with a longer "float" phase while the species that tended to be slower increased its speed by taking shorter more frequent strides (Aerts et al. 2000b). These authors suggest that the less frequent, but longer strides with an aerial phase are likely to require higher force generation (Vanhooydonck et al. 2002). Of particular interest, lizard species that differ in the capacity to perform particular behaviors are often quite similar in many aspects of their gross anatomy, suggesting that differences in underlying adaptive traits may be quite subtle (Aerts et al. 2000b; Kohlsdorf and Navas 2007).

The performance of suspensory locomotor behaviors observed in wild *Lagothrix* and *Ateles* likely vary from those observed under captive conditions for two fundamental reasons. The first is related to the physical properties of the two environments, which may impact each study species equally, the second pertains to differing levels of ability to perform certain behaviors under challenging circumstances common in their natural habitats. Each of these reasons is discussed, in turn, in the text that follows.

The distinct physical environments of captive and wild settings are likely to impact measures of performance variables for several reasons. First, movements in the wild occur on pliant supports, influencing the duration and direction of the swing and potentially resulting in displacement vectors that deviate from the direction of travel. Second, suspensory locomotion in the wild often occurs across multiple supports that vary in size, inclination, and orientation and it is reasonable to expect temporal and spatial variability in response to habitat complexity (Dickinson et al. 2000). For example, individuals may not always be able to position their supporting hands and feet optimally, resulting in asymmetrical strides and irregular patterns of hand and tail touchdowns. Lastly, locomotion in complex arboreal environments often involves sudden, even mid-stride, alterations in the course of travel which may reflect the types of supports available but may also occur in response to numerous other motivations, e.g., social interactions, opportunistic feeding, response to a perceived predation threat, etc.

Differences between *Lagothrix* and *Ateles* that emerge from an analysis of temporal gait characteristics collected in the wild are likely to reflect differing capacities for efficiently coping with the challenges mentioned in the preceding text while moving in a forelimb suspensory manner. Arboreal animals, in particular, may exhibit different strategies for coordinating movement, maneuverability, and stability in a discontinuous and highly complex environment (Stevens 2006). Differing compensating mechanisms that may be critical to safely and efficiently sustaining forelimb suspensory locomotion in the wild may or may not be apparent in the predictable environment of a laboratory setting. For example, *Lagothrix* incorporates more frequent tail support during suspensory movements in both captivity and in the wild (see Table 13.3 and Turnquist et al. 1999). Previous investigators suggested that tail use may play a critical role in minimizing lateral sway during forward progression (e.g., Jenkins et al. 1978; Lemelin 1995; Turnquist et al. 1999; Schmitt et al. 2005). This lateral displacement is likely to increase the risk that an individual will lose contact with the support and it is also likely to negatively impact both speed of travel and the ability to quickly change direction. Studies of captive subjects indicate that *Ateles* exhibits less side-to-side movements during forelimb suspension compared to *Lagothrix*, and these differences have been related to osteological and myological differences in tail and back morphology in the two species (Lemelin 1995; Schmitt et al. 2005). The more frequent tail contacts made by *Lagothrix* during suspension in both the wild and captivity may serve to compensate, to a degree, for the lack of underlying specializations (Jungers and Stern 1981; Schmitt et al. 2005), yet this compensating strategy exhibited by *Lagothrix* may not be adequate to allow for the sustained suspensory movements that are frequently performed by *Ateles*.

Previous kinematic analyses of captive individuals describe several important biomechanical features of brachiation that allow individuals to sustain this specialized locomotor behavior (Cartmill and Milton 1977; Jenkins et al. 1978; Jenkins 1981; Hollihn 1984; Jungers and Stern 1984; Preuschoft and Demes 1984; Swartz

1989; Turnquist et al. 1999; Schmitt et al. 2005). For example, the potential energy of each swing can be maximized by flexing the elbow, pumping the legs, or through use of a prehensile tail to elevate the body (e.g., Jungers and Stern 1984; Turnquist et al. 1999; Schmitt et al. 2005). Although results were nonsignificant, *Ateles* tended to use a flexed elbow more often during forelimb suspensory behavior than *Lagothrix*, suggesting that suspensory style of *Ateles* involves more energetically efficient movements. Interestingly, *Lagothrix* and *Ateles* do not appear to differ in their ability to perform overhead, "straight" arm behaviors in which the elbow is extended and the shoulder flexed or abducted, and both species performed these behaviors significantly more frequently than the quadrupedal *Alouatta* in all contexts observed (Table 13.2). These data suggest that differences in suspension between *Ateles* and *Lagothrix* is not explicitly tied to differences in elbow or joint mobility, rather, forelimb suspension performance differences are likely more closely linked to distinctions in muscle fiber orientation, insertion, and recruitment as well as neurological differences that may influence each species' capacity to sustain suspension efficiently in a complex environment.

The capacity of *Ateles* for frequent, fluid, rapid, and even acrobatic forelimb suspensory movements is widely supported by prior quantitative and qualitative studies of their behavior in the wild (e.g., Erikson 1963; Richard 1970; Mittermeier 1978; Fontaine 1990; Bergeson 1996; Defler 1999; Cant et al. 2001; 2003). Suspensory activity is broadly understood to facilitate the movements of these individuals as they travel and feed in the flexible, distal branches of trees (Napier 1967; Grand 1972; Cartmill and Milton 1977; Fleagle and Mittermeier 1980) and to increase the speed and directness of travel (Cannon and Leighton 1994; Cant et al. 2003). Although *Lagothrix* is clearly capable of incorporating forelimb suspensory movements, data presented here suggest that they may lack underlying specializations that facilitate the performance of this behavior. Differences in the performance of suspensory locomotion are likely to be intimately tied to the efficacy of these behaviors, in terms of strategies that minimize risk, travel time and the energetic costs of these movements, particularly in the wild.

Conclusion

The performance of forelimb suspension by *Ateles belzebuth* and *Lagothrix poeppigii* was variable in several measures. The tendency for so many strides to by asymmetrical and for these species to use a variety of handhold/tail hold patterns in the wild likely reflects the heterogeneous nature of an arboreal environment as well as differing capacities to respond efficiently to the challenges of pliant and unreliable supports. These findings are relevant to understanding the ecological function of forelimb suspensory locomotion in particular and positional behavior in general.

Comparative studies of performance capacity, such as the one presented here, are particularly relevant to our understanding of form/function relationships. A wide range of reliance on this behavioral mode may result in superficial morphological similarity; however, underlying neurological, physiological, or muscle fiber differences may result in distinct performance capacities that are functionally important.

It is important to emphasize that performance studies measure capacities thought to emerge from underlying adaptive complexes with presumed fitness gains. In order to understand the processes influencing the evolution of primate positional behavior, studies need to be greatly extended to examine within and between population differences in performance capacity. Although studies of other animal models have found support for the heritability and functional role of performance capacities, this has rarely been attempted for primates (Lawler 2006). Future studies of primate locomotor and postural performance should consider the following questions: 1) Do individuals vary in performance parameters?; 2) Are performance measures repeatable (low intra-individual variation) and heritable?; and 3) Can variation in these abilities and their morphological, physiological and/or neurological correlates be related to differential survival and reproductive success through direct testing? These additional data are important to understanding how novel positional behaviors evolve in primate lineages.

Acknowledgments Drs. Richard Lawler and Laura MacLatchy patiently read numerous versions of this manuscript and helped me to develop and refine my arguments. I am especially gratefully to the editors of this volume for their editorial comments, their remarkable patience, and the opportunity to contribute this chapter. Special thanks are also owed to my field assistants, Noah Scovronik and Keely Savoie, and the staff of Tiputini Biodiversity Station for their hard work and advice in the field. This research was supported by an NSF Doctoral Dissertation Improvement Award and Boston University.

References

Aerts P, Van Damme R, Van Elsacker L, Duchene V (2000a) Spatio-temporal gait characteristics of the hind-limb cycles during voluntary bipedal and quadrupedal walking in bonobos (*Pan paniscus*). Am J Phys Anthrop 111:508–517.

Aerts P, Van Damme R, Vanhooydonck B, Zaaf A, Herrel A (2000b). Lizard locomotion: how morphology meets ecology. Neth J Zool 50(2):261–277.

Aerts P, D'Aout K, Herrel A, Van Damme R (2002) Topics in functional and ecological vertebrate morphology. Shaker Publishing, Maastricht, The Netherlands.

Arnold SJ (1983) Morphology, performance and fitness. Am Zool 23(2):347–361.

Avis V (1962) Brachiation: the crucial issue of Man's ancestry. SW J Anthrop 18:119–148.

Blank S, Seiter Bruce P (2001) Resampling Stats add-in for Excel user's guide: Version 2. Resampling Stats, Inc., Arlington.

Bergeson DJ (1996) The positional behavior and prehensile tail use of *Alouatta palliata, Ateles geoffroyi* and *Cebus capucinus*. PhD dissertation, Washington University.

Byron CD, Covert HH (2004) Unexpected locomotor behaviour: brachiation by an Old World monkey (Pygathrix nemaeus) from Vietnam. J Zool Lond 263:101–106.

Cannon CH, Leighton M (1994) Comparative locomotor ecology of gibbons and macaques: Selection of canopy elements for crossing gaps. Am J Phys Anthrop 93:505–524.

Cant JGH (1987) Positional behavior of female Bornean orangutans (*Pongo pygmaeus*). Am J Primatol 12:71–90.

Cant JGH (1992) Positional behavior and body size of arboreal primates: a theoretical framework for field studies and an illustration of its applications. Am J Phys Anthrop 88:273–283.

Cant JGH, Youlatos D, Rose MD (2001) Locomotor behavior of *Lagothrix lagothricha* and *Ateles belzebuth* in Yasuni National Park, Ecuador: general patterns and nonsuspensory modes. J Hum Evol 41:141–166.

Cant JGH, Youlatos D, Rose MD (2003) Suspensory locomotion of *Lagothrix lagothricha* and *Ateles belzebuth* in Yasuni National Park, Ecuador. J Hum Evol 44:685–699.

Cartmill M, Milton K (1977) The lorisiform wrist joint and the evolution of "brachiating" adaptations in the Hominoidea. Am J Phys Anthrop 47:249–272.

Cartmill M, Lemelin P, Schmitt D (2002) Support polygons and symmetrical gaits in mammals. Zool J Linn Soc 136:401–420.

De Silva J (2007) Foot dorsiflexion during vertical climbing in chimpanzees. Am J Phys Anthrop 135(S46):89.

Defler TR (1999) Locomotion and posture in *Lagothrix lagotricha*. Folia Primatol 70:313–327.

Demes B, Larson SG, Stern JT Jr, Jungers WL, Biknevicius AR, Schmitt D (1994) The kinetics of primate quadrupedalism: "hindlimb drive" reconsidered. J Hum Evol 26:353–374.

Demes B, Jungers WL, Wunderlich R, Richmond B, Fleagle J, Lemelin P (1996) Body size and leaping kinematics in Malagsy vertical clingers and leapers. J Hum Evol 31(4): 367–388.

Demes B, Carlson KJ, Franz TM (2006) Cutting corners: the dynamics of turning behaviors in two primate species. J Exp Biol 209:927–937.

Dickinson MH, Farley CT, Full RJ, Koehl MAR, Kram R, Lehman S (2000) How animals move: an integrative view. Science 288(5463):100–106.

Di Fiore A (2004) Diet and feeding ecology of woolly monkeys in a western Amazonian rain forest. Int J Primatol 25(4):767–801.

Di Fiore A, Campbell CJ (2007) The atelines: Variation in ecology, behavior and social organization. In: Campbell CJ, Fuentes A, MacKinnon KC, Panger M, Beader SK (eds), Primates in Perspective. Oxford University Press, New York 155–185.

Erikson GE (1963) Brachiation in New World monkeys and in anthropoid apes. Symp Zool Soc Lond 10:135–164.

Fleagle JG, Mittermeier RA (1980) Locomotor behavior, body size, and comparative ecology of seven Surinam monkeys. Am J Phys Anthrop 52:301–314.

Fontaine R (1990) Positional behavior in *Saimiri boliviensis* and *Ateles geoffroyi*. Am J Phys Anthrop 82:485–508.

Garland T Jr, Hankins E, Huey RB (1990) Locomotor capacity and social dominance in male lizards. Funct Ecol 4(2):243–250.

Grand TI (1972) A mechanical interpretation of terminal branch feeding. J Mammal 53(1):199–201.

Higham TE, Davenport MS, Jayne BC (2001) Maneuvering in an arboreal habitat: the effects of turning angle on the locomotion of three sympatric ecomorphs of Anolis lizards. J Exp Biol 204:4141–4155.

Hildebrand M (1967) Symmetrical gaits of primates. Am J Phys Anthrop 26:119–130.

Hirasaki E, Kumakura H, Matano S (1993) Kinesiogical characteristics of vertical climbing in *Ateles geoffroyi* and *Macaca fuscata*. Folia Primatol 61:148–156.

Hollihn U (1984) Bimanual suspensory behavior: Morphology, selective advantages and phylogeny. In: Preuschoft H, Chivers DJ, Brockelman WY, Creel N (eds), The Lesser Apes: Evolutionary and Behavioural Biology. Edinburgh University Press, Edinburgh 85–95.

Hunt KD (1991) Mechanical implications of chimpanzee positional behavior. Am J Phys Anthrop 86:521–536.

Irschick DJ, Garland T Jr (2001) Integrating function and ecology in studies of adaptation: investigations of locomotor capacity as a model system. Annu Rev Ecol Syst 32:367–396.

Irschick DJ, Herrel A, Vanhooydonck B, Huyghe K, Van Damme R (2005) Locomotor compensation creates a mismatch between laboratory and field estimates of escape speed in lizards: a cautionary tale for performance-to-fitness studies. Evolution 59(7):1579–1587.

Isler K, Thorpe SKS (2003) Gait parameters in vertical climbing of captive, rehabilitant and wild Sumatran orang-utans (*Pongo pygmaeus abelii*). J Exp Biol 206:4081–4096.

Isler K (2005) 3D-Kinematics of vertical climbing in hominoids. Am J Phys Anthrop 126:66–81.

Jenkins J, Farish A, Dombrowski PJ, Gordon EP (1978) Analysis of the shoulder in brachiating spider monkeys. Am J Phys Anthrop 48:65–76.

Jenkins FA (1981) Wrist rotation in primates: a critical adaptation for brachiators. Symp Zool Soc Lond 48:429–451.

Jungers W.L., Stern JT Jr (1981) Preliminary electromyography of brachiation in gibbons and spider monkeys. Int J Primatol 2:19–33.

Jungers WL, Stern JT Jr (1984) Kinesiological aspects of brachiation in Lar gibbons. In: Preuschoft H, Chivers DJ, Brockelman WY, Creel N (eds), The Lesser Apes: Evolutionary and Behavioural Biology. Edinburgh University Press, Edinburgh 119–134.

Kohlsdorf T, Navas CA (2007) Evolution of jumping capacity in Tropidurinae lizards: does habitat complexity influence obstacle-crossing ability? Biol J Linn Soc 91:393–402.

Larson SG, Stern JT Jr (1987) EMG of chimpanzee shoulder muscles during knuckle-walking: problems of terrestrial locomotion in a suspensory adapted primate. J. Zool. Lond. 212:629–655.

Larson SG (1998) Parallel evolution of the hominoid trunk and forelimb. Evol Anthropol 6:87–99.

Larson SG, Schmitt D, Lemelin P, Hamrick M (2000) Uniqueness of primate forelimb posture during quadrupedal locomotion. Am J Phys Anthrop 112:87–101.

Lawler RR (2006) Sifaka positional behavior: ontogenetic and quantitative genetic approaches. Am J Phys Anthrop 131:261–271.

Lemelin P. (1995) Comparative and functional myology of the prehensile tail in New World monkeys. J Morph 224: 351–368.

Lemelin P, Schmitt D, Cartmill M (2003) Footfall patterns and interlimb co-ordination in opossums (Family Didelphidae): Evidence for the evolution of diagonal-sequence walking gaits in primates. J Zool Lond 260:423–429.

Losos JB, Irschick DJ, Schoener TW (1994) Adaptation and constraint in the evolution of specialization of Bahamian Anolis lizards. Evolution 48(6):1786–1798.

Mittermeier RA (1978) Locomotion and posture in *Ateles geoffroyi* and *Ateles paniscus*. Folia Primatol 30:161–193.

Napier JR (1967) Evolutionary aspects of primate locomotion. Am J Phys Anthrop 27:333–342.

Polk JD, Psutka SP, Demes B (2005) Sampling frequencies and measurement error for linear and temporal gait parameters in primate locomotion. J Hum Evol 49:665–679.

Preuschoft H, Demes B (1984) Biomechanics of brachiation. In: Preuschoft H, Chivers DJ, Brockelman WY, Creel N (eds), The Lesser Apes: Evolutionary and Behavioural Biology. Edinburgh University Press, Edinburgh 96–118.

Richard A (1970) A comparative study of the activity patterns and behavior of *Alouatta villosa* and *Ateles geoffroyi*. Folia Primatol 12:241–263.

Schmitt D, Rose MD, Turnquist JE, Lemelin P (2005) The role of the prehensile tail during ateline locomotion: experimental and osteological evidence. Am J Phys Anthrop 126:435–466.

Stanford CB (2004) Arboreal bipedalism in Bwindi chimpanzees: implications for models of the evolution of hominid bipedalism. Am J Phys Anthrop 123(S38):86–87.

Stevens NJ (2006) Stability, limb coordination and substrate type: the ecorelevance of gait sequence pattern in primates. J Exp Zool 305A:953–963.

Stevens NJ, Schmitt DO, Cole III TM, Chan L-K (2006) Technical note: out-of-plane angular correction based on a trigonometric function for use in two-dimensional kinematic studies. Am J of Phys Anthrop 129(3):399–402.

Strier KB (2009) Seeing the forest through the seeds: mechanisms of primate behavioral diversity from individuals to populations and beyond. Curr Anthrop 50(2):213–228.

Swartz SM (1989) Pendular mechanics and the kinematics and energetics of brachiating locomotion. Int J Primatol 10(5):387–418.

Thorpe SKS, Crompton RH (2006) Orangutan positional behavior and the nature of arboreal locomotion in Hominoidea. Am J Phys Anthrop 131:384–401.

Turnquist JE, Schmitt D, Rose MD, Cant JGH (1999) Pendular motion in the brachiation of captive *Lagothrix* and *Ateles*. Am J Primatol 48:263–281.

Van Damme R, Wilson RS, Vanhooydonck B, Aerts P (2002) Performance constraints in decathletes. Nature 415:755–756.

Vanhooydonck B, Van Damme R (2001) Evolutionary trade-offs in locomotor capacities in lacertid lizards: are splindid sprinters clumsy climbers? J Evol Biol 14:46–54.

Vanhooydonck B, Van Damme R, Aerts P (2002) Variation in speed, gait characteristics and microhabitat use in lacertid lizards. J Exp Biol 205:1037–1046.

Vereecke EE, D'Aout K, Aerts P (2006) Locomotor versatility in the white-handed gibbon (*Hylobates lar*): a spatiotemporal analysis of the bipedal, tripedal and quadrupedal gaits. J Hum Evol 50(5):552–567.

Wallace IJ, Demes B (2008) Symmetrical gaits of *Cebus apella*: implications for the functional significance of diagonal sequence gait in primates. J Hum Evol 54:783–794.

Chapter 14
Gait and Kinematics of Arboreal Quadrupedal Walking of Free-ranging Red Howlers (*Alouatta seniculus*) in French Guiana

Dionisios Youlatos and Jean-Pierre Gasc

Abstract The understanding of the adaptive significance of primate locomotor diversity requires studies under naturalistic conditions that combine frequency of use and biomechanical analyses. Here, we report on limb kinematics and gait parameters of the arboreal quadrupedal walk of adult free-ranging red howlers (*Alouatta seniculus*). The data derive from the analysis of original video recordings shot in a primary rain forest in French Guiana. Diagonal-sequence diagonal-couplets walks largely dominated, with mean speeds of 0.67 ± 0.26 m/s. Stance duration was equal for both limbs. During the forelimb swing phase, arm abduction and protraction and elbow extension were the principal movements. Arm abduction and retraction, progressive elbow extension, and forearm pronation dominated during the stance phase. During the swing phase of the hind limb, hip flexion, thigh abduction, and knee extension dominated. Hip extension, thigh abduction, and knee extension were the main movements during the stance phase. These findings appear to support preliminary laboratory observations, provide a background for biomechanical associations, and underline the evolutionary and adaptive importance of morpho-functional complexes within the primate radiation.

Keyword Field study • Limb excursions • Locomotion • New World monkeys • Primates

Abbreviations

COM	center of mass
DSDC	diagonal-sequence diagonal-couplet
p	probability level

D. Youlatos (✉)
Department of Zoology, School of Biology, Aristotle University of Thessaloniki, Greece
e-mail: dyoul@bio.auth.gr

K. D'Août and E.E. Vereecke (eds.), *Primate Locomotion: Linking Field and Laboratory Research*, Developments in Primatology: Progress and Prospects,
DOI 10.1007/978-1-4419-1420-0_14,

Introduction

Primates are unique among mammals in exhibiting a remarkable diversity of locomotor and postural modes (Napier and Walker 1967; Cant 1992). However, although many primates exhibit unique locomotor modes, such as the bipedalism of humans or the ricochetal brachiation of gibbons, the vast majority are arboreal and quadrupedal (Rose 1973; Rollinson and Martin 1981). In effect, quantitative studies of the positional behavior of primates have revealed that arboreal quadrupedal walking accounts for 30–70% of the primate locomotor repertoire (see reviews by Gebo 1989; Dagosto and Gebo 1998).

Arboreal quadrupedal walking is defined as quadrupedal locomotion on and along single continuous horizontal or subhorizontal arboreal supports, involving regular stance and swing phases of all limbs (Hunt et al. 1996). Primate quadrupedal walking has been noted to differ from that of other quadrupedal mammals in a number of features: 1) the extensive to exclusive use of diagonal-sequence diagonal-couplets (DSDC) gaits, where the footfall of a forelimb follows that of the diagonally opposite or contralateral hind limb (Hildebrand 1967; Gambaryan 1974; Rollinson and Martin 1981; Vilensky and Larson 1989; Cartmill et al. 2002, 2007; Schmitt and Lemelin 2002, 2004; Wallace and Demes 2008); 2) the frequent use of compliant gaits that involve substantial joint flexion at the stance phase and relatively long stride length and contact time (Gambaryan 1974; Alexander and Maloiy 1984; Reynolds 1985a,b; Cartmill et al. 2002, 2007; Schmitt and Lemelin 2002, 2004); 3) a significantly greater forelimb protraction at touchdown (Vilensky and Larson 1989; Vilensky and Gankiewicz 1990; Vilensky et al. 1994; Larson et al. 2000; Schmidt and Fischer 2000; D'Août et al. 2002; Cartmill et al. 2002, 2007; Fischer et al. 2002; Schmitt and Lemelin 2002, 2004; Schmitt 2003; 4) the occurrence of higher vertical substrate reaction forces in the hind limbs than in the forelimbs as the major part of the propulsive force is accomplished by the hind limbs (Kimura et al. 1979; Rollinson and Martin 1981; Vilensky and Larson 1989; Vilensky and Gankiewicz 1990; Vilensky et al. 1994; Schmitt 2003; Franz et al. 2005); and 5) rare use of symmetrical running gaits (such as a trot) and the direct passage from a walk to a gallop, instead of passing through a running trot, as most other mammals would do (Hildebrand 1967; Rollinson and Martin 1981; Vilensky and Larson 1989; Vilensky and Gankiewicz 1990; Vilensky et al. 1994; Cartmill et al. 2002, 2007). These differences have been attributed to several morphological as well as neurological factors that distinguish primates from other mammalian quadrupeds. Thus, primates, and especially the arboreal species, tend to bear more weight on their hind limbs, most likely by actively shifting their center of gravity toward the hind part of the body (Reynolds 1985b; Kimura 1992; Cartmill et al. 2002, 2007) by having longer limbs and more flexible joints (Alexander and Maloiy 1984; Reynolds 1985a), or present evolutionary changes in the neurological control of the forelimb and a more direct cortical control of limb movements (Vilensky 1989; Vilensky and Larson 1989).

However, these features of the kinematics and kinetics of primate quadrupedalism derive from observations based on studies under controlled conditions, where primates have been accustomed to walking on tread- or rope mills in the laboratory.

These settings provide an environment that allows optimal conditions for video recordings of multiple individuals and species, resulting in relatively large samples of exploitable sequences, and allowing tests for intraspecific and interindividual variability. However, a major drawback of this approach is the fact that controlled conditions are rarely able to represent the habitat complexity that primates encounter during all types of locomotor bouts in their natural environments (Stevens et al., Chapter 16). As a result, the features that have been recorded under laboratory conditions may depart from walking patterns of wild primates. Unfortunately, the scarcity of gait parameters and kinematic data from the wild does not yet lend for meaningful comparisons, although the few existing data imply the occurrence of similar gait and kinematic features in the wild and in the laboratory (Alexander and Maloiy 1984; Isler and Thorpe 2003; Isler and Gruter 2006; Stevens et al., Chapter 16). However, certain practical difficulties appear to discourage researchers from attempting similar observations in the wild. These may be the overall frequency of quadrupedal walking bouts, the occurrence of full strides on single supports, the overall reduced visibility in primary rain forests (hosting most arboreal primates), the search for good filming conditions and the relevant equipment, along with the fact that all these situations tend to gradually reduce the exploitable sample size, decreasing the statistical robustness of the whole effort. Nevertheless, such data from the wild can provide a mainly qualitative and preliminary quantitative approach of the study of gaits and kinematics of primates. This could help substantiate laboratory observations and elucidate the evolutionary and adaptive significance of these characteristics via available information on the spatial context in which they occur.

In this chapter, we analyze the arboreal quadrupedal walk of adult free-ranging red howlers (*Alouatta seniculus*) sampled and filmed in a primary rain forest in French Guiana. The purpose of this report is to provide new data on gait types, speed, stride length and duration and related gait parameters, and qualitative and quantitative kinematics of angular excursions of fore- and hind limbs of arboreal quadrupedal walking as performed by a primarily quadrupedal New World monkey in the wild.

Study Site, Subjects, and Methods

Field research was conducted at the "Station des Nouragues" (4°05′ N, 52°40′ W) in French Guiana, situated 100 km south of Cayenne, the country's main town. The study site occupies a 160-ha hilly undisturbed terra firme wet rain forest covered with trails. The site is dominated by high mature forest with a mean annual temperature of 26.1°C and annual rainfall ranging from 3000 to 3250 mm. Primate species found in the site include red howlers (*Alouatta seniculus*), black spider monkeys (*Ateles paniscus*), brown capuchins (*Cebus apella*), wedge-capped capuchins (*Cebus olivaceus*), Guianan sakis (*Pithecia pithecia*), and golden-handed tamarins (*Saguinus midas*).

During the study period (April–July 1992, July–September 1993), the site was used by two groups of red howlers. During fieldwork, the subjects were followed, sampled, and extensively video recorded from dawn to dusk. The data analyzed here derive from a total of 3 h of extensive video recordings using a Hi-8 SONY camcorder (CCD-TR705E) set on a light tripod. For better focus and zooming, additional zoom lenses of ×1.4 were used, coupled with the original ×8 lens. Recording was done at 25 frames/s, at a shutter speed of 1/1000, both providing enough light for filming and avoiding blurring during frame-to-frame analysis.

The recordings involved all locomotor and postural modes employed by both sexes in the wild. An analysis of the locomotor repertoire of the population in French Guiana (Youlatos 1994, 1998) showed that arboreal quadupedal walking was the dominant mode of locomotion (42% of all locomotor modes) and was performed primarily on horizontal supports (83% of quadrupedal walk subsample) of 5–10 cm in diameter (54% of quadrupedal walk subsample). Therefore, this mode was selected for further detailed analysis based on its high frequency of use, its importance for traveling within the canopy, and its regularity in swing and stance phases. A JVC (BR-6400TR) VCR and a JVC (TM-150PSN) monitor were used to select relatively well discerned walking sequences of at least one full stride. The trimming process provided a total of 102 walk sequences to be used for analysis of gait types and associated supports and other additional qualitative details in limb movements. The next step was to select good quality lateral views of quadrupedal walk sequences, where all joints were visible for at least one and a half stride cycles, for quantitative frame-to-frame analysis.

These criteria were filled only by 12 walking sequences. We transformed the selected video sequences into digital video format and analyzed them on a PC using a specially organized digitization program. During the frame-to-frame analysis, we calculated the following gait parameters: 1) *duty factor* (% of gait cycle period that hand or foot is in stance period), 2) *diagonality* (% of gait cycle period by which the hind footfall precedes the ipsilateral fore footfall), 3) *gait cycle, fore- and hind limb swing and stance phase duration*, 4) *speed* (distance covered as calculated by estimated body length of the animal per time unit), 5) *distance covered by the center of mass during touchdown* (stance duration times speed), and 6) *fore- and hind limb relative stride length* (length of fore- or hind limb stride over body length). Mean male or female (depending on the sequence) head-body length of Guianan red howlers were used for calibration of the video sequence (Youlatos 1994).

In addition, the joint centers of the shoulder, elbow, wrist, hip, knee, and ankle were visually estimated and digitally traced over each frame (Fig. 14.1). The angles formed by the lines that join them described the angular movements of the corresponding joints and were based on Vilensky and Gankiewicz (1990) (Fig. 14.1). In this way, angular excursions (°) in the sagittal plane of both fore- and hind limb joints were directly calculated and were plotted against gait cycle duration. Angular values and the corresponding joint movements were described as follows: in the shoulder joint, low values corresponded to retraction and higher values to protraction; in the elbow, low values corresponded to flexion and higher ones to extension; in the wrist joint, low values corresponded to extension and higher ones to flexion;

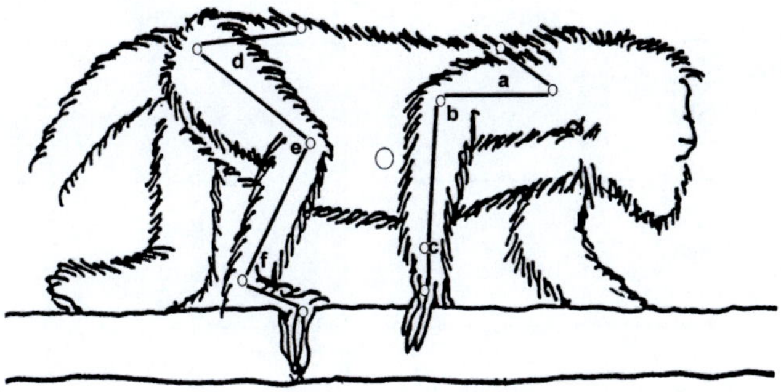

Fig. 14.1 Landmarks of the sampled joint points of red howlers for angular measurements: (**a**) shoulder; (**b**) elbow; (**c**) wrist; (**d**) hip; (**e**) knee; (**f**) ankle

in the hip and knee joints, low values corresponded to flexion, and higher ones to extension; in the ankle joint, low values corresponded to dorsal flexion and higher ones to plantarflexion. Moreover, the abduction angles of the arm and hip were indirectly calculated, using a simple trigonometric formula: cos (a) = projected segment length/estimated segment length, the latter based on external limb metrics of Guianan red howlers (Youlatos 1994). Other limb movements, such as arm and thigh medial or lateral rotations and forearm pronation or supination were only qualitatively observed and estimated during the frame-to-frame analysis. However, it must be well understood that such both quantitative and qualitative descriptions only approximate real limb excursions. Higher accuracy could be obtained only by combining simultaneous lateral, frontal, and dorsal views, which unfortunately was not feasible in the wild.

We tested differences between temporal and spatial gait parameters using Student's *t*-tests. We tested angular excursions for each joint over each phase via ANOVAs. For all tests, $p < 0.05$ were considered statistically significant.

Results

During the study period, arboreal quadrupedal walking was the main locomotor mode. The analysis of the whole sample of recorded walking sequences ($n = 102$) showed that arboreal walking was composed of exclusively symmetrical gaits, with the touchdowns of the forelimbs and hind limbs evenly spaced in time (Hildebrand 1967). These symmetrical gaits were composed of both diagonal-sequence and lateral-sequence gaits. Diagonal-sequence gaits, where the footfall of a forelimb follows that of the diagonally opposite or contralateral hind limb, largely dominated (91.9%, $n = 102$). In contrast, lateral-sequence gaits, where the footfall of a forelimb follows that of the same side or ipsilateral hind limb, were rare, and occurred primarily upon slightly downward oblique supports (66.7%, $n = 9$). Moreover, in both diagonal- and lateral-sequence gaits, it was the forefoot that initiated the stride in most occasions [63.7% in diagonal ($n = 93$) and 66.7% in lateral gaits ($n = 9$)].

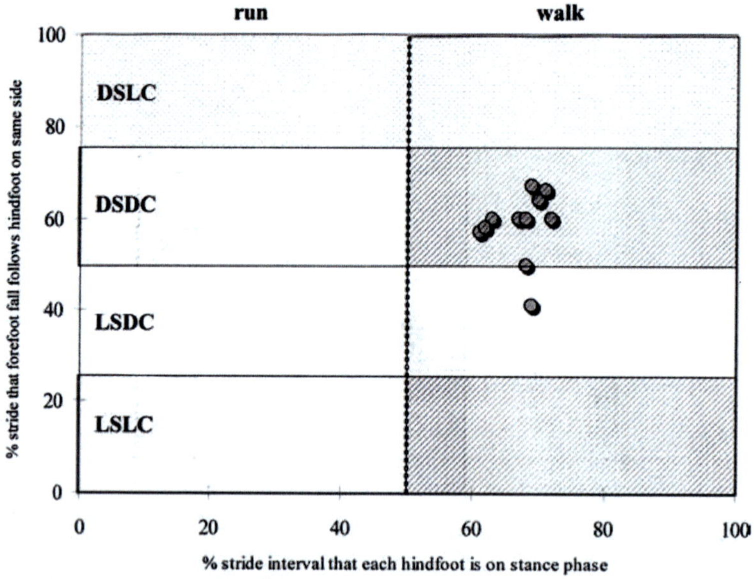

Fig. 14.2 Bivariate plot of the 12 analyzed walking gaits of red howlers. *x*-axis: hind limb duty factor; *y*-axis: diagonality. DSLC = diagonal sequence lateral couplets; DSDC = diagonal sequence diagonal couplets; LSDC = lateral sequence diagonal couplets; LSLC = lateral sequence lateral couplets

The subsequent selection of a small sample of walking sequences ($n = 12$), that were suitable for further analysis provided detailed information on several gait parameters. In this way, mean duration of the gait cycle of a quadrupedal walk of red howlers was 1.16 ± 0.24 s, and involved regular stance and swing phases of the limbs. Mean recorded speeds were 0.67 ± 0.26 m/s and mean distance covered by the center of mass was 0.53 ± 0.12 m.

Further, diagonality (mean: 58.3 ± 6.7%) and hind limb duty factor (mean: 66.7 ± 3.8%) showed that, when plotted in a modified Hildebrand graph (Fig. 14.2), the arboreal quadrupedal walk of red howlers fell within the areas of diagonal-sequence diagonal-couplets walking (DSDC). This indicates that the footfalls of the opposite or contralateral fore- and hind limb were related in time as a pair.

Metrics of the swing and stance phase of the forelimb are presented in Table 14.1. Mean forelimb stride length was particularly long, surpassing that of the animal's body length (Table 14.1) and, as expected, the stance phase durations of both fore- and hind limb were equal (stance phase forelimb/hind limb ratio: 1.04 ± 0.09).

In terms of joint angular excursions, the different joints of the forelimb showed different profiles, during both stance and swing phase. Specifically, the shoulder was characterized by a monophasic excursion, while both elbow and wrist joints exhibited biphasic excursions (Fig. 14.3). In the beginning of the stance phase, the shoulder was protracted and abducted, depicting a gradual retraction and adduction, and medial rotation of the arm throughout the rest of the phase. In this way, the

Table 14.1 Stance and swing phase metrics of fore- and hind limb of arboreal quadrupedal walking of red howlers ($n = 12$)

	Forelimb	Hind limb
Swing/stance	0.39 ± 0.12	0.44 ± 0.11
Stance duration (s)	0.83 ± 0.20	0.80 ± 0.18
Duty factor (%)	72.2 ± 5.8	66.7 ± 3.8
Stride length (body length)	1.12 ± 0.14	1.13 ± 0.13

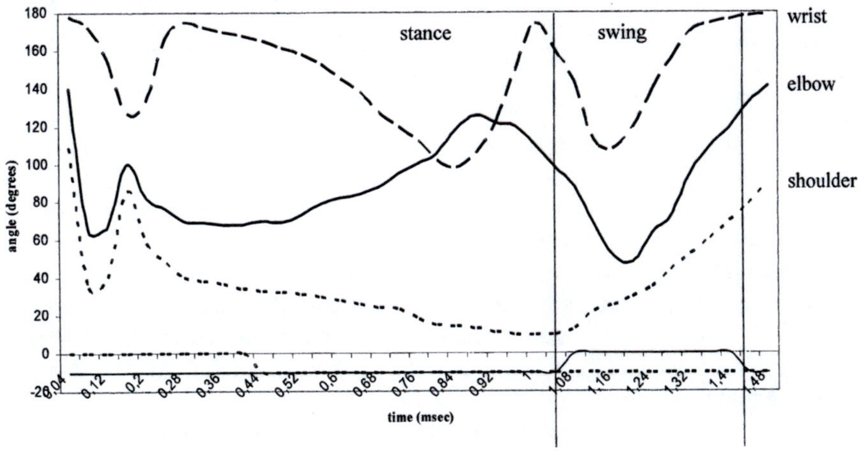

Fig. 14.3 Graph of joint angles of the forelimb; vertical lines within the graph mark lift-off (**left**) and touchdown (**right**)

shoulder was retracted and adducted at lift-off, where rapid protraction, abduction, and lateral rotation of the arm began and continued during the swing phase (Fig. 14.3). Mean arm abduction during the stance phase was only slightly greater than that recorded for the swing phase (68.7 ± 5.5° vs. 62.5 ± 11.9°, $p = 0.142$). The elbow joint was extended at initial stance, showing rapid flexion in the first third of the stance phase, a progressive extension throughout the phase, and a rapid flexion prior to lift-off that was continued in the very beginnings of the swing phase (Fig. 14.3). This was once more followed by swift extension so that the elbow was finally semiextended at touchdown. The forearm was kept mainly pronated throughout the stance phase and most of the swing phase, and shifted to semi-pronation before touchdown. The wrist joint, being extended in the beginning of the stance phase, exhibited gentle flexion during the first three quarters of the phase, only to show a quick extension at the last quarter before lift-off (Fig. 14.3). The hand grip was performed through a pincer-like position between digits II and III. The swing phase was characterized by initial swift wrist flexion that was followed by a longer extension phase finding the wrist extended at touchdown (Fig. 14.3).

When the mean highest and lowest angles of all joints during both stance and swing phases were calculated (Table 14.2), a few significant differences appeared

Table 14.2 Mean values, standard deviation (SD), and range of the minimum and maximum angles of forelimb joints during swing and stance phases (n = 12)

		Minimum angle (°)			Mamixmum angle (°)		
		Mean	SD	Range	Mean	SD	Range
Shoulder	Swing	11.5	9.8	2–26	131.8	39.2	72–176
	stance	8.3	8.2	2–28	113.9	51.9	45–178
Elbow	Swing	42.9	8.8	30–51	136.3	30.1	82–175
	stance	60.9	18.0	30–83	152.6	27.6	86–178
Wrist	Swing	95.6	25.5	47–134	174.0	5.9	164–180
	stance	85.4	36.3	24–130	162.2	31.1	88–180

to emerge. Thus, the shoulder showed significantly higher maximum angles during the swing phase than during the stance phase (p = 0.037). This is translated in a more protracted forelimb in the swing phase than in the stance phase. In the elbow joint, swing phase minimum angles were significantly lower than those of the stance phase (p = 0.009). In other words, the elbow joint appeared more flexed during the swing phase than during the stance phase.

In the hind limb, metrics of the swing and stance phase are also presented in Table 14.1. The hind limb stance phase composed almost two thirds of the gait cycle, resulting in a hind limb duty factor of 0.66. Mean hind limb stride length was also long (Table 14.1).

As in the case of the forelimb, angular excursions of the different hind limb joints during both stance and swing phase exhibited some differences. The hip joint was characterized by a monophasic excursion, while both knee and ankle joints exhibited biphasic excursions (Fig. 14.4). At the beginning of the stance phase, the hip joint was primarily flexed and slightly adducted. During most of the stance phase the hip joint extended and abducted gently, and the thigh rotated laterally progressively, to be slightly extended and abducted at lift-off (Fig. 14.4). Thus, the swing phase was characterized by rapid hip flexion and adduction, the thigh rotating fast medially (Fig. 14.4). Mean hip abduction during the stance phase was only slightly greater than that recorded for the swing phase (64.1 ± 12.3° vs. 56.2 ± 8.2°; p = 0.084). During the stance phase, the knee was initially extended, and showed a gradual flexion throughout stance, which was followed by a short and rapid extension prior to lift-off. In the beginning of the swing phase, the knee flexed again and exhibited rapid extension in the last one third of the swing, before touchdown (Fig. 14.4). Finally, the ankle joint showed a rapid and continuous plantarflexion at initial stance, followed by gradual dorsiflexion throughout the rest of the stance phase (Fig. 14.4). This movement was interrupted by a swift plantarflexion prior to lift-off that continued in the initial one third of the swing phase (Fig. 14.4). Rapid dorsiflexion of the ankle continued in the middle third of the swing phase and was again followed by rapid plantarflexion at the last third of the swing, just before touchdown (Fig. 14.4).

When the mean highest and lowest angles of all joints during both stance and swing phases were calculated, a number of significant differences emerged. The hip joint exhibited significantly smaller minimum and maximum angles at the swing phase

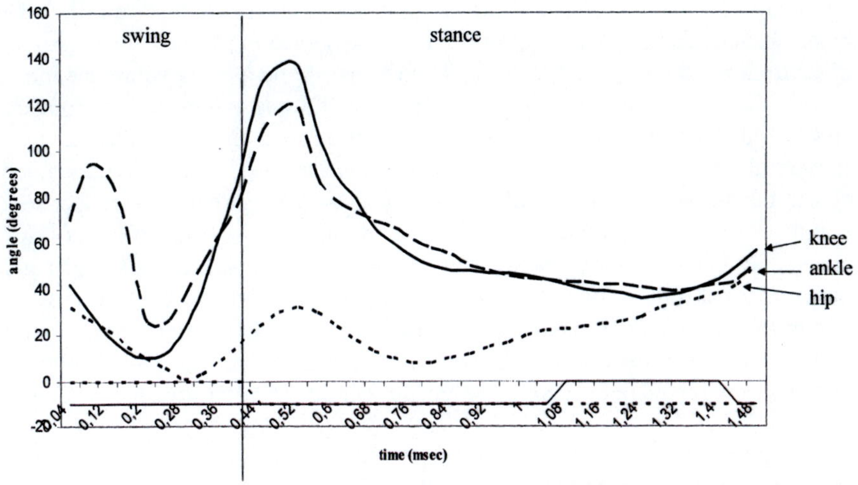

Fig. 14.4 Graph of joint angles of the hind limb; touchdown is marked by the vertical line within the graph

Table 14.3 Mean values, standard deviation (SD), and range of the minimum and maximum angles of hind limb joints during swing and stance phases ($n = 12$)

		Minimum angle (°)			Maximum angle (°)		
		Mean	SD	Range	Mean	SD	Range
Hip	Swing	12.6	13.8	2–47	81.6	39.7	42–159
	stance	22.0	15.8	2–50	98.3	24.5	55–140
Knee	Swing	43.5	21.1	10–97	140.0	19.4	90–174
	stance	73.8	25.3	30–116	134.0	14.3	115–165
Ankle	Swing	37.3	18.8	8–77	124.9	24.0	95–173
	stance	50.6	14.5	31–78	116.7	30.3	46–16

than at the stance phase (min angles: $p = 0.003$; max angles: $p = 0.016$; Table 14.3). This is translated in a more protracted and more retracted hind limb during the swing phase the stance phase. On the other hand, the knee and ankle joints were both characterized by significantly smaller minimum angles at the swing phase than at the stance phase (knee min angles: $p < 0.001$; ankle min angles: $p = 0.046$; Table 14.3). This resulted in a more flexed knee and a more dorsiflexed ankle in the swing phase than in the stance phase.

Discussion

The present report on the gait and joint movement metrics of the quadrupedal walk of free-ranging red howlers (*Alouatta seniculus*) is one of the few studies under naturalistic conditions. The data, presented and analyzed here, derived from original video recordings of individuals locomoting in their natural environment in a primary rain forest in French Guiana. In this way, these data tend to reflect the biomechanical and kinematic responses of the observed individuals to the variety of environmental constraints with which they cope, while moving within the forest canopy. Therefore, the purpose of the present analysis was to provide a preliminary understanding of the kinematic adjustments that animals utilize in order to move successfully and safely within the three-dimensional arboreal milieu. As underlined earlier, the different constraints of fieldwork lead to different means of data collection that may not be completely comparable to those used in the laboratory. This may not result in the exact, thorough, and quantitatively precise analysis of gait and joint metrics of walking patterns obtained in the laboratory. However, it can provide a comparative quantitative and qualitative background of the way a certain locomotor mode, i.e., arboreal quadrupedal walk, is performed within the natural environment of the observed individuals and eventually lead to more focused lines of future research.

In this context, our analysis of the lateral views of quadrupedal walk sequences of red howlers in a primary rain forest in French Guiana provided a set of interesting observations regarding gait parameters as well as limb joint kinematics. Thus, we believe that these data will allow a preliminary understanding of how this mode is actually performed by a relatively large-bodied anthropoid primate in the wild (Fig. 14.5).

Red howlers used diagonal-couplets diagonal-sequence walking gaits with relatively long strides extensively at low speeds. Stance phases of both limbs were equally spaced in time and composed almost two thirds of the gait cycle. In this way, both fore- and hind limb were covering the same distance with relative stride

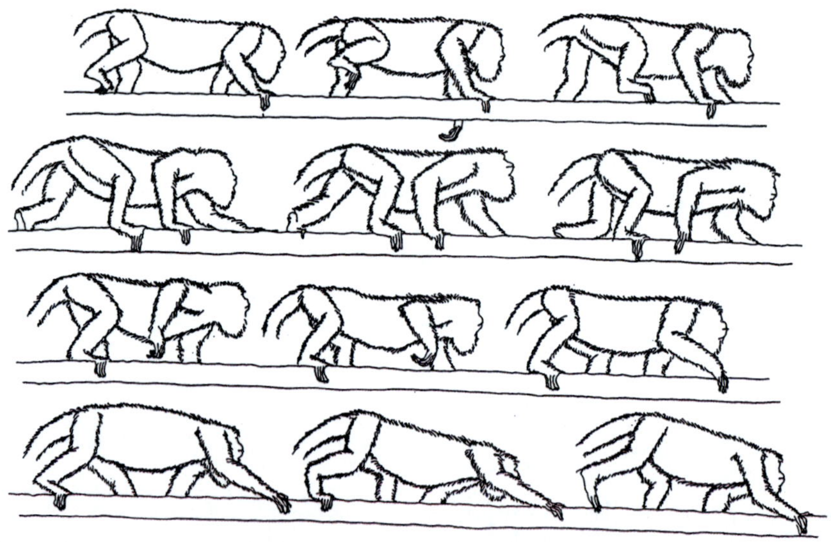

Fig. 14.5 Analyzed walking sequence of a male red howler on a medium-sized horizontal branch, representing a complete gait cycle with each image corresponding to a time difference of 120 ms

lengths superior to one, i.e., longer than body length. In terms of limb joint kinematics, the joints of the proximal segments (shoulder, hip) moved in monophasic excursions, compared to the biphasic movements of the more distal joints (elbow, knee, wrist, ankle). In the forelimb, the arm was protracted, the elbow was quite extended, and the wrist extended at touchdown, whereas at lift-off, the arm was retracted, the elbow semi-flexed, and the wrist semiextended. In the hind limb, the hip was semi-flexed, the knee extended, and the ankle plantarflexed at touchdown. At lift-off, the hip was semiextended, the knee relatively flexed, and the ankle plantarflexed.

In terms of gaits, red howlers used both diagonal and lateral gaits, with a strong preference for the former. However, the use of both types appears to reflect the overall locomotor plasticity of primates, which is ultimately related to the diversity of features of arboreal supports (Stevens 2006). The dominant walking gait of red howlers on horizontal medium-sized arboreal supports was diagonal-sequence diagonal-couplets (DSDC). This gait characterizes almost all arboreal primates (Hildebrand 1967; Rollinson and Martin 1981; Vilensky and Larson 1989; Cartmill et al. 2002, 2007; Schmitt and Lemelin 2002; Wallace and Demes 2008), as well as some other arboreal mammals such as marsupials (Lemelin et al. 2003) and carnivorans (Taylor 1970; McLearn 1992). These gaits have been hypothesized to be more stable when the center of body mass is located more posteriorly, as in the case of most primates (Kimura et al. 1979; Rollinson and Martin 1981), or to be a simple byproduct of increased supraspinal control of locomotion, especially in the forelimbs that have become particularly dexterous for manipulative activities (Vilensky and Larson 1989). More recently, Cartmill et al. (2002, 2007) suggested that DSDC gaits would be more advantageous upon small branches, enabling arboreal mammals to rely upon the grasping hind limb placed under the center of mass (COM). In this way, it could draw back and recover when the forelimb would be placed upon an unstable support, providing equal or greater stability at the point of forelimb touchdown. In red howlers, the hind limb was not in a particularly protracted position, or anchored beneath the COM of the animal at forelimb touchdown (Fig. 14.5). Instead, it was placed in a more caudal position, away from the COM, apparently increasing the base of support and more likely enhancing the dynamic stability of quadrupedal walk. However, in biomechanical terms, such a hind limb position, implying relatively extended joints, may have not favored body propulsion, but rather the capacity to withdraw by simply flexing the limb joints (see also Cartmill 2007). A similar hind limb positioning behavior has been also observed in capuchin monkeys (Wallace and Demes 2008), and may be related to the overall locomotor plasticity of primates (Stevens 2006).

The latter is also exhibited by the infrequent use of lateral sequence gaits by red howlers, occurring primarily on downward inclined supports. In effect, these gaits do occur in arboreal primates and, effectively, upon downward inclined supports (Rollinson and Martin 1981; Vilensky et al. 1994; Dunbar and Badam 2000; Schmitt 2003; Shapiro and Raichlen 2005; Nyakatura et al. 2008). On such supports, locomotion appears to emphasize the role of the forelimbs and the reduced propulsive role of the hind limbs. In this way, the forelimbs appear to generate braking effort and maintain control of velocity to counter the acceleration due to gravity (Nyakatura et al. 2008).

In the hind limb, the observed angular excursions of the different joints of red howlers did not differ importantly from those reported in studies under controlled conditions for other arboreal primates and other arboreal mammals (Jenkins and Camazine 1977; Vilensky and Gankiewicz 1990; Schilling and Fischer 1999; Fischer et al. 2002; Aerts et al. 2000; D'Août et al. 2002; Schmidt 2005). The observed differences concerned mainly joint angle values, and in some cases, some joints showed inverse or delayed movements at similar times of the gait cycle, compared to those recorded for other primates. Similar differences are difficult to explain, as they may be due to differences in the settings (laboratory vs. field), methods of recording and analysis (high-speed cameras with enough light vs. normal video camera with variable light conditions), angle of filming (completely lateral *vs.* variably lateral, although in the present work we tried to retain apparently lateral sequences for analysis), or, finally, phylogeny [cf. fore- and hind limb joint angle differences between primate families (Larson et al. 2000; Fischer et al. 2002)]. However, the overall kinematics of the hind limb in red howlers exhibited the monophasic excursion of the proximal joint (hip) in contrast to the biphasic excursions of the more distal joints (knee, ankle) typical for most mammals (Gasc 2001; Schmidt and Fischer 2000; Schmidt 2005). These movements appear to depict the major propulsive role of the hip within the hind limb, compared to the more adjusting and stabilizing role of the distal joints and segments involved in quadrupedal activities (Schmidt and Fischer 2000; Schmidt 2005).

The movements of the forelimb joints of red howlers observed upon medium-sized branches in the wild appear to be more or less similar to those reported for other primates as well as other arboreal and ambulatory mammals (Jenkins 1971; Jouffroy et al. 1983; Larson et al. 2000; Schmidt and Fischer 2000; Fischer et al. 2002; Schmidt 2005). As in the case of the hind limb, some minor differences were detected in angular values and the sequence of joint kinematics, which may be also due to the factors analyzed above. Red howlers, as all quadrupedal primates, appeared to initiate the step by a relatively protracted forelimb and retracted it gradually as the step proceeded. The protracted position contributes to a more obtuse scapulo-humeral angle that further increases the relative long stride length of the animal (Larson et al. 2000). This is promoted by a relatively greater scapular mobility. The latter contributes to the complex movements of the forelimb in primates, which appears to be primordial for quadrupedal propulsion in mammals (Schmidt and Fischer 2000). In effect, the anatomy of the shoulder joint of red howlers, which facilitates considerable abduction and substantial protraction (Grand 1968; Schön 1968; Stern et al. 1977, 1980a, b; Fleagle et al. 1981; Larson and Stern 1989, 1992; Schön Ybarra 1998; Youlatos 1994, 2000), helps the animal move forward using longer strides. The increase of stride length instead of stride frequency is a pattern that is encountered in most primates (Alexander and Maloiy 1984; Demes et al. 1994) and, very likely, contributes to the lowering of the body avoiding branch sway and disequilibrium that would result in toppling over the support (Schmitt 1999, 2003; Cartmill et al. 2002). This may explain the relatively long forelimb stride of red howlers, insignificantly lower to that of the hind limb. In this way, the animal appears to increase contact time with the unstable arboreal support, an action that contributes further to a reduction of unnecessary substrate reaction forces that could be quite harmful for forelimb joints and segments (Demes et al. 1994; Schmitt 1994, 1999, 2003).

The observed long forelimb stride length is also highlighted by the ample elbow excursions that were observed during the stance phase in red howlers. The recorded movements, similar to those of other primates (Jouffroy et al. 1983; Larson et al. 2000; Schmidt and Fischer 2000; Fischer et al. 2002; Schmidt 2005), reveal substantial elbow yield that provides further stability to the walking animal. This contributes to a crouched position, increased contact time, and reduction of vertical oscillations of the center of mass that appear to further enhance balance over the support used (Schmitt 1999; Larson et al. 2000). The dynamic stability at this level is further assisted by the pronated forearm, as is the case for other quadrupedal primates (Larson and Stern 2006), and the grasp between digits II and III that position the limb and provide a secure grip over the support (Youlatos 1999), respectively. This underscores the adjusting role of the distal limb elements in contrast to the mainly propulsive role of the proximal limb segments (Schmidt and Fischer 2000; Schmidt 2005). These movements at the level of the elbow joint appear to be facilitated by the skeletal and muscular anatomy of the region, contributing to enhanced stability during elbow flexion and forearm pronation (Grand 1968; Schön 1968; Schön Ybarra 1998; Youlatos 1994, 2000). This anatomical arrangement, coupled with complex muscle recruitment during quadrupedal locomotion, contributes to the compliant walk that characterizes red howlers, and most arboreal primates upon arboreal supports (Schmitt 1999). In effect, this was achieved behaviorally through an increase in elbow joint flexion, step length, and step duration through enhanced elbow yield, likely reducing peak stresses, vertical oscillations of the COM, and maintaining balance over the support (Larson et al. 2000). Therefore, elbow yield along with a protracted forelimb contributes to a crouched posture that may be critical for stability, balance, reduction of substrate reaction forces, and branch sway.

Enhancing stability and reducing unwanted substrate reaction may demand less energy expenditure during quadrupedal activities. In this way, the behavioral performance of the major locomotor mode of red howlers in French Guiana appears to correspond with the overall energy minimizing/saving strategy of howlers (Milton 1980). Howlers appear to have evolved early in the phylogenetic history of atelines (Rosenberger and Strier 1989; Strier 1992; Jones 2008), and might have probably simultaneously adopted a limited locomotor repertoire that was mainly restricted to above-branch quadrupedalism. The latter was probably associated with the selection of single horizontal or subhorizontal supports, whereupon quadrupedal walking with regular swing and stance phases could be executed (Grand 1984). Therefore, red howlers could have opted for the frequent utilization of a locomotor mode that was associated with overall stability, balance, reduction of peak stresses, and a safer way to increase relative speed upon single arboreal supports. This relatively conservative way of moving within the canopy could have further contributed to their novel shifting toward an energy-minimizing strategy compared to early atelines (Jones 2008). This may have further required some major rearrangements of the musculoskeletal anatomy of the forelimb to increase forelimb yield and protraction, albeit retaining an important propulsive role, analogous to the ancestral primary role in the flexible locomotor repertoire of early atelines (Jones 2008).

In conclusion, the present study showed that the analysis of major and/or critical locomotor modes of primates in the wild can provide clues to their actual performance in the natural environment with which they constantly interact. In this way, similar analyses can contribute to the elucidation of the evolutionary and adaptive significance of these modes within the respective radiations. In addition, it highlighted the significance of gait and kinematic studies in the wild in helping us comprehend the evolutionary importance of locomotion in the primate radiation, as well as in mammalian evolutionary history. Given that modern technology has reduced the gap between the laboratory and the field, we believe that future research in this domain should be directed to similar studies in order to obtain a comparable bulk of data to that from laboratory-oriented studies.

Acknowledgments We are particularly grateful to Evie Vereecke and Kris D'Août for organizing such an exciting symposium and inviting us to participate. Data collection and analysis were feasible through grants from the Action Spécifique Guyane of the Muséum National d'Histoire Naturelle, Paris, and from the Centre National de la Recherche Scientifique, France. Many thanks also go to the anonymous referees who greatly improved earlier versions of the manuscript.

References

Aerts P, van Damme R, van Elsacker L, Duchene V (2000) Spatio-temporal gait characteristics of the hindlimb cycles during voluntary bipedal and quadrupedal walking in bonobos *(Pan paniscus)*. Am J Phys Anthropol 111:503–517.

Alexander RMcN, Maloiy GMO (1984) Stride lengths and stride frequencies of primates. J Zool Lond 202:577–582.

Cant JGH (1992) Positional behavior and body size of arboreal primates: a theoretical framework and an illustration of its application. Am J Phys Anthropol 88:273–283.

Cartmill M, Lemelin P, Schmitt D (2002) Support polygons and symmetrical gaits in mammals. Zool J Linn Soc 136: 401–420.

Cartmill M, Lemelin P, Schmitt D (2007) Understanding the adaptive value of diagonal-sequence gaits in primates: a comment on Shapiro and Raichlen, 2005. Am J Phys Anthrop 133:822–825.

Dagosto M, Gebo DL (1998) Methodological issues in studying positional behavior. In: Strasser E, Fleagle J, Rosenberger A, McHenry H, (eds), Primate Locomotion: Recent Advances. Plenum Press, New York, pp 5–29.

D'Août K, Aerts P, Clercq de D, De Meester K, Van Elsacker L (2002) Segment and joint angles of the hind limb during bipedal and quadrupedal walking of the bonobo (*Pan paniscus*). Am J Phys Anthropol 119:37–51.

Demes B, Larson SG, Stern JT Jr, et al (1994) The kinetics of primate quadrupedalism: "hind limb drive' reconsidered. J Hum Evol 26:353–374.

Dunbar DC, Badam GL (2000) Locomotion and posture during terminal branch feeding. Int J Primatol 21:649–669.

Fischer MS, Schilling N, Schmidt M, Haarhaus D, Witte H (2002) Basic limb kinematics of small therian mammals. J Exp Biol 205:1315–1338

Fleagle JG, Stern JT, Jungers WL, Susman RL, Vangor AK, Wells JP (1981) Climbing: a biomechanical link with brachiation and bipedalism. Symp Zool Soc Lond 48:359–375.

Franz TM, Demes B, Carlson KJ (2005) Gait mechanics of lemurid primates on terrestrial and arboreal substrates. J Hum Evol 48:199–217.

Gambaryan PP (1974) How Mammals Run. Keter Publishing House, Jerusalem.

Gasc JP (2001) Comparative aspects of gait, scaling and mechanics in mammals. Comp Biochem Physiol A Mol Integr Physiol 131:135–144.

Gebo DL (1989) Locomotor and phylogenetic considerations in anthropoid evolution. J Hum Evol 18:201–233.

Grand TI (1968) Functional anatomy of the upper limb. In: Malinow MR (ed), Biology of the Howler Monkey (*Alouatta caraya*). Karger, Basel, pp 104–125.

Grand TI (1984) Motion economy within the canopy: four strategies for mobility. In: Rodman PS, Cant JGH (eds), Adaptations for Foraging in Nonhuman Primates: Contributions to an Organismal Biology of Prosimians, Monkeys, and Apes. Columbia University Press, New York, pp 54–72.

Hildebrand M (1967) Symmetrical gaits of primates. Am J Phys Anthropol 26:119–130.

Hunt KD, Cant JGH, Gebo DL, Rose MD, Walker SE, Youlatos D (1996) Standardized descriptions of primate locomotor and postural modes. Primates 37:363–387.

Isler K, Thorpe SKS (2003) Gait parameters in vertical climbing of captive, rehabilitant and wild Sumatran orang-utans *(Pongo pygmaeus abelii)*. J Exp Biol 206:4081–4096.

Isler K, Gruter CC (2006) Arboreal locomotion in wild black-and-white snub-nosed monkeys (*Rhinopithecus bieti*). Folia Primatol 77:195–211.

Jenkins FA Jr (1971) Limb posture and locomotion in the Virginia opossum (*Didelphis marsupialis*) and in other non-cursorial mammals. J Zool Lond 165:303–315.

Jenkins FA Jr, Camazine SM (1977) Hip structure and locomotion in ambulatory and cursorial carnivores. J Zool Lond 181:351–370.

Jones AL (2008) The evolution of brachiation in ateline primates, ancestral character states and history. Am J Phys Anthropol 137:123–144.

Jouffroy FK, Renous S, Gasc JP (1983) Etude cinéradiographique des déplacements du membre antérieur du potto du Bosman (*Perodicticus potto* Muller, 1766) au cours de la marche quadrupède sur une branche horizontale. Ann Sci Nat Zool 5:75–87.

Kimura T, Okada M, Ishida H (1979) Kinesiological characteristics of primate walking: its significance in human walking. In: Morbeck ME, Preuschoft H, Gomberg N (eds), Environment, Behavior, and Morphology: Dynamic Interactions in Primates. Gustav Fischer, New York, pp 297–311.

Kimura T (1992) Hind limb dominance during primate high-speed locomotion. Primates 33:465–474.

Larson SG, Stern JT (1989) The use of supraspinatus in the quadrupedal locomotion of vervets (*Cercopithecus aethiops*): implications for interpretation of humeral morphology. Am J Phys Anthropol 79:369–377.

Larson SG, Stern JT (1992) Further evidence for the role of supraspinatus in quadrupedal monkeys. Am J Phys Anthropol 87:359–363.

Larson SG, Schmitt D, Lemelin P, Hamrick M (2000) Uniqueness of primate forelimb posture during quadrupedal locomotion. Am J Phys Anthropol 112:87–101.

Larson SG, Stern JT (2006) Maintenance of above-branch balance during primate arboreal quadrupedalism: coordinated use of forearm rotators and tail motion. Am J Phys Anthropol 129:71–81.

Lemelin P, Schmitt D, Cartmill M (2003) Footfall patterns and interlimb coordination in opossums (Family Didelphidae): evidence for the evolution of diagonal-sequence walking gaits in primates. J Zool 260:423–429.

McLearn D (1992) Locomotion, posture and feeding behavior of kinkajous, coatis and racoons. J Mammal 73:245–261.

Milton K (1980) The Foraging Strategy of Howler Monkeys. Columbia University Press, New York.

Napier JR, Walker AC (1967) Vertcial clinging and leaping: a newly recognized category of locomotor behavior of primates. Folia Primatol 6:204–219.

Nyakatura JA, Fischer MS, Schmidt M (2008) Gait parameter adjustments of cotton-top tamarins (*Saguinus oedipus*, Callitrichidae) to locomotion on inclined arboreal substrates. Am J Phys Anthropol 135:13–26.

Reynolds TR (1985a) Mechanics of increased support of weight by the hind limbs in primates. Am J Phys Anthropol 67:335–349.

Reynolds TR (1985b) Stresses on the limbs of quadrupedal primates. Am J Phys Anthropol 67:351–362.

Rollinson J, Martin RD (1981) Comparative aspects of primate locomotion, with special reference to arboreal cercopithecines. Symp Zool Soc Lond 48:377–427.

Rose MD (1973) Quadrupedalism in primates. Primates 14:337–357.

Rosenberger AL, Strier KB (1989) Adaptive radiation of the ateline primates. J Hum Evol 18:717–750.

Schilling N, Fischer MS (1999) Kinematic analysis of treadmill locomotion of tree shrews, *Tupaia glis* (Scandentia: Tupaiidae). Z Saugetierk 64:129–153.

Schmidt M, Fischer MS (2000) Cineradiographic study of forelimb movements during quadrupedal walking in the brown lemur (*Eulemur fulvus*, Primates: Lemuridae). Am J Phys Anthropol 111:245–262.

Schmidt M (2005) Quadrupedal locomotion in squirrel monkeys (Cebidae: Saimiri sciureus): a cineradiographic study of limb kinematics and related substrate reaction forces. Am J Phys Anthropol 128:359–370.

Schmitt D (1994) Forelimb mechanics as a function of substrate type during quadrupedalism in two anthropoid primates. J Hum Evol 26:441–458.

Schmitt D (1999) Compliant walking in primates. J Zool 248:149–160.

Schmitt D, Lemelin P (2002) Origins of primate locomotion: gait mechanics of the woolly opossum. Am J Phys Anthropol 118:231–238.

Schmitt D (2003) Evolutionary implications of the unusual walking mechanics of the common marmoset (*C. jacchus*). Am J Phys Anthropol 122:28–37.

Schmitt D, Lemelin P (2004) Locomotor mechanics of the slender loris (*Loris tardigradus*). J Hum Evol 47(1–2): 85–94.

Schön MA (1968) The muscular system of the red howling monkey. US Natl Mus Bull 273:1–185.

Schön Ybarra MA (1998) Arboreal quadrupedalism and forelimb articular anatomy of red howlers. Int J Primatol 19:599–613.

Shapiro LJ, Raichlen DA (2005) Lateral sequence walking in infant *Papio cynocephalus*: implications for the evolution of diagonal sequence walking in primates. Am J Phys Anthropol 126:205–213.

Stern JT, Wells JP, Vangor AK, Fleagle JG (1977) EMG of some muscles of the upper limb in *Ateles* and *Lagothrix*. Yrbk Phys Anthropol 20:498–507.

Stern JT, Wells JP, Jungers WL, Vangor AK (1980a) An EMG study of the serratus anterior in atelines and *Alouatta*: implications for hominoid evolution. Am J Phys Anthropol 52:323–334.

Stern JT, Wells JP, Jungers WL, Vangor AK, Fleagle JG (1980b) An EMG study of the pectoralis major in atelines and *Hylobates*, with special reference to the evolution of a pars clavicularis. Am J Phys Anthropol 52:13–25.

Stevens NJ (2006) Stability, limb coordination and substrate type: the ecorelevance of gait sequence pattern in primates. J Exp Zool 305A:953–963.

Strier KB (1992) Atelinae adaptations: behavioral strategies and ecological constraints. Am J Phys Anthropol 88:515–524.

Taylor ME (1970) Locomotion in some East African viverrids. J Mammal 51:42–51.

Vilensky JA (1989) Primate quadrupedalism: how and why does it differ from that of typical quadrupeds? Brain Behav Evol 34:357–364.

Vilensky JA, Larson SG (1989) Primate locomotion: utilization and control of symmetrical gaits. Ann Rev Anthropol 18:17–35.

Vilensky JA, Gankiewicz E (1990) Effects of speed on forelimb joint angular displacement patterns in vervet monkeys (*Cercopithecus aethiops*). Am J Phys Anthropol 83:203–210.

Vilensky JA, Moore AM, Libii JN (1994) Squirrel monkey locomotion on an inclined treadmill: implications for the evolution of gaits. J Hum Evol 26:375–386.

Wallace I, Demes B (2008) Symmetrical gaits of *Cebus apella*: implications for the functional significance of diagonal sequence gait in primates. J Hum Evol 54:783–794.

Youlatos D (1994) Maitrîse de l'espace et accès aux ressources chez le singe hurleur roux (*Alouatta seniculus*) de la Guyane Française: étude morpho-fonctionnelle. PhD dissertation, Muséum National d'Histoire Naturelle, Paris.

Youlatos D (1998) Seasonal variation in the positional behavior of red howling monkeys. Primates 39:447–455.

Youlatos D (1999) The schizodactylous grasp of the howling monkey. Z Morph Anthrop 82:187–198.

Youlatos D (2000) Functional anatomy of forelimb muscles in Guianan Atelines (Platyrrhini: Primates). Ann Sci Nat Zool Biol Anim 21:137–151.

Chapter 15
From Treadmill to Tropics: Calculating Ranging Cost in Chimpanzees

Herman Pontzer, David A. Raichlen, and Michael D. Sockol

Abstract Ecological studies of wild primates often use travel time or distance as a measure of ranging and foraging cost. Recent laboratory studies of walking and climbing cost in primates and other mammals can greatly improve the accuracy of these estimates. Here, we review recent studies of climbing cost in primates and walking costs in chimpanzees, and apply this work to calculating daily ranging cost for wild chimpanzees. Laboratory investigations of locomotor cost suggest that daily locomotor cost is best calculated by multiplying the distance walked or climbed by the cost per meter of each activity. We discuss the reliability of different variables in predicting the cost per meter traveled, including body mass, hip height, and step length. We calculate daily ranging cost using these variables for different populations and sex-age classes of wild chimpanzees, and propose a hierarchical approach to estimating walking cost in wild populations, preferring hip height to body mass.

Keywords Biomechanics • Energetics • Foraging ecology • Primate locomotion

Abbreviations

α	angle of take-off, relative to horizontal, for a leaping primate
BMR	basal metabolic rate (kJ/day)
COT	cost of transport (J m^{-1}); the metabolic cost of traveling a meter during terrestrial travel
COL	cost of locomotion (J s^{-1}); the rate of metabolic energy use during terrestrial locomotion
d	step length; the horizontal distance traveled during contact time for a limb

H. Pontzer (✉)
Department of Anthropology, Washington University, USA
e-mail: hpontzer@artsci.wustl.edu

K. D'Août and E.E. Vereecke (eds.), *Primate Locomotion: Linking Field and Laboratory Research*, Developments in Primatology: Progress and Prospects,
DOI 10.1007/978-1-4419-1420-0_15,

DEE	daily energy expenditure (kJ day^{-1}); the total amount of energy used each day
EMA	effective mechanical advantage; the ratio of the muscle's moment arm to the moment arm of the ground reaction force
E_{leap}	metabolic cost of a leap (J)
E_{climb}	metabolic cost of a climbing bout (J)
E_k	kinetic energy (J)
fore	percentage of vertical ground force born by the forelimbs during terrestrial locomotion
g	gravitational acceleration (9.81 m s^{-2})
GRF	ground reaction force (N)
hind	percentage of vertical ground force born by the hindlimbs during terrestrial locomotion
h	vertical distance traveled during a climbing bout (m)
M	body mass (kg)
L_{skel}	skeletal limb length (cm); the summed lengths of a limb's long bones, e.g., femur + tibia
L_E	effective limb length (cm); the length of a limb as a strut, typically measured as hip height while standing (Fig. 15.2).
l_{fasc}	fascicle length (m); the mean fascicle length for an extensor muscle group
v	forward velocity (m s^{-1}) during terrestrial locomotion
σ	force generated per cm^2 of active muscle
t_c	contact time (s); the period of foot-ground contact for one limb during one stride
V_{act}	volume of muscle (cm^3) activated to support bodyweight during terrestrial locomotion

Introduction

Ecological studies of primates are inherently multidimensional, involving comparisons among different activities, habitat qualities, and behaviors. It is useful, and often necessary, to convert these disparate variables to a common currency, the most popular being *time* and *energy*. With a common currency such as energy, measures of habitat quality (e.g., food patch size and distribution) can be compared directly to behavioral options (e.g., travel or continue feeding), and an accurate accounting of the economic decisions facing a primate can be made. In this chapter, we examine recent experimental work investigating the energy cost of walking and climbing in primates, and discuss the application of these studies to estimates of ranging cost in wild populations. We focus on chimpanzees, but our approach is applicable to studies of other primates and terrestrial mammals in general.

Why focus on energy? Ideally, studies seeking to understand the evolutionary costs and benefits of a behavioral or morphological strategy would measure lifetime reproductive success (e.g., Arnold 1983), the ultimate measure of evolutionary fitness. However, since fitness outcomes are difficult to measure for long-living species such as primates, measuring lifetime reproductive success is rarely possible. Studies attempting such ambitious measurements are typically limited to cross-sectional demographic data or reproductive success over a portion of the lifespan (cf. Altmann 1991). By comparison, measuring energy intake and expenditure is much easier, and can be done over a shorter timeframe. Further, there is good evidence that energy balance is causally linked to survival and reproduction. Experimental and comparative studies suggest that somatic maintenance and reproductive investment are directly correlated with energy balance (Wiersma and Verhulst 2005; Charnov and Ernest 2006).

Locomotor cost is of particular importance in the primate energy budget for two reasons. First, locomotion is energetically costly relative to other activities. Depending on travel speed, the rate of energy use during terrestrial travel is approximately two to ten times greater than basal metabolic rate (BMR) defined as the rate of energy used while sleeping (Fig. 15.1). In contrast, modeling studies (e.g., Leonard and Robertson 1997) estimate other typical primate activities, such as resting (1.25× BMR), feeding (1.38 × BMR), and play (2.35 × BMR) as relatively inexpensive energetically. Second, because all primates must travel to

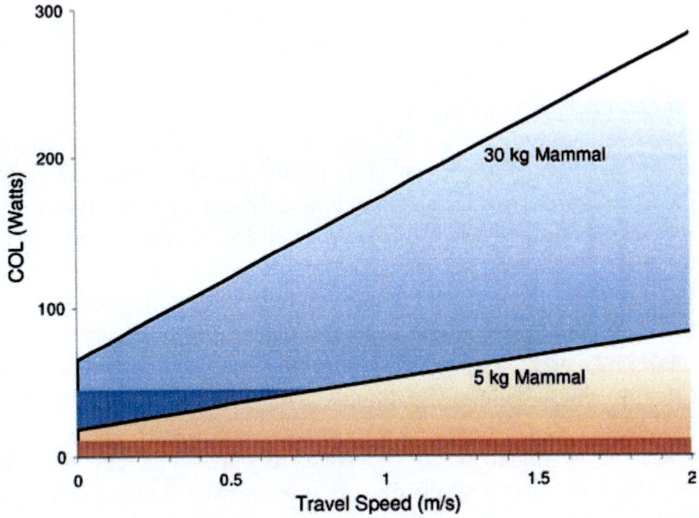

Fig. 15.1 COL (solid line) and BMR (hashed area) versus travel speed for a 5-kg and 30-kg mammal. COL is calculated using Eq. 15.1, the general endotherm regression in Taylor et al. (1982). BMR is calculated using Eq. 15.15, taken from Leonard and Robertson (1997). The y-intercept for COL is the postural cost of locomotion; the percentage by which the postural cost exceeds BMR is greater for smaller animals

acquire food, energy use during locomotion is fundamental to cost-benefit analyses all primates make in their foraging behavior. Locomotion is one of the few activities that provide a net energy return (i.e., energy expended is offset by energy gained from foraging), and determining the locomotor costs for individuals is therefore essential for understanding the economic choices faced by a foraging primate.

Despite the importance of locomotor cost in primate ecology, calculating the daily cost of walking, climbing, and leaping for wild primates has been hampered by a general lack of laboratory studies investigating locomotor cost in primates, and a lack of consensus on the best method for applying insights from the laboratory to animals in the wild. Following seminal work by Taylor and colleagues (1970, 1982) investigating the scaling of locomotor cost in terrestrial animals, most estimates of locomotor cost have used body mass to calculate travel cost for wild primates (e.g., Garland 1983; Altmann and Samuels 1992; Steudel 2000; Pontzer and Wrangham 2004). This approach has two critical shortcomings. First, body mass estimates used in these studies are typically species means, which introduce error and hamper comparisons of cost between different size- or age-classes of uncertain mass. Second, there is considerable variation in locomotor cost among mammals even when controlling for body mass (Taylor et al. 1982). This variation in cost is primarily due to differences in locomotor anatomy (Kram and Taylor 1990; Roberts et al. 1998a, b; Pontzer 2007), which are not considered in mass-based estimates of cost.

Estimating ranging costs for wild primates is further hampered by a lack of data on the cost of non-terrestrial locomotion, such as leaping and climbing. Prior to the recent work by Hanna and colleagues (Hanna et al. 2008), there were no studies of the metabolic cost of climbing in nonhuman primates, and consequently studies of ranging energetics in wild primates have used mechanical work and ballistics calculations (e.g. Warren and Crompton 1998) or human rock-climbing studies (e.g., Pontzer and Wrangham 2004) to estimate leaping and climbing costs. Similarly, a lack of data on daily energy expenditure, DEE, the total energy expended per day in the wild (also called *field metabolic rate*; see Nagy et al. 1999), for primates has led researchers to use human-based factorial models in estimating the size of the daily energy budget. Factorial models sum estimates of individual behavioral costs and each layer of estimation, including estimation of body mass, walking cost, climbing cost, and DEE, introduces additional error into the calculation of ranging cost and its proportion of the daily energy budget.

In this chapter, we review recent experimental work on the cost of walking and climbing in primates, as well as recent work investigating the determinants of locomotor cost for terrestrial animals. We then discuss the application of these insights to field studies, and assess the accuracy and reliability of different methods for estimating ranging cost in wild primates. Finally, we apply these methods to estimating ranging costs in wild chimpanzees.

Laboratory Studies of Energy Expenditure

Walking, Trotting, and Galloping

Allometric Studies

Empirical studies of locomotor cost have established that, per gram of body mass, large animals use less energy than small animals to walk or run at a given speed, or over a given distance. Taylor and colleagues (1982) established the nature of this relationship in a series of pioneering comparative studies measuring oxygen consumption during treadmill trials, culminating in a large allometric study of the cost of terrestrial locomotion in 65 species of birds and mammals. The allometric relationship for the rate of energy expenditure during locomotion, also called the *cost of locomotion* (COL) emerged as

$$\mathrm{COL} = 10.7M^{0.68}v + 6.07M^{0.70} \qquad \text{(Eq. 15.1)}$$

where COL is given in Watts, v is velocity of locomotion (meters/second), and body mass M is in kg. Taylor and colleagues also reported allometric relationships for separate phylogenetic groups. The relationship for primates was given as

$$\mathrm{COL}_{\text{primates}} = 10.5M^{0.70}v + 6.07M^{0.84} \qquad \text{(Eq. 15.2)}$$

As noted by Steudel-Numbers (2003), this "Primates" relationship is problematic, because it includes tree shrews and is derived largely from measurements of juveniles.

Both equations above give the whole-body cost of locomotion, rather than the mass-specific cost often reported, since ecological studies typically calculate whole-body costs. In both equations, the whole-body cost of terrestrial locomotion increases linearly with speed. The slope is a linear function of speed, and increases with body mass$^{0.68}$. The intercept for this function also increases with body size, as mass$^{0.70}$ (Fig. 15.1). In many studies of ranging cost, the intercept cost, also called the *postural cost* of locomotion (Taylor 1977), is disregarded, and only the slope-term is used (e.g., Garland 1983). This greatly simplifies the estimation of ranging cost, because once the intercept term is eliminated, dividing each side of the equation by travel speed gives the cost per distance, also called the *cost of transport* (COT; J m^{-1}):

$$\mathrm{COT} = 10.7M^{0.68} \qquad \text{(Eq. 15.3)}$$

Or, for the Primate equation,

$$\mathrm{COT}_{\text{primates}} = 10.5M^{0.70} \qquad \text{(Eq. 15.4)}$$

COT values for the primates included in Taylor et al. (1982) are not significantly different from those of other quadrupedal mammals of similar size. Calculating the cost of travel this way implicitly assumes the postural cost is equivalent to the energy used when the individual was not moving, and can therefore be disregarded. In fact, the postural cost is ca. 35–80% greater than BMR (Fig. 15.1), whereas the cost of resting while awake is estimated to be 25% greater than BMR (Leonard and Robertson 1997). Thus, eliminating the postural cost will lead to a small underestimation of total locomotor cost. However, this underestimation is typically viewed as preferable to the error introduced in estimating travel speed, which is needed to calculate ranging cost using Eqs. 15.1 or 15.2. This may be especially important because travel speeds likely vary *during* foraging bouts, further complicating ecologically relevant estimates of speed-dependent COL.

Equations 15.3 and 15.4 indicate that the energy cost to travel a given distance is a function of animal size, but not travel speed. In fact, a U-shaped relationship between speed and COT has been reported during walking in some species, including humans (Margaria et al. 1968), elephants (Langman et al. 1995), caribou (Luick and White 1986), and horses (Hoyt and Taylor 1981), and during trotting and galloping in horses (Hoyt and Taylor 1981) and caribou (Luick and White 1986). Nonetheless, locomotor cost data reported for the majority of terrestrial animals indicate a linear relationship between COL and speed and an independence of COT and speed (e.g., Taylor et al. 1982). For species with a U-shaped relationship between COT and speed within walking or running gaits, the change in COT with speed is small, particularly over the limited range of speeds habitually chosen by free-ranging individuals (e.g., Hoyt and Taylor 1981; Pennycuick 1975). The independence (or near-independence) of COT and speed are useful for ecological studies, as daily travel cost can be calculated by multiplying COT by the distance traveled without requiring data on travel speed.

Determinants of COL and COT

The dependence of COT on animal size has been the focus of investigation for several decades. Intriguingly, the mechanical work done to move the body and swing the limbs is a poor predictor of COT (Cavagna and Kaneko 1977; Heglund et al. 1982; Willems et al. 1995; Minetti et al. 1999). Though the mechanical work done on the body during locomotion must derive from metabolic energy, the apparent efficiency with which chemical energy is converted to mechanical energy changes with travel speed and between species (Cavagna and Kaneko 1977; Heglund et al. 1982; Willems et al. 1995; Minetti et al. 1999). In addition, muscles use energy even when they produce no work, such as during isometric contractions (Kram and Taylor 1990; Taylor 1994). As a result, the scaling of COT with body mass is not explained by differences in mechanical work performed (Heglund et al. 1982).

To account for the scaling of COT, Kram and Taylor (1990) proposed that locomotor cost derives primarily from the muscle force generated to support body weight. While walking or running at a steady speed over level ground, the limb muscles must generate force each step to counteract the force of gravity. The *average* vertical ground force generated by the muscles must equal body weight, but whereas the force of gravity is constant, the magnitude of ground force generation fluctuates. For example, during running or galloping, no ground force is generated during the aerial phase, (i.e., when all limbs are off the ground), while ground force during stance phase (i.e., when one or more limb is in contact with the ground), may exceed two to three times bodyweight. Kram and Taylor (1990) noted that as stance duration decreases, the magnitude of ground force generation must increase so that average ground force, calculated over an entire stride cycle, equals bodyweight: shorter stance duration, which they termed *contact time*, required an animal to generate larger ground forces over a shorter amount of time. Their key insight (Kram and Taylor 1990) was that COL is directly linked to this rate of force generation, such that

$$\mathrm{COL} = cMgt_c^{-1} \qquad \text{(Eq. 15.5)}$$

where c is a constant relating the rate of force production to the rate of energy use (COL), Mg is bodyweight (g is gravitational acceleration: 9.8 ms^{-2}), and t_c is contact time, the duration of foot-ground contact per step for a given foot. As with the Eqs. 15.3 and 15.4, the postural cost is disregarded. By dividing both sides by travel speed, Equation 15.5 gives the cost per distance as a function of step length, d, the horizontal distance traveled during contact time:

$$\mathrm{COT} = cMgd^{-1} \qquad \text{(Eq. 15.6)}$$

Notably, contact time and step length outperform body mass and work in predicting COL and COT (Pontzer et al. 2009). This force-production approach also correctly predicts the scaling of COT and the independence of COT and speed noted in the preceding text. Since larger animals will generally have longer limbs and take longer steps (Kram and Taylor 1990; Hoyt et al. 2000), d increases with body size such that COT is lower, per gram of body mass, for larger animals. Further, because terrestrial animals generally increase speed by increasing step frequency and stride length rather than step length (Heglund and Taylor 1988), d changes little with speed (Kram and Taylor 1990), and thus COT remains independent of travel speed.

Step length d is generally a function of animal size, particularly limb length: animals with longer limbs take longer steps (Kram and Taylor 1990; Hoyt et al. 2000). The critical measure of limb length is *effective* limb length, L_E, the length of the limb as a strut, typically measured as the distance from the hip joint to the ground while standing (Pontzer 2007; Fig. 15.2). In an analysis of 28 species including arthropods, reptiles, birds, and mammals, Pontzer (2007) found that L_E explained 98% of the variance in mass-specific COT, and that after controlling for L_E, body mass had no effect on mass-specific COT. That is, using L_E to predict mass-specific COT was more accurate than using body mass, e.g., Eq. 15.3.

Fig. 15.2 Effective limb length, L_E. This limb length, also termed hip height, is the distance from the hip joint (or greater trochantor) to the ground while standing

Multiplying both sides of the mass-specific regression equation given in Pontzer (2007) by body mass *M* gives the whole-body cost of transport as

$$\mathrm{COT} = 90.284\, M\, L_{\mathrm{E}}^{-0.77} \qquad \text{(Eq. 15.7)}$$

Notably, *skeletal* limb length, L_{skel}, the summed lengths of a limb's long bones, is not a reliable substitute for L_{E}. Most animals habitually use crouched postures, and small animals generally crouch more than large animals (Biewener 1989). Therefore, the lengths of the long bones do not accurately reflect L_{E}, and skeletal limb length is no more reliable than body mass in predicting COT (Steudel and Beattie 1995; Pontzer 2007).

A simplifying assumption of the force-production approach (including the use of L_{E}) to predict COT is that the ratio of ground force produced to metabolic energy consumed (*c* in Eqs. 15.5 and 15.6) is constant across species. This assumption is probably justified in most large-scale comparisons of cost (e.g., Kram and Taylor 1990; Pontzer 2007), because the volume of muscle activated to produce a Newton of ground force (V_{act}) is largely invariant across terrestrial animals. V_{act} is the product of muscle fascicle length (l_{fasc}) and muscle force, and therefore increases linearly with l_{fasc}, but decreases linearly with the muscle's effective mechanical advantage (EMA), the ratio of the muscle's moment arm to the moment arm of the ground reaction force (GRF; Biewener 1989).[1] Because l_{fasc}

[1] As EMA increases, the moment arm of the muscle is relatively large compared to the moment arm of the GsRF and thus, the muscle produces less force to balance the opposing moment from the GRF.

and EMA both scale with body mass$^{0.26}$ (Biewener 1990), the ratio of l_{fasc} to EMA generally remains constant as body size increases, supporting the assumption that the ratio of force to energy cost is independent of size (Kram and Taylor 1990; Taylor 1994).

More recent work has examined energy cost in species which deviate from the typical scaling relationships for l_{fasc} and EMA. Roberts and colleagues (1998a, b) first examined this issue in a comparison of running cost in ground birds and mammals, and demonstrated that birds use more energy to produce ground force, and therefore have higher COT for a given step length, because of their longer l_{fasc}. We have extended this approach to primates (Sockol et al. 2007; Pontzer et al. 2009), and have shown that the difference in walking cost between humans and chimpanzees, and between bipedal and quadrupedal walking in chimpanzees, can be predicted by considering differences in l_{fasc} and EMA along with differences in step length. For a given body mass or a given speed, primates in general differ from other terrestrial mammals in having longer l_{fasc} (Alexander 1991), lower EMA (Schmitt 1999; Polk 2004; Pontzer et al. 2009), and in using longer step lengths (Schmitt 1999). Our numerical model for predicting cost while accounting explicitly for these variables takes the form:

$$\mathrm{COT} = kMgd^{-1}\sigma^{-1}\left[\mathrm{fore}\left(\frac{l_{fasc,a}}{EMA_a}+\frac{l_{fasc,b}}{EMA_b}+\frac{l_{fasc,c}}{EMA_c}\right)+\mathrm{hind}\left(\frac{l_{fasc,d}}{EMA_d}+\frac{l_{fasc,e}}{EMA_e}+\frac{l_{fasc,f}}{EMA_f}\right)\right] \quad \text{(Eq. 15.8)}$$

where k is a constant relating the volume of activated muscle to the amount of energy consumed (J/cm^3), g is gravity, M is body mass, σ is the force generated per cross-sectional area of active muscle (typically 20 N/cm^2; Biewener et al. 2004), d is step length, *fore* is the proportion of bodyweight borne by the forelimbs and *hind* is the proportion of bodyweight borne by the hind limbs, and l_{fasc} and EMA are the mean fascicle lengths and effective mechanical advantages for each of the three forelimb joints (a: shoulder, b: elbow, c: wrist) and three hind limb joints (d: hip, e: knee, f: ankle). By explicitly incorporating l_{fasc} and EMA, this model (Eq. 15.8) is able to account for the long muscle fibers and crouched postures typical of primates. This approach predicted over 90% of the variation in COT for a diverse group terrestrial species, including ground birds, dogs, walking and running humans, and bipedal and quadrupedal chimpanzees, outperforming other common predictors of COT (Pontzer et al. 2009).

Climbing and Leaping

Primates are arboreal, and walking, trotting, and galloping make up only a portion of their daily locomotor repertoire. For most primates, the cost of climbing and leaping must be considered in estimates of daily locomotor energy cost. Until recently, a lack of reliable empirical data on the cost of climbing or leaping has led researchers to estimate these costs using data from human climbers (e.g., Pontzer and Wrangham 2004) or standard equations for mechanical work (e.g., Warren and Crompton 1998).

Hanna and colleagues (2008) recently reported climbing costs for a taxonomically diverse sample of five primates ranging in body mass from 0.17 to 1.40 kg, collected while the primates climbed on a rope-mill enclosed in a metabolic chamber. Climbing on a rope-mill differs from climbing on static structures in that the potential energy of the climber remains constant. However, the work done while climbing on the rope-mill to maintain position while opposing gravity is equal to the potential energy that would be gained climbing. Thus, while rope-mill climbing may differ somewhat from climbing on static structures, this experimental design provides a useful means of measuring climbing cost.

In contrast to studies of terrestrial locomotor cost, the metabolic cost of climbing was reliably predicted by the mechanical work performed in lifting the body's mass against the force of gravity (Hanna et al. 2008). Moreover, the efficiency with which this mechanical work was performed was nearly independent of body mass, with 0.17 kg lorises using nearly the same amount of energy per kilogram to climb a meter as 62.2 kg human rock climbers (Hanna et al. 2008). The allometric equation for climbing efficiency reported by Hanna and colleagues is:

$$\%\text{Efficiency} = 9.50M^{0.109} \qquad \text{(Eq. 15.9)}$$

This near independence of efficiency and body size, and the apparent independence of climbing cost and locomotor anatomy, is convenient for ecologists estimating climbing costs. Since %Efficiency (Eq. 15.9) is the ratio of mechanical work performed to the amount of metabolic energy consumed, the metabolic energy spent climbing, E_{climb} (J) can be calculated by dividing the mechanical work done by the estimated climbing efficiency:

$$E_{climb} = \frac{Mgh}{\%\text{Efficiency}} \qquad \text{(Eq. 15.10)}$$

where M is body mass (kg), g is gravity, h is the height climbed (meters), and %Efficiency is calculated by Eq. 15.9. Note that %Efficiency must be entered as a decimal, e.g., 10 % as 0.10.

Equation 15.10 may also be useful for estimating the cost of leaping and other ballistic movements. Because these activities are of short duration and fueled largely by anaerobic metabolism, their metabolic cost cannot be measured using standard oxygen-consumption techniques. However, in the absence of empirical data on leaping cost, Eq. 15.10 provides an estimate for cost which assumes that the mechanical work performed by the leaping primate is similar to the efficiency seen in climbing, ca. 10%. To use this approach, the work term in Eq. 15.10 (Mgh) must be replaced with the equivalent term for leaping. The mechanical work performed in a jump is equivalent to the kinetic energy (E_K) of the individual at take off.

If take-off velocity, v, is known, E_K can be calculated directly as $0.5Mv^2$. Alternatively, E_K can be calculated from the distance jumped and the take-off angle. In the simplest case, where a primate takes-off and lands at the same height and travels a horizontal distance x, the potential energy term can be ignored (since the change in height is zero) and the kinetic energy at take-off can be calculated following Warren and Crompton (1998) as:

$$E_K = \frac{Mgx}{2\sin 2\alpha} \qquad \text{(Eq. 15.11)}$$

where α is the take-off angle measured relative to the horizontal and x is the horizontal distance traveled. The metabolic energy consumed during a leap, E_{leap}, could then be calculated as

$$E_{leap} = \frac{Mgx}{\%\text{Efficiency}\left(2\sin 2\alpha\right)} \qquad \text{(Eq. 15.12)}$$

Warren and Crompton (1998) discuss alternative forms of Eq. 15.11 that account for gain or loss of height from take-off to landing, as well as methods for estimating α from jump height.

Daily Energy Expenditure (DEE)

Ranging costs are often placed in the context of total daily energy expenditure in order to compare relative ranging costs among different species or to determine the proportion of daily energy expenditure (DEE; kJ/day) spent on locomotion (e.g., Garland 1983; Altmann 1987; Warren and Crompton 1998; Pontzer and Wrangham 2004). For primates, DEE is typically estimated using a factorial approach, in which each behavior in the daily activity budget is assigned a metabolic rate, and the daily cost of each activity is summed to give DEE (Coehlo et al. 1979; Leonard and Robertson 1997; Key and Ross 1999, Steudel-Numbers 2006). Leonard and Robertson (1997) used this approach to model DEE for 18 primate species, including humans, and developed the following allometric regression:

$$\text{DEE}_{\text{factorial}} = 359.82M^{0.792} \qquad \text{(Eq. 15.13)}$$

This factorial approach has the advantage of being easily applied to the large number of species with published activity budgets, but it may underestimate true DEE. Direct measures of DEE, using the doubly labeled water method (Nagy et al. 1999), produce energy budgets substantially greater than those estimated by factorial modeling. In *Alouatta palliata*, measured DEE exceeds modeled DEE by 60%. Similarly, the allometric regression for DEE measured in wild populations of 79 mammal species (Nagy et al. 1999) is:

$$\text{DEE}_{\text{DLW}} = 873.6M^{0.772} \qquad \text{(Eq. 15.14)}$$

As is evident in comparing Eqs. 15.13 and 15.14, direct measures of DEE in mammals exceed factorial-based estimates by approximately twofold. Clearly, more direct measures of DEE in primates are needed to determine whether primates do in fact use smaller energy budgets than other mammals of similar body mass, or if the factorial method consistently underestimates DEE. In our case-study of ranging costs in chimpanzees, we examine the effect of these different equations (Eqs. 15.13 and 15.14) for estimating DEE and the proportion of the energy budget spent on locomotion.

Case Study: Kanyawara Chimpanzees

The preceding equations above provide several methods for calculating ranging costs and DEE for wild primates. In the following discussion, we examine the accuracy and reliability of each of these methods, and the ease with which each method might be employed in the field. We focus on chimpanzees living in the Kanyawara community in Kibale National Park, Uganda. Daily climbing and travel distances are available for this population (Pontzer and Wrangham 2004), as are long-bone lengths and skeletally based body mass estimates (Carter et al. 2008), making each of the approaches described above feasible. We compare estimated COT for these chimpanzees to published values for laboratory studies of chimpanzee walking cost (Taylor and Rowntree 1973; Sockol et al. 2007) and examine the proportion of the daily energy budget spent on walking and climbing under the different approaches proposed.

Daily Walking Cost

We estimated daily walking cost (kJ/day) by multiplying the distance walked each day (km/day) by COT. COT values for adult chimpanzees were taken from Sockol et al. (2007), and COT for juveniles was taken from Taylor and Rowntree (1973), the only empirical data available for chimpanzee locomotor cost. Because COT is dependent on body size, calculating COT for Kanyawara chimpanzees using data from captive chimpanzees assumes similarity in body size among these populations. This assumption is supported by the similarity in hip height, L_E, between Kanyawara adults (Carter et al. 2008) and the sample in Sockol et al. (2007); as noted earlier, hip height is a better predictor of mass-specific COT than body mass (Pontzer 2007). Hip height estimates are not available for juveniles at Kanyawara, and so body mass for the juveniles was matched to the juvenile chimpanzees in Taylor and Rowntree (1973). Body mass estimates and hip heights for all age-sex classes are given in Table 15.1.

Table 15.1 Estimated COT, climbing cost, and DEE for Kanyawara chimpanzees

	Mass	L_{skel}[b]	L_E[d]	Day range[e]	Climbing[e]	Walking cost[f]			Climbing cost		DEE (kJ/day)	
Class	(kg)[a]	(cm)	(cm)	km/day	m/day	J kg^{-1}m^{-1}	J m^{-1}	kJ/day	J/m (Eq. 15.10)	kJ/day	Eq. 15.13	Eq. 15.14
Males	43.0	54.5	53.7	2.4	104	3.8	164.2	394.1	2944	306.2	7076	15 935
Females	36.9	54.5	53.7	2.0	117	3.8	140.9	281.8	2569	300.5	6269	14 160
Mothers	36.9 + infant	54.5	53.7	1.9	96	3.8	160.0	304.0	2917	280.0	6269	14 160
Juveniles	17.5	41.6[c]	41.0	2.0	134	5.0	87.9	175.9	1321	177.1	3472	7 960

[a] Adult mass from Carter et al. (2008). Juvenile mass adopted from Taylor and Rowntree (1973) sample. Clinging infant mass: 5 kg.

[b] Mean (femur + tibia) length for combined sex sample from Carter et al. (2008).

[c] Estimated from the ratio of L_{skel} to body mass in adult males and females, assuming geometric similarity.

[d] Estimated from the ratio of (thigh + shank) length to L_E for chimpanzees in Sockol et al. (2007).

[e] From Pontzer and Wrangham (2004).

[f] Mass-specific cost from Sockol et al. (2007) for adults and Taylor and Rowntree (1973) for juveniles.

We also estimated COT for adult males, adult females, mothers carrying infants, and juveniles using Eqs. 15.3, 15.4, and 15.7, to compare these estimates to empirical measures of walking cost. Estimates of body mass (Carter et al. 2008) were used to estimate COT using Eqs. 15.3 and 15.4. Hip height, L_E, for Eq. 15.7 was calculated by summing average femur and tibia lengths for adult chimpanzees from Kibale National Park (Carter et al. 2008), using the ratio of L_{skel} to L_E for chimpanzees in Sockol et al. (2007); L_E for juveniles was estimated assuming geometric similarity.

For mothers carrying infants, maternal body mass, 36.9 kg, was used to calculate COT; this estimate of COT was then multiplied by 1.14, to account for the 14% increase in gross weight incurred by carrying a 5-kg infant. Note that for Eq. 15.7, COT can simply be calculated by substituting combined (mother + infant) mass for M. This approach assumes that the additional cost of carrying is simply proportional to the increase in total mass, as has been shown for several species in load-carrying experiments (Taylor et al. 1982; Marsh et al. 2006).

Equations 15.6 and 15.8 may be more accurate predictors of COT than the anatomical measures used here for Kanyawara chimpanzees (Pontzer et al. 2009). However, Eqs. 15.6 and 15.8 require data on step length and posture that are not available for wild chimpanzees. We examine the practicality and utility of obtaining these measures in the wild, and compare the advantages of each approach (Eqs. 15.3, 15.4, 15.6, 15.7, and 15.8) for estimates of ranging cost in wild populations.

Daily Climbing Cost

As with walking, we estimated the daily cost of climbing by multiplying the distance climbed per day (Table 15.1) by the estimated cost per meter (Eq. 15.10). Daily climbing distance for each age-sex class was taken from Pontzer and Wrangham (2004). For mothers, the mechanical work performed (Mgh) was calculated using the combined (mother + infant) mass of 41.9 kg.

Daily Energy Expenditure

We calculated daily energy expenditure (DEE, kJ/day) using both Eqs. 15.13 and 15.14. Equation 15.13 adopts the trendline reported by Leonard and Robertson (1997), which used the factorial method for calculating DEE. Species-specific estimates of chimpanzee DEE in Leonard and Robertson (1997) fall within 10% of the DEE trendline using the factorial method, suggesting that their primate regression (Eq. 15.13) may be reliably applied to the Kanyawara population. Eq. 15.14 adopts the trendline reported by Nagy et al. (1999), based on doubly labeled water measurements of DEE in wild populations of 79 mammals.

Results and Discussion

Daily Ranging Costs and DEE for Kanyawara Chimpanzees

The daily cost of ranging for Kanyawara chimpanzees is low, relative to DEE, across all age-sex classes. When COT from similarly sized, i.e., similar hip-height adults, similar body mass juveniles, chimpanzees is used to calculate daily ranging cost (Sockol et al. 2007; Taylor and Rowntree 1973), the daily cost of walking for adult males is 394.1 kJ/day, equal to 94.2 kcal/day using the standard ratio of 0.239 calories/joule. Estimated daily walking cost was 281.8 kJ/day (67.4 kcal/day) for adult females, 304.0 kJ/day (72.7 kcal/day) for mothers carrying infants, and 175.9 kJ/day (42.0 kcal/day) for juveniles. Daily travel distances and walking costs are given in Table 15.1.

Daily climbing costs were also low relative to DEE. The %Efficiency during climbing (Eq. 15.9) was 14.3% for adult males, 14.1% for adult females and mothers, and 13.0% for juveniles. These efficiencies, when combined with the mechanical work done each day during climbing for these sex-age classes (Eq. 15.10), produced estimated daily climbing costs of 306.2 kJ/day (73.2 kcal/day) for adult males, 300.5 kJ/day (71.8 kcal/day) for adult females, 280.0 kJ/day (66.9 kCal/day) for mothers with clinging infants, and 177.1 kJ/day (42.3 kcal/day) for juveniles. Daily climbing distance and estimated cost for each sex-age class is given in Table 15.1.

Estimates of DEE ranged widely, depending on the method used for estimation. Using Eq. 15.13, the trendline derived from the factorial model of DEE presented in Leonard and Robertson (1997), produced DEE estimates of 7076 kJ/day (1691 kcal/day) for adult males, 6,269 kJ/day (1498 kcal/day) for adult females and mothers, and 3471 kJ/day (829 kcal/day) for juveniles. Using Eq. 15.14, the trendline for field metabolic rate in 79 mammal species (Nagy et al. 1999), produced DEE estimates of 15,935 kJ/day (3808 kcal/day) for adult males, 14,160 kJ/day (3384 kcal/day) for adult females and mothers, and 7960 kJ/day (1902 kcal/day) for juveniles. If mothers are assumed to be nursing, their DEE estimates would be higher. Key and Ross (1999) estimate that DEE for lactating mothers is ca. 40% greater than adult females without nursing infants. Such an increase would produce DEE estimates of 8776 kJ/day (2098 kcal/day) and 19,823 kJ/day (4737 kcal/day) using Eqs. 15.13 and 15.14, respectively.

The estimated proportion of DEE spent on ranging, also called the *ecological cost of transport* (Garland 1983), in Kanyawara chimpanzees is similar to that reported for other terrestrial mammals (Garland 1983). Together, walking and climbing account for ca. 4–10% of total daily energy use depending on the method used to estimate DEE. When Eq. 15.13 is used to estimate DEE, walking accounts for 4.9–5.6% of the daily energy budget, while climbing accounts for 4.3–5.1%. The percentage of DEE spent on travel drops considerably for both walking (2.0–2.5%) and climbing (1.9–2.2%) when Eq. 15.14 is used to estimate DEE.

These estimates for the proportion of DEE spent on walking and climbing differ somewhat from previous estimates for this population (Pontzer and Wrangham 2004). Daily walking cost estimates here disregard the postural cost of locomotion, whereas walking cost in Pontzer and Wrangham (2004) was estimated using Eqs. 15.1 and

15.2, which include the postural cost. As a result, walking cost estimates here are lower. In contrast, climbing costs estimated using Eq. 15.10 are nearly twice those reported in Pontzer and Wrangham (2004). This difference is a result of the improved, empirically based methods for estimating cost that were not previously available, as well as improved body mass estimates. As a result, cost estimates here for walking and climbing, which we consider more reliable than those reported by Pontzer and Wrangham (2004), are more similar than previously suggested.

Estimating Ranging Costs for Wild Primates

Laboratory measurements of walking cost are available for both adult and juvenile chimpanzees (Taylor and Rowntree 1973; Sockol et al. 2007), but this is rare among primate species. Direct measurements of COT are available for only eight species of primate; many of these measurements are from juveniles; and only chimpanzees have accompanying kinematic, kinetic, and hip height measurements (Steudel-Numbers 2003; Sockol et al. 2007). In the absence of direct measures of COT, what is the most reliable, and most accurate, predictor of COT?

We estimated COT using the general endotherm regression from Taylor et al. 1982 (Eq. 15.3), the Primates regression from Taylor et al. (1982), and the regression for COT and L_E reported in Pontzer (2007; Eq. 15.7). Results are shown in Table 15.2. Mass-based estimates were consistently low, with the general endotherm regression (Eq. 15.3) producing COT estimates 11.7–15.9% below laboratory measures of COT, and the Primates regression (Eq. 15.4) producing COT estimates 7.8–13.1% below laboratory measures. Hip height proved more accurate, with estimates from Eq. 15.7 ranging from 3.4–11.3% above laboratory measures of COT. Overestimation is likely preferable in this instance, since overground travel in the wild is likely less efficient than treadmill walking. These results, as well as recent work demonstrating that hip height is generally a better predictor of COT than body mass (Pontzer 2007; Pontzer et al. 2009) suggest field researchers should employ hip height (Eq. 15.7) in estimates of daily walking cost when possible. Digital image-based approaches (e.g., Caillaud et al. 2008) may even be used to measure hip height for different individuals or sex-age classes in the field.

The methods for calculating ranging cost advocated here disregard the postural cost of locomotion. This simplifies the calculation of ranging cost, avoids the error induced in estimating travel speed, and essentially subtracts BMR from gross estimates of locomotor cost, e.g., Eqs. 15.1 and 15.2. However, as discussed in the preceding text and by others (Altmann 1987; Steudel 2000), disregarding the postural cost will also underestimate true COT, because the postural cost is somewhat greater than BMR (Fig. 15.1). To determine the magnitude of this under-estimation, we calculated the difference between the postural cost term in Eq. 15.2 ($6.03M^{0.70}$) and estimated BMR for each age-sex class, using the BMR regression given in Leonard and Robertson (1997):

$$\text{BMR} = 3.462M^{0.762} \qquad \text{(Eq. 15.15)}$$

Table 15.2 Predictors of COT for Kanyawara chimpanzees

	Lab. based COT		COT Predictor					
	Mass		Body mass: Eq. 15.3		Body mass: Eq. 15.4		Hip height: Eq. 15.7	
Class	(kg)	(J/m)	J/m	Difference (%)	J/m	Difference (%)	J/m	Difference (%)
Males	43.0	164.2	138.1	–15.9	146.1	–13.1	180.8	11.3
Females	36.9	140.9	124.4	–11.7	131.3	–7.8	155.1	10.8
Mothers	36.9	160.0	141.3	–11.7	149.0	–7.8	176.2	10.8
Juveniles	17.5	87.9	74.9	–14.8	77.9	–11.5	90.6	3.4

where BMR is in watts. We then multiplied this difference—the magnitude by which the postural cost exceeds BMR—by the estimated time spent walking. We estimated the time spent walking by dividing daily travel distance by the average walking speed for chimpanzees (0.8 m/s, average across all age-sex classes; see Pontzer and Wrangham 2004). This additional daily cost was 72.6 kJ/day (17.4 kcal/day) for adult males, 55.5 kJ/day (13.3 kcal/day) for adult females, 52.7 kJ/day (12.6 kcal/day) for mothers, and 36.8 kJ/day (8.8 kcal/day) for juveniles. These results suggest that the underestimation of daily walking cost incurred by disregarding the postural cost is ca. 15–21% of daily walking cost. This underestimation will generally be larger for smaller species because of the differential scaling of postural cost and BMR. Estimates of time spent walking, or, if possible, of mean travel speed, may allow field researchers to account for this additional cost by subtracting BMR from the postural cost as shown here.

Recent work on the cost of climbing in primates (Hanna et al. 2008) suggests that climbing costs can be reliably estimated for primates in the field (Eq. 15.10), provided the distance climbed per day is known. Methods for measuring the distance climbed each day are discussed elsewhere (Pontzer and Wrangham 2004; see also Blanchard et al., Chapter 10). Notably, the similarity of climbing efficiency across a wide range of body size and locomotor anatomy (Hanna et al. 2008) suggests that anatomical differences do not need to be considered when estimating climbing costs.

The nearly twofold difference in DEE generated by Eqs. 15.13 and 15.14 underscores how little is known about daily energy expenditure in wild primates. Factorial methods (e.g., Leonard and Robertson 1997) produce DEE estimates approximately 50% lower than direct measures of energy expenditure in free-ranging mammals. Either a large source of energy expenditure is not captured by factorial models, or primates have, as a group, significantly lower DEE than is seen in other mammals. The difference in factorial- and doubly labeled water-based measures of DEE in howlers, discussed earlier, suggests that factorial models do in fact underestimate cost. However, there is substantial variation in DEE among wild populations (Nagy et al. 1999), and it is possible that many primates do in fact have low DEE. Direct measures of DEE in primates, using doubly labeled water or similar methods, are needed to investigate species differences in DEE and to appreciate differences in the proportion of DEE spent on ranging.

Integrating Laboratory and Field Investigation of Primate Ranging Cost

Though locomotion accounts for a small portion of DEE (Table 15.1), climbing and walking are nonetheless among the costliest activities for wild primates (Leonard and Robertson 1997). Understanding the relative costs of walking, trotting, climbing, leaping, and other activities can shed significant light on the evolutionary pressures shaping locomotor anatomy and ranging strategies. For example, while the improved estimates of daily walking and climbing cost differ somewhat from those reported previously (Pontzer and Wrangham 2004), results here confirm that chimpanzee anatomy results in a lowered walking efficiency that is not offset by improved climbing efficiency. Short hind limbs, as well as longer l_{fasc} and relatively poor EMA, result in a high COT for chimpanzees relative to other, more terrestrially adapted mammals of similar body mass (Taylor et al. 1982; Pontzer 2007; Sockol et al. 2007). These costly anatomical traits do not appear to affect climbing efficiency; climbing efficiency is essentially constant across primates, regardless of anatomical differences (Hanna et al. 2008), including humans. Thus, as suggested by Pontzer and Wrangham (2004), the locomotor anatomy of chimpanzees appears to be energetically inefficient, and may instead be selected to minimize the risk of falling from the canopy. Integrating laboratory and field studies of primates is necessary to shape and test hypotheses regarding anatomical form and ecological function.

In this study, we examine ranging cost for chimpanzees in Kibale National Park, Uganda, but the methods discussed are applicable for all primates. Direct measures of locomotor cost for size-matched individuals of a given species are ideal for estimating ranging costs in wild primates, but such data are rarely available. To estimate the daily cost of terrestrial travel, results here and recent work on the determinants of locomotor cost in terrestrial animals suggest that hip height (Eq. 15.10) is the most accurate predictor of COT. Failing reliable measures of hip height, body mass can be used (Eqs. 15.3 and 15.4), although errors will likely increase (see Pontzer 2007). Climbing cost can be estimated reliably (Hanna et al. 2008) using the mechanical work performed in lifting the body's mass against gravity (Eq. 15.10); this approach may be more broadly applied to activities such as leaping, but this has not been validated with direct measures of cost. DEE can be estimated using either factorial modeling (Eq. 15.13) or direct measures of field metabolic rate (Eq. 15.14). For all primates, understanding of the evolutionary pressures shaping morphology and behavior will be improved by direct measures of daily ranging demands and locomotor performance.

Acknowledgments We thank Evie Vereecke and Kristiaan D'Août for inviting us to join this volume.

References

Alexander RMcN (1991) Elastic mechanisms in primate locomotion. Z Morphol Anthropol 78:315–320.

Altmann SA (1987) The impact of locomotor energetics on mammalian foraging. J Zool, Lond 211:215–225.

Altmann SA (1991) Diets of yearling female primates (*Papio cynocephalus*) predict lifetime fitness. Proc Natl Acad Sci USA 88:420–423.

Altmann J, Samuels A (1992) Costs of maternal care: infant-carrying in baboons. Behav Ecol Sociobiol 29:391–398.

Arnold SJ (1983) Morphology, performance and fitness. Am Zool 23:347–361.

Biewener AA (1989) Scaling body support in mammals: limb posture and muscle mechanics. Science 245:45–48.
Biewener AA (1990) Biomechanics of mammalian terrestrial locomotion. Science 250:1097–1103.
Biewener AA, Farley CT, Roberts TJ, Temaner M (2004) Muscle mechanical advantage of human walking and running: implications of energy cost J Appl Physiol 97:2266–2274.
Caillaud D, Levrero F, Gatti S, Menard N, Raymond M (2008) Influence of male morphology on male mating status and behavior during interunit encounters in Western lowland gorillas. Am J Phys Anthropol 135:379–389.
Carter ML, Pontzer H, Wrangham RW, Kerbis Peterhans J (2008) Skeletal pathology in *Pan troglodytes schweinfurthii* in Kibale National Park, Uganda. Am J Phys Anthropol 135:389–403.
Cavagna GA, Kaneko M (1977) Mechanical work and efficiency in level walking and running. J Physiol 268:467–481.
Charnov EL, Ernest SKM (2006) The offspring-size/clutch-size trade-off in mammals. Am Nat 167:578–582.
Coelho AM, Bramblett CA, Quick LB (1979) Activity patterns in howler and spider monkeys: an application of socio-bioenergetic methods In: Bernstein IS, Smith EO (eds), Primate Ecology and Human Origins. Gartland, New York, pp 175–199.
Garland T (1983) Scaling the ecological cost of transport to body mass in terrestrial mammals. Am Nat 121:571–87.
Hanna J, Schmitt D, Griffin TM (2008) The energetic cost of climbing in primates. Science 320:898.
Heglund NC, Fedak MA, Taylor CR, Cavagna GA (1982) Energetics and mechanics of terrestrial locomotion IV: total mechanical energy changes as a function of speed and body size in birds and mammals. J Exp Biol 79:57–66.
Heglund NC, Taylor CR (1988) Speed, stride frequency and energy-cost per stride – how do they change with body size and gait. J Exp Biol 138:301–318.
Hoyt DF, Taylor CR (1981) Gait and energetics of locomotion in horses. Nature 292:239–240.
Hoyt DF, Wickler SJ, Cogger EA (2000) Time of contact and step length: the effect of limb length, running speed, load carrying and incline. J Exp Biol 203:221–227.
Key CA, Ross C (1999) Sex differences in energy expenditure in non-human primates. Proc R Soc Lond B 266:2479–2485.
Kram R, Taylor CR (1990) Energetics of running: a new perspective. Nature 346:265–267.
Langman VA, Roberts TJ, Black J, Maloiy GMO, Heglund NC, Weber JM, Kram R, Taylor CR (1995) Moving cheaply: energetics of walking in the African elephant. J Exp Biol 198:629–632.
Leonard WR, Robertson ML (1997) Comparative primate energetics and hominoid evolution. Am J Phys Anthropol 102:265–281.
Luick BR, White RG (1986) Oxygen consumption for locomotion by caribou calves. J Wildl Manage 50:148–152.
Marsh RL, Ellerby DJ, Henry HT, Rubenson J (2006) The energetic costs of trunk and distal-limb loading during walking and running in guinea fowl *Numida meleagris*: I. Organismal metabolism and biomechanics. J Exp Biol 209:2050–2063.
Minetti AE, Ardigo LP, Reinach E, Saibene F (1999) The relationship between mechanical work and energy of locomotion in horses. J Exp Biol 202:2329–2338.
Nagy KA, Girard IA, Brown TK (1999) Energetics of free-ranging mammals, reptiles, and birds. Annu Rev Nutr 19:247–277.
Pennycuick CJ (1975) On the running of the gnu (*Connochaetes taurinus*) and other animals. J Exp Biol 63:775–799.
Polk JD (2004) Influences of limb proportions and body size on locomotor kinematics in terrestrial primates and fossil hominins J Hum Evol 47:237–252.
Pontzer H (2007) Limb length and the scaling of locomotor cost in terrestrial animals. Journal of Experimental Biology. 210:1752–1761.
Pontzer H, Wrangham RW (2004) Climbing and the daily energy cost of locomotion in wild chimpanzees: implications for hominoid locomotor evolution. J Hum Evol 46:315–333.

Pontzer H, Raichlen DA, Sockol MD (2009) The metabolic cost of walking in humans, chimpanzees, and early hominins. J Hum Evol 56:43–54.

Roberts TJ, Kram R, Weyand PG, Taylor CR (1998a) Energetics of bipedal running: I. metabolic cost of generating force J Exp Biol 201:2745–2751.

Roberts TJ, Chen MS, Taylor CR (1998b) Energetics of bipedal running: II. Limb design and running mechanics J Exp Biol 201:2753–2762.

Schmitt D (1999) Compliant walking in primates. J Zool Lond 248:149–160.

Sockol MD, Raichlen DA, Pontzer H (2007) Chimpanzee locomotor energetics and the origin of human bipedalism. Proc Natl Acad Sci USA 30:12265–12269.

Steudel K, Beattie J (1995) Does limb length predict the relative energetic cost of locomotion in mammals? J Zool 235:501–514.

Steudel K (2000) The physiology and energetics of movement: effects on individuals and groups In: Boinski S, Garber P (eds), On the Move: How and Why Animals Travel in Groups. University of Chicago, Chicago, pp 9–23.

Steudel-Numbers K (2003) The energetic cost of locomotion: humans and primates compared to generalized endotherms. J Hum Evol 44:255–262.

Steudel-Numbers K (2006) Energetics in Homo erectus and other early hominins: the consequences of increased lower limb length. J Hum Evol 51:445–453.

Taylor CR, Schmidt-Nielsen K, Raab JL (1970) Scaling of energetic cost of running to body size in mammals. Am J Physiol 319:1104–1107.

Taylor CR, Rowntree VJ (1973) Running on two or on four legs: which consumes more energy? Science 179:186–187.

Taylor CR (1977) The energetics of terrestrial locomotion and body size in vertebrates. In: Pedley TJ (ed), Scale Effects in Animal Locomotion. Academic Press, New York, pp 127–114.

Taylor CR, Heglund NC, Maloiy GMO (1982) Energetics and mechanics of terrestrial locomotion: I. Metabolic energy consumption as a function of speed and body size in birds and mammals. J Exp Biol 97:1–21.

Taylor CR (1994) Relating mechanics and energetics during exercise. Comparative vertebrate exercise physiology: unifying physiological principles. Adv Vet Sci Comp Med A 38:181–215.

Warren DR, Crompton RH (1998) Diet, body size and the energy costs of locomotion in saltatory primates. Folia Primatol 69 (Suppl1):86–100.

Wiersma P, Verhulst S (2005) Effects of intake rate on energy expenditure, somatic repair and reproduction of zebra finches. J Exp Biol 208:4091–4098.

Willems PA, Cavagna GA, Heglund NC (1995) External, internal and total work in human locomotion. J Exp Biol 198:379–393.

Chapter 16
Linking Field and Laboratory Approaches for Studying Primate Locomotor Responses to Support Orientation

Nancy J. Stevens, Jonah H. Ratsimbazafy, and Fidy Ralainasolo

Abstract Studies of primate locomotor kinematics typically focus on data conducted in the laboratory setting, with cameras carefully positioned to afford clear views of the focal subjects, and strict control of a range of other variables ranging from light levels, to the travel path and even locomotor velocity of the focal subject. Such studies permit the manipulation of support types, facilitating the collection of a large number of data points relating to specific aspects of animal locomotion, and detailing how an individual responds to differences in arboreal support types. Studies of primate behavior in the field setting provide a completely different window into locomotor behavior. These approaches lend insights into the choices that animals make with regard to support use, providing information on the frequencies of different locomotor behaviors on different locomotor substrates. In this study, we explore the relationship between arboreal support type and forelimb and hind limb kinematics using locomotor data gathered in both the laboratory and natural settings. In the laboratory, we test a biomechanical model generated to explain limb kinematic response to support orientation using a large number of strides obtained from *Eulemur* individuals negotiating simulated arboreal supports. Next, using techniques that expand laboratory analyses into the field setting, we examine limb movements in *Eulemur cinereiceps* at the Manombo Special Reserve in southeastern Madagascar. Results suggest that animals tend to protract forelimbs more at touchdown on declines, and retract hind limbs more at lift-off on declines, patterns that generally maintain the line of gravity between the points of contact with oblique substrates. Focal individuals flex the elbow and knee joints more at midsupport on inclined and declined branches, bringing the center of mass closer to oblique supports. Patterns observed in the natural setting accord well with strides collected in the laboratory. This study both documents the effects of support orientation upon strepsirhine primate quadrupedal locomotion and represents the first record of limb kinematics in the natural setting for the critically endangered *Eulemur cinereiceps*.

N.J. Stevens (✉)
Department of Biomedical Sciences, Ohio University, Ohio, USA
e-mail: stevensn@ohio.edu

K. D'Aoû t and E.E. Vereecke (eds.), *Primate Locomotion: Linking Field and Laboratory Research*, Developments in Primatology: Progress and Prospects,
DOI 10.1007/978-1-4419-1420-0_16, © Springer Science+Business Media, LLC 2011

Keywords Arboreal locomotion • *Eulemur* • Substrate orientation

Introduction

Numerous studies have identified primate behavioral and morphological specializations correlated with the use of an arboreal habitat (e.g., Grand 1972; Morbeck 1976, 1979; Oxnard 1976; Fleagle 1979; Ripley 1979; Crompton 1984; Garber 1984; Cant 1988, 1992; Boinski 1989; Larson and Stern 1992; Cannon and Leighton 1994; Larson 1995; Hunt et al. 1996). Studies conducted in the wild have reported on primate strategies for moving on arboreal supports, for example, grasping multiple branches and/or adopting a crouching posture (i.e., flexed and abducted limbs), to move along narrow supports, presumably to improve balance by lowering the center of mass (Fleagle 1977a,b; Fleagle and Mittermeier 1980; Grand 1984; 1988; Dunbar and Badam 2000). Recent decades have seen increased interest in exploring specializations for arboreality, yielding a wealth of experimental kinematic studies aimed at identifying specific locomotor accommodations to different aspects of the arboreal habitat (Rollinson and Martin 1981; Meldrum 1991; Hirasaki et al. 1993; Demes et al. 1995; Schmitt and Larson 1995; Hamrick 1996; Lemelin and Schmitt 1997; Jayne and Irschick 1999; Stevens and Larson 1999; Turnquist et al. 1999; Wunderlich 1999; Schmidt and Fischer 2000; Krakauer et al. 2002; Stevens 2003, 2008).

The Problem of Branch Inclination: Limb Kinematics on Oblique Supports

Angular orientation of supports constitutes one of the most obvious challenges that primates must overcome in negotiating the arboreal habitat. This study focuses strictly on kinematic accommodations to support inclination, using data gathered in the laboratory and natural settings. Oblique arboreal substrates create two primary types of challenges to maintaining balance, namely, increased chances of sliding down or toppling off a support (Fig. 16.1). Sliding and toppling relate to the shear forces that exist between the contact point and the substrate (Hirasaki et al. 1993). A number of animals exhibit well-developed strategies for resisting shear forces, by generating "strong bonds" with the support itself. For example, bats are capable of generating negative pressure along specialized contact surfaces, producing enough suction to remain firmly attached to smooth surfaces (Cartmill 1979). Anurans use both capillary surface tension and viscous adhesion, in order to maintain contact with inclined supports (Barnes 1997). Many animals that use suction and/or adhesive tactics possess an ability to secrete fluids from specialized sweat or mucous glands located in textured toe pads to assist in this process, e.g., tree frogs (Barnes 1997). Those that do not produce

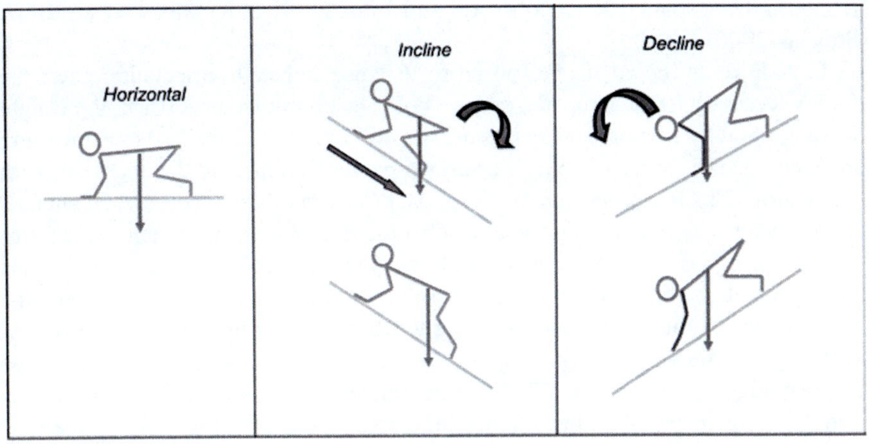

Fig. 16.1 On horizontal branches, the gravity vector is perpendicular to the support. On oblique supports, a surface parallel component of the gravity vector increases with support inclination, increasing the tendency to slide or topple from a support. This component is in the direction of movement on declines, and against the direction of movement on inclines. One way that primates may maintain balance on oblique supports is by altering limb joint angles and protracting and retracting limbs to keep the line of gravity between the points of contact with the substrate

secretions may possess specialized features such as intricate internal tendons that can raise and lower textured surfaces of the contact pads, thereby increasing intermolecular forces between the animal and the substrate via dry adhesion (Cartmill 1979; Haffner 1996).

Another strategy for resisting shear forces on arboreal supports is to find some way to interlock with a branch. One common method employed by small animals is to resist the shear forces via manual and/or pedal claws, which they can imbed into irregularities in tree bark or other surfaces (Cartmill 1974; Haffner 1996; Giannoni et al. 1999). Some rodents that habitually use claws for this purpose have tendon-locking mechanisms that reduce energy expenditure associated with muscle flexion (Haffner 1996). Sloths, dermopterans, and bats also employ tendon-locking mechanisms to interlock with substrates by hooking entire hands and/or feet around the support (Quinn and Baumel 1993; Simmons and Quinn 1994).

Most primates lack well-developed claws for interlocking with supports and/or specialized manual and pedal glands for adhesion. Nonetheless they remain adept at arboreality. Primates are known to make numerous kinematic adjustments in order to counteract balance challenges on inclined branches. Hamrick (1996) and Lemelin (1996) have shown that wrist joint angles and hand placement respectively change with differences in support orientation, and Stevens and Larson (1999) reported that limb joint angles are altered, perhaps to shift the center of mass closer to the substrate. Primates moving along oblique supports exhibit myriad other changes in angular and support patterns

(Prost and Sussman 1969; Rollinson and Martin 1981; Krakauer et al. 2002; Stevens 2006).

In addition to the challenges posed for balance, substrate orientation alters the forces necessary for quadrupedal progression. On a horizontal substrate, the weight force or gravity vector acts perpendicular to the surface. When substrates are inclined, the weight force includes a surface-parallel component that increases with inclination. This is a propulsive force (in the direction of movement) when animals walk downhill, and a braking force (against the direction of movement) when animals walk uphill. Therefore, when walking down an inclined branch, the limbs must generate higher braking forces than they would on a horizontal to overcome the tendency to slide down, and conversely, when walking up, the limbs must generate higher propulsive forces. Aligning the limbs with the gravity vector may assist in generating these impulses. Preliminary studies of possums navigating inclines and declines suggest that this is indeed the case (Lammers et al. 2002). One way that a clawless animal may maintain balance while applying necessary braking and propulsive forces is to make accommodations in limb protraction, retraction and flexion. At this point, it is convenient to consider walking up and walking down oblique supports separately.

Incline Predictions

In addition to the use of grasping hands and feet to resist shear forces on oblique supports, it is predicted that elbow and knee joints may become more flexed at midstance in order to bring the center of mass closer to oblique substrates. At lift-off, hind limbs should be more retracted and knees should be more extended on inclines than they are on the horizontal (Fig. 16.1). Because the line of gravity falls more posteriorly along an inclined support, this angular excursion pattern can act to reduce the animal's tendency to topple off of a support by keeping the gravity vector within the line of support defined by the points of contact with the substrate.

Decline Predictions

As an animal walks head-first down declined branches, it is also expected that the elbows and knees will be more flexed at midstance to bring the center of mass closer to the substrate. But in this case, the forelimb should be more protracted and the elbow more extended at touchdown. As the line of gravity falls more anteriorly on a declined support, this excursion pattern should help to maintain the gravity vector near the line of support defined by the points of contact with the substrate, and to control acceleration on the decline to avoid toppling over head-first (Fig. 16.1). Individuals that *are* able to reverse their hind feet, e.g., ruffed lemurs, may do so, in order to achieve a better grasp on the substrate by placing the hind limb in tension (Laborde 1986; Meldrum et al. 1997). This case may actually result in high hind limb angular excursions on declines. Those that *cannot* reverse their hind feet may take shorter steps, flex their knees in order to lower their center of mass, and be unwilling to walk headfirst down the steepest declines.

Locomotor Kinematics in *Eulemur*: A Case Study for Naturalistic Experimental Research

This study examines a subset of these hypotheses by generating a large sample of strides on horizontal, inclined and declined simulated arboreal supports in a laboratory assembled at the Duke Lemur Center. Importantly, it also extends these hypotheses into the natural setting to examine whether the critically endangered gray-headed lemur, *Eulemur cinereiceps* (until recently known as the white-collared lemur, *E. albocollaris*—see Johnson et al. 2008 for a taxonomic overview), exhibits similar responses to support angulation as the baseline pattern established for its congeners in the laboratory setting.

Eulemur represents a good choice for understanding kinematic accommodations to support orientation, representing an adept arboreal quadrupedal primate that prefers to travel in the continuous canopy (e.g., Sussman 1976), and with well-documented morphology and positional behavior (e.g. Sussman 1976; Jouffroy and Lessertisseur 1978; Ward and Sussman 1979; Ganzhorn 1985; Jungers 1985; Dagosto 1995; Hamrick 1996; Overdorff 1996). *Eulemur cinereiceps*, restricted to the low-altitude southeastern coastal rain forests of Madagascar, finds protection only within Manombo Special Reserve, a plot containing 8000 ha of forest located at 22°57′S, 23°08′S latitude and 47°36′E, 47°48′E longitude. In recent years, the Manombo forest habitat has drastically reduced as it has withstood a major cyclone, a natural forest fire, and increasing human pressure through logging and hunting (Ratsimbazafy 2002). Although resilient, this confluence of challenges has rendered *Eulemur cinereiceps* one of the 25 most critically endangered primates in the world, as defined by the IUCN (Mittermeier et al. 2006, 2007). The need for more baseline data on positional behavior and habitat use of this taxon to inform the conservation effort sets the context for its inclusion in this study.

Methods

Focal Subjects

This study adhered to the principles of ethical treatment of nonhuman primates, using noninvasive kinematic data collection protocols approved by the Institutional

Animal Care and Use Committees (IACUC) at Stony Brook University, Duke University, and Ohio University. The study sample included one adult male and one adult female of both *Eulemur collaris* and *Eulemur rubriventer* filmed by N.J. Stevens in the laboratory setting, and one adult male and one adult female *Eulemur cinereiceps* filmed in the natural setting of Manombo Forest, Madagascar. Laboratory subjects had regular access to naturalistic supports in their large enclosures. The body mass for *Eulemur* species is ca. 2 kg (Smith and Jungers 1997), and *Eulemur* intermembral indices fall between 68 and 72 (Fleagle 1999). Of the *Eulemur* species examined herein, only *E. rubriventer* has been the subject of long-term research on habitat use (Overdorff 1996). However, all *Eulemur* species with published field data are agile arboreal quadrupeds, similar in body mass and proportions, that travel and forage in the arboreal canopy (e.g., Sussman 1976; Jouffroy and Lessertisseur 1978; Ward and Sussman 1979; Ganzhorn 1985; Jungers 1985; Dagosto 1995; Hamrick 1996; Overdorff 1996; Smith and Jungers 1997; Fleagle 1999), justifying intrageneric comparisons made in this study.

Measuring Locomotor Kinematics in the Laboratory and in the Field

In the laboratory setting, subjects moved upon simulated branches constructed from 2.44 m sections of polyvinyl chloride (PVC) pipes, 1.25 cm in diameter, coated with a nonslip surface and oriented horizontally and at 30° and 60° angles from the horizontal. Individuals acclimated to these substrates in their naturalistic enclosures before data collection, allowing them to become accustomed to the experimental support orientations. Stevens (2003) provides a more comprehensive description of the laboratory experimental setup. In the natural setting, we filmed individuals on oblique supports ranging between 30° and 60° in inclination, and between 1 and 4 cm in diameter. Although some of these branches were flexible, strides analyzed in this study were restricted to supports that oscillated less than 2 cm in the superoinferior plane during the stride cycle. Using standard 2D kinematic techniques in both the laboratory and in Manombo Forest, we positioned cameras on tripods to capture lateral views of the study subjects, placing recording devices at a sufficient distance to reduce parallax, 5 m from the path of movement of the focal subjects. In the laboratory setting, N.J. Stevens used two Panasonic AG-195 VHS professional video cameras, and in the field setting, we used a Sony DCR-HC42 NTSC digital camcorder. In the laboratory, a shutter speed of 1/1000 s was used to reduce motion blur; in the field we used the highest shutter speeds possible in the variably lit settings. Polk et al. (2005) note that error associated with sampling at lower frequencies is most pronounced in animals moving at very high velocity. For both camera types, we optimized frame rates to reduce motion blur by splitting interlaced video fields to achieve 60 Hz, a sampling rate adequate to capture the angular kinematics, shoulder and hip heights, and stride lengths used by *Eulemur* subjects at the velocities that they traveled in this study.

Using Peak Motus (version 9.1) to import video clips, N.J. Stevens collected 15 strides per individual per substrate that provided unobstructed views of *all* of the anatomical points of interest, with the total number of individuals (n = 4) and support (n = 5) combinations yielding 240 strides in the laboratory setting (notably *Eulemur rubriventer* chose not to walk on 60° declines in the lab). Laboratory-derived data formed a baseline for comparison with a smaller number of strides (n = 19) collected in the natural setting, following field kinematic methods outlined in Stevens et al. (2006). Data collected at Manombo included 10 strides for the adult male (horizontal: 3, 30° incline: 3, 60° incline: 2, 30° decline: 1, 60° decline: 1), and nine strides for the adult female (horizontal: 1, 30° incline: 1, 60° incline: 2, 30° decline: 3, 60° decline: 2). In frame-by-frame analysis, we collected kinematic variables by digitizing forelimb and hind limb positions at touchdown and lift-off events. Variables are summarized in Table 16.1.

For the purposes of this study, forelimb angle refers to the angle between the substrate and a line segment connecting the greater tubercle of the humerus with the lateral aspect of the distal end of the fifth metacarpal, such that a larger forelimb angle reflects greater forelimb *protraction*. Hind limb angle refers to the angle between the substrate and a line segment connecting the greater trochanter of the femur with the lateral aspect of the distal end of the fifth metatarsal, such

Table 16.1 Kinematic variables considered in this study

Parameter	Abbreviation	Description
Forelimb angle[a]	FL	Angle made by a line connecting shoulder and hand markers and the substrate
Forelimb excursion	FLEXC	Difference in forelimb angle between touchdown and lift-off
Hind limb angle[a]	HL	Angle made by a line connecting hip and foot markers and the substrate
Hind limb excursion	HLEXC	Difference in hind limb angle between touchdown and lift-off
Elbow angle[a]	Elb	Angle made by the proximal and distal forelimb segments
Knee angle[a]	Knee	Angle made by the proximal and distal hind limb segments
Shoulder height[a]	ShoHT	Perpendicular distance of the shoulder marker from the substrate
Hip height[a]	HipHT	Perpendicular distance of the hip marker from the substrate
Forelimb stride length	FSTRLEN	Distance traveled by the hand marker from one touchdown to the next
Hind limb stride length	HSTRLEN	Distance traveled by the foot marker from one touchdown to the next
Velocity	SPEED	Distance covered in one stride cycle divided by the number of frames

[a]Individual angles measured at touchdown (TD, the first frame in which the limb is in contact with the support), midstance (MS, the frame in which the shoulder marker passes over the hand marker/hip marker passes over the foot marker), and lift off (LO, the last frame in which the limb is in contact with the support).

that a larger hind limb angle reflects greater hind limb *retraction*. Elbow angle refers to the angle made by the proximal and distal forelimb segments, whereas knee angle refers to the angle made by the proximal and distal hind limb segments, such that for both the elbow and the knee, a smaller angle indicates a more *flexed* limb posture. To account for out-of-plane positions of the elbow and knee joints, 2D angles collected from lateral view were subjected to a trigonometric correction (following Stevens et al. 2006). We measured shoulder and hip heights as the perpendicular distance between the shoulder or hip marker and the substrate. We calculated stride lengths as the distance traveled by the hand/foot marker between consecutive touchdown events for the limb, and velocity as the distance traveled in one stride cycle divided by the number of frames. In the laboratory setting N.J. Stevens placed reflective tape markers on anatomical landmarks on the study subjects, facilitating the collection of a large laboratory data set for each substrate. For data collected in the natural setting, we visually identified anatomical landmarks only for strides that preserved clear and well-lit views of the study subjects moving along supports that had obvious landmarks measured for calibration purposes.

Owing to the smaller sample of strides obtained in the field setting, we statistically analyzed only the laboratory data set for this study, incorporating field data in more qualitative comparisons. As kinematic variables do not always follow a normal distribution, we rank-transformed laboratory data prior to analysis and replaced original variates with the ranks, breaking ties by assigning mean ranks to tied cases (SPSS version 13.0). Rank transformations permit two-way analysis of variance without loss of power in data sets that are not normally distributed (Iman 1974; Conover and Iman 1981).

Because differences in velocity may themselves constitute a response to branch orientation, we did not attempt to constrain velocity in either the laboratory or the natural settings. To explore substrate-related differences in limb kinematics while taking into account differences in velocity, we subjected data to an analysis of covariance (ANCOVA) (Sokal and Rohlf 1981) with velocity (m/s) as the covariate. ANCOVAs compare kinematic variables collected on two substrates at the mean velocity observed on both substrates. When we found no significant relationships with velocity, we analyzed data using analyses of variance for substrates of differing orientations.

Results

Means and standard deviations for all kinematic variables by individual and substrate type are presented in Table 16.2. Significance levels for ANOVA/ANCOVA analyses of strides collected in the lab are summarized in Table 16.3. Males and females exhibited similar locomotor responses to support orientation in the variables discussed herein, although not all differences in limb kinematics reached significance in both individuals (Table 16.3).

Table 16.2a Mean (AVG) and standard deviations (SD) for kinematic variables of *Eulemur collaris* (laboratory setting)

	Female										Male									
	H		I 30		I 60		D 30		D 60		H		I 30		I 60		D 30		D 60	
	Avg.	SD	Avg.	SD	Avg.	SD	Avg.	SD	Avg.	SD	Avg.	SD	Avg.	SD	Avg.	SD	Avg.	SD	Avg.	SD
FLTD[a]	127.73	1.10	124.21	3.10	127.72	2.39	137.94	1.53	143.16	1.60	127.84	2.11	121.31	3.06	123.72	3.87	138.31	1.49	136.88	3.18
FLTO[a]	50.73	1.73	45.18	4.53	46.08	3.39	63.83	3.47	97.58	5.88	54.54	2.52	47.61	3.02	45.42	3.35	59.89	3.67	71.95	6.87
elb MS[a,b]	116.04	4.503	99.14	5.64	93.67	4.13	102.08	3.01	87.56	44.57	122.77	4.01	93.34	3.61	90.81	4.92	106.17	5.77	74.36	7.82
HLTD[a]	58.61	0.79	66.37	2.27	76.30	3.03	43.08	1.63	39.27	2.12	63.57	2.85	70.52	2.79	74.26	4.52	43.73	1.65	39.10	10.07
HLTO[a]	123.66	1.65	130.73	1.40	133.82	2.38	110.88	2.48	129.79	6.49	128.30	2.30	134.12	1.91	136.97	2.12	124.71	3.43	118.45	9.87
knee MS[a,b]	105.11	2.81	93.09	3.70	83.20	2.99	67.48	3.76	56.19	8.26	99.98	3.64	86.21	2.30	81.14	3.17	71.39	2.95	60.49	5.24
f str leng[c]	0.49	2.42	0.60	6.61	0.53	4.63	0.46	3.16	0.32	4.64	0.58	2.27	0.59	4.51	0.56	4.52	0.50	2.86	0.38	5.95
h str leng[c]	0.49	2.55	0.60	6.48	0.51	3.52	0.46	2.89	0.64	7.76	0.57	2.29	0.58	3.75	0.53	7.05	0.50	3.03	0.51	7.69
ShoHt TD[c]	0.15	0.51	0.16	1.06	0.14	0.56	0.13	0.46	0.12	1.17	0.16	0.74	0.15	0.74	0.17	0.67	0.14	0.44	0.12	1.01
ShoHt MS[c]	0.16	0.61	0.16	0.66	0.16	0.68	0.14	0.51	0.14	2.24	0.18	0.76	0.16	0.82	0.18	0.69	0.16	0.39	0.14	1.14
ShoHt LO[c]	0.14	0.70	0.14	0.88	0.14	0.85	0.14	0.56	0.14	1.53	0.17	0.40	0.15	0.82	0.16	0.70	0.15	0.71	0.14	1.05
HipHt TD[c]	0.23	0.63	0.22	1.10	0.20	0.24	0.18	0.47	0.17	1.75	0.23	0.86	0.21	0.36	0.21	0.55	0.19	0.53	0.17	1.31
HipHt MS[c]	0.23	0.97	0.22	1.17	0.21	0.49	0.18	0.71	0.16	2.03	0.23	0.73	0.21	0.45	0.21	0.51	0.18	0.66	0.18	1.10

n = 15 for all variables collected in laboratory setting. Abbreviations follow those of Table 16.1.

[a]Units in degrees.

[b]2D.

[c]Units in meters.

Table 16.2b Mean (AVG) and standard deviations (SD) for kinematic variables of *Eulemur rubiventer* (laboratory setting)

	Female								Male							
	H		I 30		I 60		D 30		H		I 30		I 60		D 30	
	Avg.	SD	Avg.	SD	Avg.	SD	Avg.	SD	Avg.	SD	Avg.	SD	Avg.	SD	Avg.	SD
FLTD[a]	128.20	1.79	127.08	3.29	118.09	12.66	138.38	2.26	127.17	1.96	120.42	4.99	124.97	3.28	137.46	3.42
FLTO[a]	52.75	2.59	49.64	3.62	36.06	1.85	61.22	6.50	54.02	2.97	49.05	3.99	39.84	4.35	62.58	8.43
Elb MS[a,b]	125.97	2.60	109.16	5.95	88.86	7.53	104.05	4.66	126.74	3.91	106.00	8.16	90.98	7.57	108.38	6.78
HLTD[a]	57.68	2.71	74.24	4.17	89.11	3.20	37.40	5.14	62.32	2.40	81.61	2.61	82.03	3.51	40.93	2.54
HLTO[a]	125.54	1.82	131.62	1.81	142.16	2.17	111.65	6.80	125.92	1.93	133.85	2.21	141.12	1.72	107.69	9.96
Knee MS[a,b]	103.02	2.62	100.07	4.36	87.08	5.92	72.03	6.06	107.70	2.84	90.20	4.60	77.69	5.30	74.67	8.81
f str leng[c]	0.47	1.82	0.54	3.21	0.56	5.43	0.41	3.36	0.47	2.17	0.52	5.51	0.51	7.67	0.40	5.18
h str leng[c]	0.47	1.63	0.54	3.50	0.53	4.55	0.44	7.87	0.47	2.11	0.51	5.05	0.53	7.54	0.40	4.82
ShoHt TD[c]	0.16	0.47	0.14	0.63	0.14	0.71	0.13	0.65	0.15	0.46	0.13	0.88	0.13	0.90	0.13	0.72
ShoHt MS~	0.17	0.47	0.15	0.54	0.14	0.81	0.14	0.52	0.17	0.72	0.14	0.96	0.13	0.65	0.14	1.07
ShoHt LO[c]	0.15	0.73	0.14	0.69	0.11	0.60	0.14	0.82	0.15	0.70	0.14	0.92	0.11	1.03	0.14	1.37
HipHt TD[c]	0.22	0.61	0.21	0.77	0.21	1.28	0.16	1.55	0.23	0.46	0.20	0.57	0.19	0.63	0.17	0.70
HipHt MS[c]	0.24	0.80	0.21	0.71	0.20	5.59	0.18	1.47	0.24	0.48	0.20	0.54	0.18	0.49	0.17	1.31
HipHt LO[c]	0.22	0.80	0.20	0.69	0.17	0.73	0.16	1.92	0.22	0.66	0.20	0.57	0.18	0.86	0.16	0.85

$n = 15$ for all variables collected in laboratory setting. Abbreviations follow those of Table 16.1.

[a] Units in degrees.

[b] 2D.

[c] Units in meters.

Table 16.2c Mean (Avg.) and standard deviations (SD) for kinematic variables of *Eulemur cinereiceps* (natural setting)

	H		I 30		I 60		D 30		D 60	
n =	4		4		4		4		3	
	Avg.	SD	Avg.	SD	Avg.	SD	Avg.	SD	Avg.	SD
FLTD	127.19	4.24	123.85	3.02	123.68	2.65	141.01	2.06	142.94	4.48
FLTO	52.79	6.38	43.55	0.66	43.78	3.47	53.68	5.68	98.46	22.99
elb MS	109.89	4.64	98.68	3.70	89.83	4.75	100.82	3.67	92.93	7.25
HLTD	57.98	3.84	69.38	1.29	77.58	3.55	37.14	3.50	40.20	5.52
HLTO	125.04	2.03	132.82	2.38	136.48	1.69	114.69	5.28	125.78	8.02
knee MS	103.63	5.62	87.94	2.66	81.08	3.04	62.81	3.99	53.41	5.67
f str leng	0.55	0.06	0.56	0.03	0.57	0.03	0.47	0.02	0.36	0.17
H str leng	0.55	0.06	0.55	0.03	0.58	0.08	0.47	0.02	0.59	0.10
ShoHt TD	0.17	0.01	0.15	0.01	0.16	0.01	0.13	0.01	0.12	0.01
ShoHt MS	0.17	0.01	0.16	0.01	0.16	0.01	0.16	0.00	0.13	0.01
ShoHt LO	0.16	0.01	0.14	0.01	0.14	0.00	0.14	0.01	0.14	0.01
HipHt TD	0.24	0.00	0.22	0.00	0.20	0.01	0.17	0.01	0.16	0.00
HipHt MS	0.24	0.01	0.22	0.01	0.21	0.01	0.18	0.01	0.16	0.02
HipHt LO	0.23	0.01	0.21	0.00	0.20	0.01	0.17	0.01	0.16	0.02

Abbreviations follow those of Table 16.1.

Table 16.3 ANOVA/ANCOVA significance levels for kinematic variables on oblique supports

	Eulemur collaris				*Eulemur rubiventer*				*Eulemur collaris*				*Eulemur rubiventer*	
	Female		Male		Female		Male		Female		Male		Female	Male
	H-I 30	H-I 60	H-I 30	H-I 60	H-I 30	H-I 60	H-I 30	H-I 60	H-D 30	H-D 60	H-D 30	H-D 60	H-D 30	H-D 30
FLTD	0.000	–	0.000	0.001	–	0.007	0.000	0.044	0.000	–	0.000	***0.000***	0.000	0.000
FLTO	0.001	0.000	0.000	***0.000***	–	0.000	***0.000***	***0.000***	***0.000***	0.000	0.000	0.000	0.000	0.000
HLTD	0.000	0.000	0.000	0.000	0.000	0.000	0.000	0.000	***0.000***	0.000	0.000	0.000	0.000	0.000
HLTO	0.000	0.000	0.000	0.000	0.000	0.000	0.000	0.000	0.000	–	0.001	–	0.000	***0.000***
elb MS	–	0.014	0.000	0.000	0.000	0.000	0.000	0.000	0.001	***0.001***	***0.000***	0.000	0.000	0.000
knee MS	0.000	0.000	0.000	0.000	0.000	0.000	0.000	0.000	0.000	0.000	0.000	0.000	0.000	***0.000***
FL str leng	–	–	–	–	–	–	–	–	***0.000***	***0.000***	***0.009***	***0.000***	***0.000***	***0.000***
HL str leng	–	–	–	–	–	–	–	–	***0.000***	–	–	***0.003***	***0.000***	***0.000***
ShoHt TD	–	0.000	0.000	–	0.000	0.000	***0.000***	***0.000***	0.000	0.000	0.000	0.000	0.000	***0.000***
ShoHt MS	–	–	0.000	–	0.000	0.000	0.000	0.000	0.000	***0.013***	0.000	0.000	0.000	0.000
ShoHt LO	–	0.014	0.000	0.000	0.000	0.000	***0.000***	***0.000***	–	–	0.000	0.000	0.000	0.011
HipHt TD	–	0.000	0.000	0.000	0.047	***0.006***	***0.000***	***0.000***	0.000	0.000	0.000	0.000	0.000	***0.000***
HipHt MS	–	0.000	0.000	0.000	0.000	0.000	0.000	0.000	***0.000***	0.000	0.000	0.000	0.000	***0.000***
HipHt LO	–	0.000	0.000	0.000	0.000	0.000	0.000	0.000	0.000	0.000	0.000	0.000	0.000	0.000

All significant differences in table indicate lower values for oblique support relative to horizontal *EXCEPT* in HLTD and HLTO comparisons on inclines (H-I 30, H-I 60), *AND* in FLTD and FLTO comparisons on declines (H-D 30, H-D 60); regular text indicates ANOVA results; bold italics indicate relationship with speed, thus ANCOVA results; .000 denotes $p << 0.001$; blanks indicate nonsignificant differences.
Kinematic abbreviations follow those in Table 16.1.

In the laboratory setting, both *Eulemur collaris* and *E. rubriventer* exhibited significantly more protracted forelimbs (larger forelimb angles) at touchdown on declined supports than either did on the horizontal branches (Fig. 16.2a). These patterns were present throughout the stride cycle, with larger forelimb angles also observed at lift-off on declines. Hind limb angles were generally smaller, reflecting in this case that hind limbs too, tended to be more protracted at touchdown and lift-off on declined supports (Table 16.3).

On inclines, both *Eulemur* species examined in the laboratory utilized more retracted hind limbs (larger hind limb angles) at both touchdown lift-off events (e.g., Fig. 16.2b). Similar patterns were observed for the forelimbs, typically less protracted on inclines at touchdown and lift-off events (Table 16.3). *Eulemur cinereiceps* in Manombo Forest used similar kinematic patterns, e.g., exhibiting larger forelimb angles at touchdown on declines (Fig. 16.3a), and larger hind limb angles at lift-off on inclines (Fig. 16.3b).

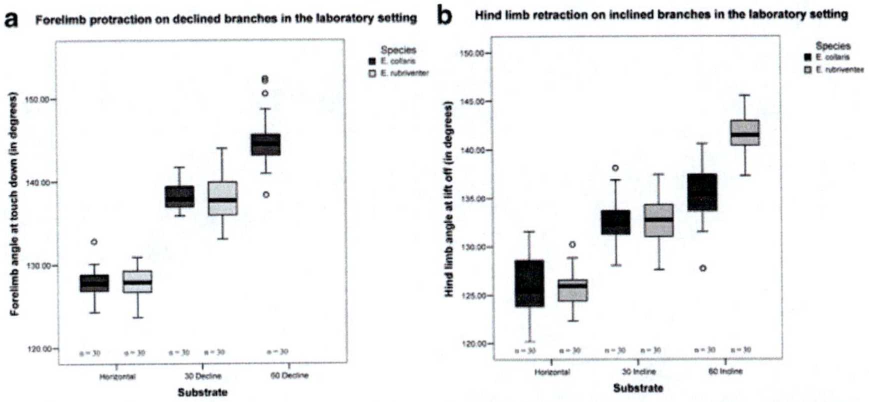

Fig. 16.2 Limb protraction and retraction on oblique supports in the laboratory. **(a)** Forelimb protraction at touchdown on declines. **(b)** Hind limb retraction at lift-off on inclines. Heavy lines in these and subsequent boxes represent median values

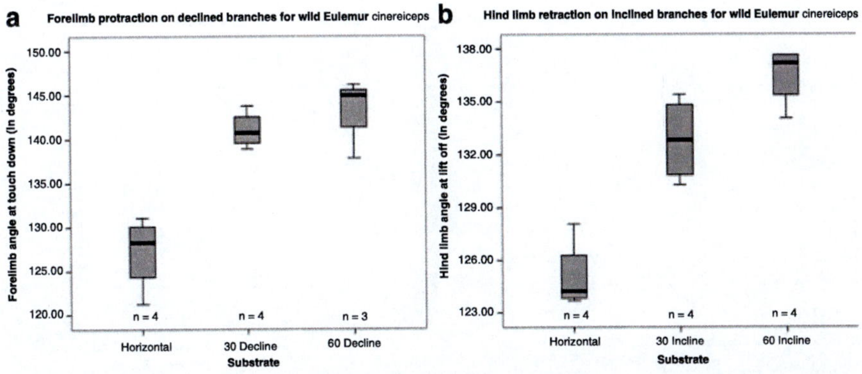

Fig. 16.3 Limb protraction and retraction on oblique supports in the field. **(a)** Forelimb protraction at touchdown on declines. **(b)** Hind limb retraction at lift-off on inclines

With regard to limb flexion during locomotion, in the laboratory setting, *Eulemur* species tended to exhibit smaller elbow and knee angles at midsupport on oblique substrates (Fig. 16.4a, b), a pattern also observed for *Eulemur cinereiceps* in the field (Fig. 16.5a, b). On simulated arboreal supports, focal subjects tended to take relatively shorter forelimb strides on declines (Fig. 16.6a), and carried their center of mass closer to oblique branches, as evidenced by lower shoulder and hip heights. For example, hind limb kinematics reflected lower hip heights at throughout the stride on both inclines and declines (Table 16.3). Although the Manombo rain forest habitat provided a greater variability in support types than was present in the laboratory, *Eulemur cinereiceps* exhibited kinematic responses to substrate orientation consistent with its congeners (e.g., Table 16.2; Fig. 16.6b).

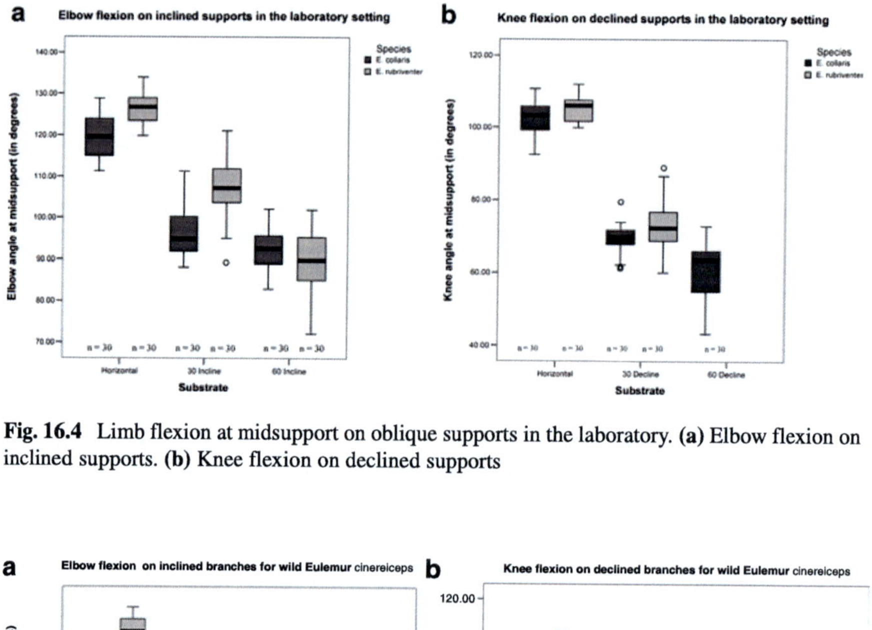

Fig. 16.4 Limb flexion at midsupport on oblique supports in the laboratory. **(a)** Elbow flexion on inclined supports. **(b)** Knee flexion on declined supports

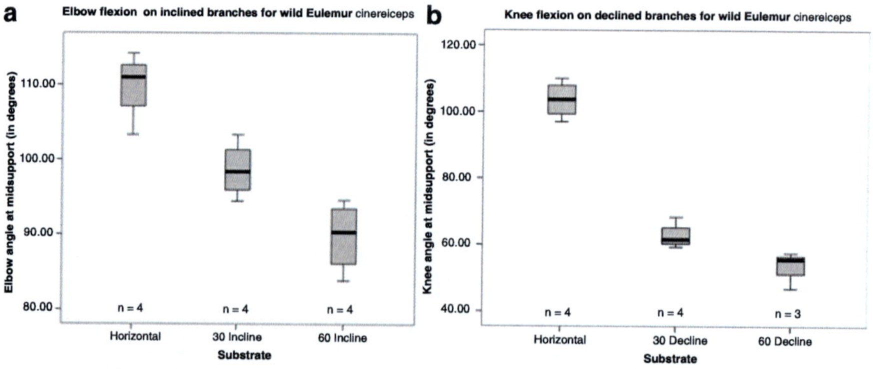

Fig. 16.5 Limb flexion at midsupport on oblique supports in the field. **(a)** Elbow flexion on inclined supports. **(b)** Knee flexion on declined supports

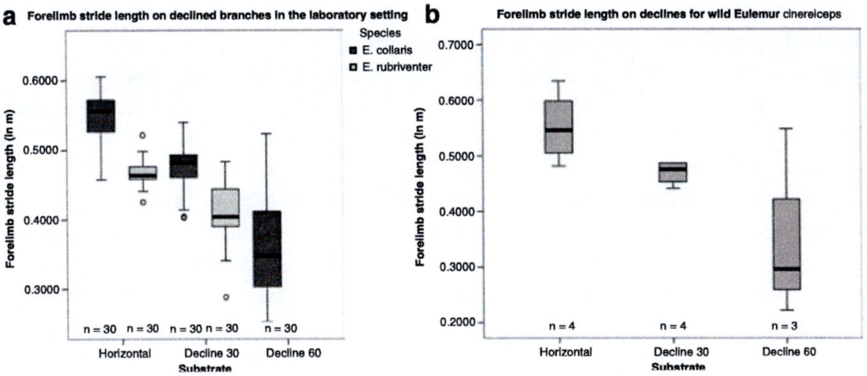

Fig. 16.6 Forelimb stride lengths on declined branches. **(a)** *Eulemur collaris* and *E. rubriventer* in the laboratory setting. **(b)** *Eulemur cinereiceps* in the natural setting

Discussion

Laboratory studies offer clear benefits for the collection of kinematic data, utilizing sophisticated camera setups and/or cineradiographic approaches in well lit spaces specifically designed to capture and accurately document limb movements and postural adjustments with large samples of strides collected in a repeatable manner (e.g., Schmidt and Fischer 2000; Polk 2001). In addition, lab based studies allow us to isolate different characteristics of arboreal substrates, permitting the examination of their separate and combined effects upon locomotion (e.g., Stevens 2003, 2006, 2008).

Yet, the laboratory setting necessarily limits locomotor choices available to study subjects. By design, substrate complexity in experimental studies rarely approaches that available in the natural environment. Captive animals may never be inspired to move as rapidly as they do when faced with a predator, competitor, or potential mate in the wild; they may not ever reach so far to select that perfectly ripe fruit (Stevens and Carlson 2008). An even more fundamental problem arises from the fact that laboratory experimental subjects are often housed in small cages, and often lack regular access to natural supports and enclosures of sufficient size to enjoy the daily path lengths and overall activity levels reflective of their wild counterparts (Chang et al. 1999; Stevens and Carlson 2008). Food provided at regular intervals obviates the need for physically demanding travel necessitated by foraging, and often renders captive individuals less able to be active and exhibit behaviors similar to individuals of their species in the wild.

Hence the collection of detailed kinematic data in the wild can offer some clear advantages for capturing postural and locomotor strategies that quadrupedal primates actually employ to maintain balance on arboreal supports. Yet, field studies are not without their challenges. Visibility of moving subjects in the leafy arboreal setting restricts the sample sizes of strides available for detailed kinematic analyses. Sunny days provide filming challenges in variable lighting and backlighting of study subjects, rainy days for maintaining lens clarity and protecting delicate electronic equipment from the weather. Navigation of natural terrain with cumbersome cameras and tripods is complicated by additional limitations imposed by camera battery life. Whereas some kinematic variables remain fairly robust to the challenges of field data collection, even in the best of natural conditions, object-observer positioning is expected to be more variable than in the laboratory, contributing to a realistic expectation of lower precision and/or accuracy for field kinematic data (Stevens et al. 2006). Linking approaches *ex situ*, capable of assembling large sample sizes of strides with strategic kinematic data collection in the field provides a compromise to explore the potential and actual locomotor niches in concert (Stevens and Carlson 2008). In this way, one can take advantage of the best of both laboratory and field data collection worlds, altering one substrate parameter at a time in the laboratory and recording precise kinematic responses. Lab-derived biomechanical models can be tested in the wild, to explore whether larger and more diverse sample sizes of animals make similar accommodations in their more complex natural habitats.

Exploring Kinematic Patterns in Experimental and Natural Conditions

This study revealed that the highly arboreal genus *Eulemur* exhibits limb kinematic responses to support orientation that are consistent between the laboratory setting and the rain forest habitat. Such observations are significant for documenting the utility of laboratory studies for documenting ecologically relevant behaviors. Given the greater variability in substrate orientations observed in the natural setting, taken together with variability in substrate diameter, flexibility, texture, and connectivity, these results are particularly compelling. Had we found different limb kinematic responses in *Eulemur* species between the laboratory and natural settings, additional data would be required to determine whether discrepancies might reflect differences between the substrate environments or whether they might reflect distinctive aspects of the species themselves.

To counteract shear forces, subjects traveling along oblique arboreal supports were expected to make a host of kinematic accommodations. For example, because the line of gravity falls more posteriorly along an inclined support, *Eulemur* was predicted to exhibit more retracted hind limbs at lift-off (e.g., Stevens and Larson 1999; Stevens 2003; Lammers et al. 2006). Conversely, as the line of gravity falls more anteriorly along a declined support, it was predicted that *Eulemur* would exhibit more protracted forelimbs at touchdown on these supports (e.g., Stevens and Larson 1999; Stevens 2003; Lammers et al. 2006). These excursion patterns were expected to improve balance on inclines and declines by keeping the gravity vector within the line of support defined by the placement of the hands and feet on the branch. Such adjustments may also be beneficial in generating greater propulsive forces to overcome gravity during forward progression on inclines, and enabling the application of greater braking forces to limit the tendency to accelerate out of control on declines. Finally, it was predicted that *Eulemur* would flex elbows and knees more on oblique supports to bring the center of mass closer to the support.

Each of these predictions was upheld in both the laboratory setting and in the wild. *Eulemur* individuals exhibited more protracted forelimbs at touchdown on declines (Figs. 16.2a, 16.3a), more retracted hind limbs at lift-off on inclines (Figs 16.2b, 16.3b), and more flexed elbows and knees at midsupport (Figs. 16.4, 16.5), resulting in significantly lower shoulder and hip heights on oblique supports (Table 16.3). Lorisids and cheirogaleids have been shown to exhibit similar limb excursion patterns on arboreal inclines and declines in the laboratory setting (e.g., Stevens 2003). Moreover, consistent patterns have been observed in the laboratory setting in New World monkeys (inclined treadmills: Vilensky et al. 1994; inclined poles: Stevens and Larson 1999; rope treadmills: Nyakatura et al. 2007), Old World Monkeys (inclined poles: Rollinson and Martin 1981), as well as in nonprimate mammals (*Monodelphis* on trackways and simulated arboreal supports: Lammers et al. 2002). Taken together, these observations suggest a common solution in limb alignment among quadrupedal mammals faced with the problem of an increased tendency to topple off of or slide down an oblique support. This study is the first to document kinematics of *Eulemur* in the natural setting, revealing consistent patterns in forelimb and hind limb touchdown and lift-off angles in the laboratory and the wild. This correspondence suggests that laboratory-based approaches can prove useful in the formulation of kinematic hypotheses about locomotion in the wild.

Why is this important? Clearly, primates have a myriad of possible responses to mechanical challenges, not limited to those related to limb flexion and excursion measured for this study, such as by altering locomotor mode altogether, e.g., descending feet-first, hopping or leaping, or by refusing to negotiate steep supports entirely as did *Eulemur rubriventer* on the steepest declines. Body mass, relative limb segment lengths, and joint configurations may each play a role in defining the range of potential options an individual may choose among in responding to locomotor challenges. In short, collecting basic kinematic data on *Eulemur* made it possible to test more explicitly whether locomotion observed in the simplified laboratory setting offers a valid window into natural behavior in the face of a frequently encountered mechanical challenge. Of course, comparisons in this study were necessarily restricted to reflect measurable kinematic variables on a particular set of support types, and likely many significant differences in oblique support kinematics remain between the laboratory and field settings. In short, this study focused on measuring variables relating to clear mechanical predictions for coping with inclines and declines. Should lab and wild observations differ greatly for those variables, there would be no reason for expecting this particular laboratory setting to effectively simulate natural locomotor conditions for the study of kinematics on oblique arboreal supports.

The Role of Kinematic Approaches in the Wild

The application of kinematic questions in the natural setting has become increasingly important for understanding locomotion in an ecologically relevant context. This study has demonstrated consistency between laboratory and field results for a number of forelimb and hind limb kinematic parameters. Other studies demonstrate a more complex reality. For example, whereas theoretical and laboratory biomechanical studies predict an increase in the use of diagonal sequence gait patterns in the terminal branch setting (e.g., Cartmill et al. 2002), Dunbar and Badam (2000) found quite the opposite in the wild, in that juvenile bonnet macaques in the natural setting actually utilized a *lower* proportion of diagonal sequence gaits in the terminal branch milieu than they did on more stable supports closer to the tree trunks. Interestingly, Isler and Grüter (2006) also reported the incorporation of lateral sequence gaits during vertical climbing in wild snub-nosed monkeys. Whereas many studies assume that animals optimize travel distances to minimize costs of locomotion, Dunbar and Badam (2000) note that given the choice, bonnet macaques may travel greater distances rather than crossing directly among terminal branches to avoid utilizing unstable supports. The use of unstable supports has been modeled in the laboratory as more energetically costly (Alexander 1991; Demes et al. 1995), yet, Thorpe et al. (2007) suggest that in the wild at least some primates increase energetic efficiency of locomotion through the use of compliant branches. In addition, although most laboratory studies utilize cylindrical simulated branches, *Eulemur cinereiceps* individuals, like the bonnet macaques examined by Dunbar and Badam (2000), often grasped small supports perpendicular to the path of motion on oblique supports, improving balance by spreading their body support over a broader area, and suggesting that more information about balance strategies may be gleaned in a more complex setting. Studies of red slender lorises in the natural setting have also documented a surprising array of unpredicted behaviors including the use of rapid arboreal quadrupedal locomotion at velocities far outstripping those collected for lorises in the laboratory setting (Nekaris and Stevens 2007). Whereas laboratory studies offer the opportunity to break down the natural environment into individual components for study, it is important to recognize that this simplification necessarily limits the choices available to focal subjects. In the laboratory setting, it may not be possible or even desirable to replicate the complexity of locomotor options available to animals in the wild. Field kinematic approaches provide complementary approaches to the study of primate locomotion, pivotal for addressing questions related to how individuals select among and use arboreal supports for locomotion in an ecologically relevant setting.

Challenges and Solutions for Integrating In Situ *and* Ex Situ *Data*

Clearly, it is far simpler to configure a kinematic data collection system in the laboratory setting than in the wild. As in any field study of positional behavior, visibility of the focal subjects in a forest setting often presents an issue. Individuals can move great distances over the course of just a few hours, sometimes appearing only for a moment in a particular camera view before moving on. To maximize the chance of collecting usable locomotor sequences, we restricted kinematic data collection to habituated individuals at Manombo Forest in Madagascar, with the assistance of graduate students and expert field guides well versed in their travel patterns. This provided the additional advantage of having data on the morphometrics of focal individuals, recorded in previous captures for behavioral studies. To minimize the impact of growth-related changes, we considered only subjects recorded as adults during previous captures in this study. This, along with the fact that these primates are critically endangered and quietly cryptic at Manombo, necessarily limited the sample size of strides in this analysis. We compensated for these challenges by maximizing the number of hours spent in the field each day, and by streamlining as much as possible the kinematic data collection setup.

Traditional kinematic setups can be heavy and cumbersome to move about within the field, not to mention the delicate and expensive nature of electronic equipment in terms of transport/repair. Designing a portable setup that could record locomotor kinematics in a remote setting required taking a large supply of lithium ion batteries and sufficient protection of the camera from rain. The solution outlined here utilizes a single rugged, lightweight and inexpensive camcorder/tripod setup for the field, with the potential for collecting 3D kinematic data using a simple trigonometric correction described in Stevens et al. (2006).

Future Directions for Linking Laboratory and Field Techniques

Despite a growing number of laboratory kinematic studies examining different aspects of primate arboreality, we still have relatively little information about the kinematic solutions primates employ to navigate their complex three-dimensional natural arboreal habitats. Merging data on "how often" primates employ different locomotor behaviors with specifically "how" they utilize natural substrates promises to be a significant avenue of research for biological anthropologists and primatologists. Melding laboratory and field approaches offers the hope of unraveling specific contributions of life history to both potential and actual locomotor behavior, e.g., in exploring the effects of ontogeny/aging, dimorphism, and gestation upon the solutions that primates have for solving problems posed by their ever-shrinking habitats. Quantifying the ways that habitat alteration can affect substrate use and locomotor energetics of different species will be particularly critical for informing conservation efforts. Preliminary work in rainforests of southeastern Madagascar suggests abundant opportunities exist for clear filming of habituated primates moving at will and at their own pace during their daily activities. Although animal visibility varies from site to site and season to season, it becomes increasingly evident that prospects are quite good for developing a larger bank of detailed, ecologically-relevant, kinematic data for primates engaging in their daily activities in the wild.

Conclusions

This study documents the effects of support orientation upon quadrupedal locomotion in *Eulemur*, and represents the first quantitative record of limb kinematics in the natural setting for the critically endangered *E. cinereiceps*. In both the laboratory and natural settings, forelimbs tend to be more protracted at touchdown on declines and hind limbs tend to be more retracted at lift-off on inclines. Moreover, individuals flex the elbows and knees more at midsupport on oblique branches, bringing the center of mass closer to the support, as evidenced by lower shoulder and hip distances from the substrate. This study represents a good test case for the integration of experimental and naturalistic data collection in the study of arboreal primate locomotor kinematics.

Acknowledgments We thank E. Vereecke and K. D'Août for overseeing the organization of this volume. S. Larson, B. Demes, J. Fleagle, W. Jungers, J. Stern, P. Wright, P. O'Connor, L. Jolley, S. Williams, K. Carlson, P. Lemelin, M. Cartmill, A. Biknevicius, and S. Reilly provided useful discussions on functional morphology. T. Cole, L-K. Chan, and D. Schmitt collaborated on developing a trigonometric correction for out-of-plane angles. Special thanks go to staff at the Duke Lemur Center, D. Brewer, W. Hess, C. Stahala, and B. Grossi. We especially thank field guides that habituated study animals as a part of a larger study of the feeding behavior and habitat use of *Eulemur cinereiceps*. We also gratefully acknowledge ANGAP staff and guides at Manombo Forest for logistical assistance. The following sources provided funding for this project: National Science Foundation (BCS-0086835), LSB Leakey Foundation, Sigma Xi Foundation, Stony Brook-IDPAS, and Ohio University College of Osteopathic Medicine.

References

Alexander R McN 1991 Elastic mechanisms in primate locomotion. Z Morphol Anthropol 78:315–320.

Barnes WJP 1997 Mechanisms of adhesion in tree frogs: an allometric study. J Morph 232:232.

Boinski S 1989 The positional behavior and substrate use of squirrel monkeys: ecological implications. J Hum Evol 18:659–677.

Cannon CH, Leighton M 1994 Comparative locomotor ecology of gibbons and macaques: selection of canopy elements for crossing gaps. Am J Phys Anthropol 94:505–524.

Cant JGH 1988 Positional behavior of long-tailed macaques (*Macaca fascicularis*) in northern Sumatra. Am J Phys Anthropol 76:29–37.

Cant JGH 1992 Positional behavior and body size of arboreal primates: a theoretical framework for field studies and an illustration of its application. Am J Phys Anthrop 88:273–283.

Cartmill M 1974 Pads and claws in arboreal locomotion. In: Jenkins FA (ed), Primate Locomotion. Academic Press, New York 45–83.

Cartmill M 1979 The volar skin of primates: its frictional characteristics and their functional significance. Am J Phys Anthropol 50:497–510.

Cartmill M, Lemelin P, Schmitt D 2002 Support polygons and symmetrical gaits in primates. Zool J Linn Soc 136:401–420.

Chang TR, Forthman DL, Maple TL 1999 Comparison of confined mandrill behavior (*Mandrillus sphinx*) in traditional and ecologically representative exhibits. Zoo Biol 18:163–176.

Conover WJ, Iman RI 1981 Rank transformations as a bridge between parametric and nonparametric statistics. Am Stat 35:124–133.

Crompton RH 1984 Foraging, habitat structure, and locomotion in two species of *Galago*. In: Adaptations for Foraging in Nonhuman Primates. Columbia University Press, New York 73–111.

Dagosto M 1995 Seasonal variation in positional behavior of Malagasy lemurs. Int J Primatol 16:807–833

Demes B, Jungers WL, Gross TS, Fleagle JG 1995 Kinetics of leaping primates: influence of substrate orientation and compliance. Am J Phys Anthropol 96:419–429.

Dunbar DC, Badam GL 2000 Locomotion and posture during terminal branch feeding. Int J Primatol 21:649–669.

Fleagle JG 1977a Locomotor behavior and muscular anatomy of sympatric Malaysian leaf monkeys (*Presbytis obscura* and *Presbytis melalophos*). Am J Phys Anthropol 46:297–308.

Fleagle JG 1977b Locomotor behavior and skeletal anatomy of sympatric Malaysian leaf monkeys *Presbytis obscura* and *Presbytis melalophos*. Yrbk Phys Anthropol 20:440–453.

Fleagle JG 1979 Primate positional behavior and anatomy: naturalistic and experimental approaches. In: Morbeck ME, Preuschoft H, Gomberg N (eds), Environment, Behaviour and Morphology: Dynamic Interactions in Primates. Gustav Fischer, New York 313–325.

Fleagle JG, Mittermeier RA 1980 Locomotor behavior, body size, and comparative ecology of seven Surinam monkeys. Am J Phys Anthropol 52:301–14.

Fleagle JG 1999 Primate Adaptation and Evolution. Academic Press, New York.

Ganzhorn JU 1985 Habitat separation of semifree-ranging *Lemur catta* and *Lemur fulvus*. Folia Primatol 45:76–88.

Garber PA 1984 Use of habitat and positional behavior in a neotropical primate, *Saguinus oedipus*. In: Adaptations for Foraging in Nonhuman Primates. Columbia University Press, New York 113–133.

Giannoni SM, Borghi CE, Laconi MR 1999 Climbing ability of *Microtus (Terricola) duodecimcostatus* and *M. (T.) gerbei*. J Mammal 63:41–49.

Grand TI 1972 A mechanical interpretation of terminal branch feeding. J Mammal 53:198–201.

Grand TI 1984 Motion economy within the canopy: four strategies for mobility. In: Adaptations for Foraging in Nonhuman Primates. Columbia University Press, New York 54–72.

Haffner M 1996 A comparison of the gross morphology and micro-anatomy of the foot pads in two fossorial and two climbing rodents (Mammalia). J Zool 244:287–294.

Hamrick MW 1996 Articular size and curvature as determinants of carpal joint mobility and stability in strepsirhine primates. J Morphol 230(2):113–127.

Hirasaki E, Kumakura H, Matano S 1993 Kinesiological characteristics of vertical climbing in *Ateles geoffroyi* and *Macaca fuscata*. Folia Primatol 61:148–156.

Hunt KD, Cant J, Gebo DL, Rose MD, Walker SE, Youlatos D 1996 Standardized descriptions of primate locomotor and postural modes. Primates 37:363–387.

Iman RL 1974 A power study of a rank transform for the two-way classification model when interaction may be present. Can J Sci (C) 2:227–239.

Isler K, Grüter CC 2006 Arboreal locomotion in wild black-and-white snub-nosed monkeys (*Rhinopithecus bieti*). Folia Primatol 77:195–211.

Jayne BC, Irschick DJ 1999 Effects of incline and speed on the three-dimensional hind limb kinematics of a generalized iguanian lizard (*Dipsosaurus dorsalis*). J Exp Biol 202:143–159.

Johnson S, Lei R, Martin S, Irwin M, Louis E 2008 Does *Eulemur cinereiceps* exist? Preliminary evidence from genetics and ground surveys in southeastern Madagascar. Am J Primatol 70:372–385.

Jouffroy FK, Lessertisseur J 1978 Eco-morphological study of limb proportions among primates, and especially among prosimians. Ann Sciences Nat Zool Biol Anim 20:99–128.

Jungers WL 1985 Body size and scaling of limb proportions in primates. In: Size and Scaling in Primate Biology. Plenum Press, New York 345–381.

Krakauer E, Lemelin P, Schmitt D 2002 Hand and body position during locomotor behavior in the Aye-Aye (*Daubentonia madagascariensis*). Am J Primatol 57:105–118.

Laborde C 1986 Description de la locomotion arboricole de *Cryptoprocta ferox* (Carnivore: Viverride, Malagache). Mammal 50:369–378.

Lammers AR, Parchman AJ, Earls KD, Biknevicius AR, Reilly SM, Stevens NJ 2002 The effects of incline on limb forces in a quadrupedal mammal. Int Comp Biol 42(6):1261.

Lammers AR, Earls KD, Biknevicius AR 2006 Locomotor kinetics and kinematics on inclines and declines in the gray short-tailed opossum *Monodelphis domestica.* J Exp Biol 209:4154–4166.
Larson SG, Stern JT 1992 Further evidence for the role of supraspinatus in quadrupedal monkeys. Am J Phys Anthropol 87:359–363.
Larson SG 1995 New characters for the functional interpretation of primate scapulae and proximal humeri. Am J Phys Anthropol 98:13–35.
Lemelin P 1996 The evolution of manual prehensility in primates: a comparative study of prosimians and didelphid marsupials. PhD dissertation, SUNY-Stony Brook.
Lemelin P, Schmitt D 1997 The relation between hand morphology and quadrupedalism in primates. Am J Phys Anthropol 105:185–197.
Meldrum DJ 1991 Kinematics of the cercopithecine foot on arboreal and terrestrial substrates with implications for the interpretation of hominid terrestrial adaptations. Am J Phys Anthropol 84:273–289.
Meldrum DJ, Dagosto M, White J 1997 Hind limb suspension and hind foot reversal in *Varecia variegata* and other arboreal mammals. Am J Phys Anthropol 103:85–102.
Mittermeier RA, Konstant WR, Hawkins F, Louis EE et al. 2006 Lemurs of Madagascar. Conservation International, Washington, D.C.
Mittermeier RA, Ratsimbazafy J et al. 2007 Primates in Peril: the world's 25 most endangered primates, 2006–2008. Prim Conserv 22:1–40.
Morbeck ME 1976 Leaping, bounding, and bipedalism in *Colobus guereza*: a spectrum of positional behavior. Yrbk Phys Anthropol 20:408–420.
Morbeck ME 1979 Forelimb use and positional behavior in *Colobus guereza*: integration of behavioral, ecological and anatomical data. In: Morbeck ME, Preuschoft H, Gomberg N (eds), Environment, Behaviour and Morphology: Dynamic Interactions in Primates. Gustav Fischer, New York 95–117.
Nekaris KAI, Stevens NJ 2007 Not all lorises are slow: rapid arboreal locomotion in *Loris tardigradus* of southwestern Sri Lanka. Am J Primatol 69:112–120.
Nyakatura JA, Fischer MS, Schmidt M 2007 Gait parameter adjustments of cotton-top tamarins (*Saguinus oedipus*, Callitrichidae) to locomotion on inclined substrates. Am J Phys Anthropol 135:13–26.
Overdorff DJ 1996 Ecological correlates to activity and habitat use of two prosimian primates: *Eulemur rubriventer* and *Eulemur fulvus rufus* in Madagascar. Am J Primatol 40:327–342.
Oxnard CE 1976 Primate quadrupedalism: some subtle structural correlates. Yrbk Phys Anthropol 20:538–553.
Polk JD 2001 The influence of body size and body proportions on primate quadrupedal locomotion. PhD dissertation, SUNY-Stony Brook.
Polk JD, Psutka SP, Demes B 2005 Sampling frequencies and measurement error for linear and temporal gait parameters in primate locomotion. J Hum Evol 49:665–679.
Prost J, Sussman JH 1969 Monkey locomotion on inclined surfaces. Am J Phys Anthropol 31:53–58.
Quinn TH, Baumel JJ 1993 Chiropteran tendon locking mechanism. J Morphol 216:197–208.
Ratsimbazafy JH 2002 On the brink of extinction and the process of recovery: *Varecia variegata variegata* in Manombo Forest. PhD dissertation, SUNY-Stony Brook.
Ripley S 1979 Environmental grain, niche diversification, and positional behavior in Neogene primates: An evolutionary hypothesis. In: Morbeck ME, Preuschoft H, Gomberg N (eds), Environment, Behavior, and Morphology: Dynamic Interactions in Primates. Gustav Fischer, New York 37–74.
Rollinson J, Martin RD 1981 Comparative aspects of primate locomotion with special reference to arboreal cercopithecines. Symp Zool Soc Lond 48:377–427.
Schmidt M, Fischer MS 2000 Cineradiographic study of the forelimb movements during quadrupedal walking in the brown lemur (*Eulemur fulvus*, Primates: Lemuridae). Am J Phys Anthropol 111:245–262.
Schmitt D, Larson SG 1995 Heel contact as a function of substrate type and speed in primates. Am J Phys Anthropol 96:39–50.

Simmons NB, Quinn TH 1994 Evolution of the digital tendon locking mechanism in bats and dermopterans: a phylogenetic perspective. J Mamm Evol 2:231–254.

Smith RJ, Jungers WL 1997 Body mass in comparative primatology. J Hum Evol 32:523–559.

Sokal R, Rohlf F 1981 Biometry, 3rd ed. WH Freeman, San Francisco.

Stevens NJ, Larson SG 1999 The influence of substrate orientation on shoulder and hip angular excursions in *Aotus*. Am J Phys Anthropol Suppl 28:257–258.

Stevens NJ 2003 The influence of substrate size, orientation and compliance upon prosimian arboreal quadrupedalism. PhD dissertation, SUNY-Stony Brook.

Stevens NJ 2006 Stability, limb coordination and substrate type: the ecorelevance of gait sequence pattern in primates. J Exp Zool 305A:953–963.

Stevens NJ, Schmitt DO, Cole TM, Chan LK 2006 Technical note: out-of-plane angular correction based on a trigonometric function for use in two-dimensional kinematic studies. Am J Phys Anthropol 129:399–402.

Stevens NJ 2008 The effect of branch diameter on primate gait sequence pattern. Am J Primatol 70:356–362.

Stevens N, Carlson K 2008 Bridging gaps between experimental and naturalistic approaches in the study of primate behavior. Int J Primatol 29(6): 1395–1399.

Sussman RW 1976 Ecological distinctions in sympatric species of *Lemur*. In: Prosimian Behavior. Martin RD, Doyle GA & Walker AC, eds. pp 75–108. Duckworth, London.

Thorpe SKS, Crompton RH, Alexander RMcN 2007 Orangutans use compliant branches to lower the energetic cost of locomotion. Biol Lett 3(3):253–256.

Turnquist JE, Schmitt D, Rose MD, Cant JGH 1999 Pendular motion in the brachiation of captive *Lagothrix* and *Ateles*. Am J Primatol 48:263–281.

Tuttle RH 1969 Quantitative and functional studies on the hands of the anthropoidea: I. The Hominoidea. J Morph 128:309–363.

Vilensky JA, Moore AM, Libii AN 1994 Squirrel monkey locomotion on an inclined treadmill: implications for the evolution of gaits. J Hum Evol 26:375–386.

Ward C, Sussman RW 1979 Correlates between locomotor anatomy and behavior in two sympatric species of lemur. Am J Phys Anthropol 50:575–590.

Wunderlich RE 1999 Pedal form and plantar pressure distribution in anthropoid primates. PhD dissertation, SUNY-Stony Brook.

Chapter 17
Quadrupedal Locomotion of *Saimiri boliviensis*: A Comparison of Field and Laboratory-based Kinematic Data

Liza J. Shapiro, Jesse W. Young, and Art Souther

Abstract As a result of a plethora of lab-based studies focusing on primate quadrupedalism, it is well known that compared to most other mammals, primates exhibit distinctive quadrupedal kinematics when moving on artificial "terrestrial" or "arboreal" substrates. However, we have little knowledge of how quadrupedal kinematics are impacted by the complexity of natural habitats, in which pathways may be obstructed, unstable, or vary dramatically in size, orientation, shape, or texture. In this study, we compared data on the quadrupedal kinematics of *Saimiri boliviensis* in both laboratory and field settings by comparing kinematic responses across laboratory substrates (pole, floor) and natural substrates (branches that varied in size and orientation). Field results indicate that *Saimiri boliviensis* adjusted to larger branches by increasing limb duty factors, but used a wide variety of gait types (as measured by limb phase) across all branch sizes and orientations, rather than fine tuning limb phase to these aspects of substrate. Lab poles elicited similar average limb phases and duty factors, but reduced gait flexibility compared to branches. Lab studies would benefit from greater complexity of simulated arboreal substrates, and field studies should strive to measure numerous substrate characteristics to most effectively test hypotheses about the adaptive nature of primate locomotion.

Keywords Gait • Quadrupedalism • New World monkeys • Squirrel monkeys

Abbreviations

DS	diagonal sequence
DSDC	diagonal-sequence, diagonal-couplets
LS	lateral sequence

L.J. Shapiro (✉)
University of Texas at Austin, USA
e-mail: liza.shapiro@mail.utexas.edu

K. D'Aoû̂t and E.E. Vereecke (eds.), *Primate Locomotion: Linking Field and Laboratory Research*, Developments in Primatology: Progress and Prospects,
DOI 10.1007/978-1-4419-1420-0_17,

LSDC	lateral-sequence, diagonal-couplets
LSLC	lateral-sequence, lateral-couplets
p	probability level
r	Pearson product-moment correlation
rho (ρ)	Spearman rank-order correlation

Introduction

Primates exhibit a highly diverse and specialized repertoire of locomotor behaviors, indicating that locomotion has played a key role in their evolutionary adaptive strategies. Accordingly, much research has been devoted to understanding the morphological, biomechanical, and ecological factors associated with locomotor variation and evolution across the primate order. The accomplishment of these research goals requires both field and laboratory data. Field studies are critical for providing the ecological context for primate locomotion, documenting the relative frequencies of positional behaviors used by a particular species, the context in which they are used, e.g., travel versus feeding, and the types of substrates on which certain behaviors are preferred. Laboratory studies benefit from the ability to isolate and measure aspects of locomotor biomechanics or morphology and to test specific functional hypotheses about muscle function, bone structure, and locomotor kinematics and kinetics. While field studies have been limited by the inability to use complex equipment to measure locomotor biomechanical variables directly on primate subjects, laboratory studies suffer from the isolation of primate locomotion from the wide variety and complexity of substrates to which primates have become adapted in their natural habitat. These two approaches are complementary; the data provided by one can and should be used to provide key insights into the other. Moreover, recent attempts at *overlapping* the two methods hold much promise toward providing a more complete analysis of primate locomotion. For example, laboratory studies have worked toward increasing the complexity of substrates on which primates are tested (e.g., Stevens 2006, 2007, chapter 16; Nyakatura et al. 2008) or examining nonstereotypical movements such as turning (e.g., Demes et al. 2006). Conversely, field-based studies of locomotion are experimenting with methods to measure detailed aspects of locomotor morphology and energetics from a distance (Sellers and Crompton 2004; Rothman et al. 2008; Blanchard et al., chapter 10; Pontzer et al., chapter 15).

One type of primate locomotor behavior that has been very well studied is quadrupedalism. Based nearly exclusively on data collected in the lab, the kinematics and kinetics of primate quadrupedalism have been shown to be unusual among mammals, suggesting an adaptive advantage to this form of locomotion early in the evolution of primates. Unlike the quadrupedal walking of most other mammals, primate quadrupedalism is characterized by a preference for diagonal-sequence, diagonal-couplets (DSDC) gait, increased forelimb protraction, reduced vertical ground reaction forces on forelimbs relative to hind limbs, compliant gait, greater limb excursion angles, long stride lengths, and low stride frequencies (Hildebrand 1967; Kimura et al. 1979; Alexander and Maloiy 1984; Reynolds 1985; Demes et al. 1994; Larson et al. 2000, 2001; Li 2000; Cartmill et al. 2007b).

Given the (not unreasonable) assumption that quadrupedal biomechanics measured in the laboratory reflects similar behavior in the wild, evolutionary explanations for the distinctive aspects of primate quadrupedalism have focused on the importance of substrate type. Current consensus states that the unusual aspects of primate quadrupedalism are a biomechanical complex that gave early primates a selective advantage over their mammalian competitors by allowing them exclusive access to resources available in the "fine branch niche" (Cartmill 1972; Larson 1998; Cartmill et al. 2002; Schmitt and Lemelin 2002; Lemelin et al. 2003). Primate locomotor features are viewed as adaptations to movement on branches of narrow diameter and/or nonhorizontal orientation because they are theorized to increase balance and stability, reduce branch oscillations, and enhance the forelimb's manipulatory abilities (Prost and Sussman 1969; Rollinson and Martin 1981; Demes et al. 1994; Vilensky et al. 1994; Larson 1998; Schmitt 1999; Larson et al. 2000, 2001; Cartmill et al. 2002; Schmitt and Lemelin 2002; Lemelin et al. 2003 Schmitt 2003a, c; Stevens 2003). The convergent expression of these kinematic features, along with the presence of grasping hands and feet in some arboreal marsupials (and their absence in more terrestrial mammals), has provided further support for the importance of small branches in the evolution of primate quadrupedal locomotion (Hildebrand 1976; White 1990; Pridmore 1994; Larson et al. 2000; Schmitt and Lemelin 2002; Lemelin et al. 2003).

Our current view of the adaptive advantage of primate quadrupedalism has benefited greatly from numerous laboratory studies demonstrating that primates exhibit distinctive quadrupedal biomechanics when moving on artificial "terrestrial" versus "arboreal" substrates (Schmitt 1994, 1998, 1999; Schmitt and Hanna 2004; Franz et al. 2005; Wallace and Demes 2008; Young 2009). Researchers have also attempted to analyze the impact of more detailed arboreal environments on primate quadrupedalism by varying the size and/or inclination of simulated branches (usually continuous, smooth, stable poles; Schmitt 2003c; Stevens 2007; Nyakatura et al. 2008; cf. Stevens 2003, 2006). However, we have little knowledge of how quadrupedal kinematics are impacted by the complexity of natural habitats, in which pathways may be obstructed, unstable, or vary dramatically in size, shape, texture, or inclination. To progress toward a better understanding of primate quadrupedalism from an adaptive and evolutionary perspective, it is imperative that we get a broader picture of the variability in substrate use in natural habitats and how aspects of those substrates, e.g., size and inclination, affect quadrupedal kinematics.

In this study, we compare data on the quadrupedal kinematics of Bolivian squirrel monkeys (*Saimiri boliviensis*) in both laboratory and field settings, i.e., Cocha Cashu Biological Station, Manu National Park, Peru. In many ways, *Saimiri boliviensis* is an excellent species with which to explore the adaptive significance of primate gait kinematics. First, squirrel monkeys at Manu are exceedingly active, frequently traveling 2–5 km per day in order to evade predators and gain access to distributed foraging resources (Terborgh 1983; Mitchell 1990). In fact, squirrel monkeys are the most itinerant primates at Manu, with home ranges more than twice as large as similarly-sized primates at the site, e.g., *Cebus* and *Saguinus* (Terborgh 1983; Mitchell 1990). Second, squirrel monkeys frequently travel and forage on a variety of substrates that vary widely in diameter and orientation (Terborgh 1983; Boinski 1989; Fontaine 1990; Mitchell 1990; Arms et al. 2002). Finally, previous laboratory studies have provided conflicting data on the predominant pattern of interlimb coordination in squirrel monkeys. Prost and Sussman (1969) and Vilensky and colleagues (Vilensky and Patrick 1985; Vilensky et al. 1994) found that squirrel monkeys primarily used lateral sequence gaits when walking on declined and level substrates, but diagonal sequence gaits on inclined surfaces. In contrast, more recent observations indicate that diagonal sequence gaits predominate on all substrates (Arms et al. 2002; Schmidt 2005; see also Youlatos, chapter 14, on howlers).

Our objectives are to:

1) Provide additional field-based data documenting the range of variation of substrate size and orientation utilized by squirrel monkeys during quadrupedal walking and running in a natural habitat;
2) Provide data on footfall patterns and interlimb timing utilized by *Saimiri* for comparison to previous laboratory studies;
3) Assess whether artificial arboreal substrates capture similar quadrupedal behavior when compared to locomotion in natural habitats;
4) Evaluate the degree to which quadrupedal kinematics are "fine-tuned" to substrate characteristics; i.e., is kinematic variation across substrates greater than that within substrates?

Materials and Methods

Laboratory Data

J. Young collected laboratory data at the Center for Neotropical Primate Research and Resources (CNPRR, Mobile, AL). All procedures were approved by the CNPRR Institutional Animal Care and Use Committee (IACUC). The sample consisted of five female squirrel monkeys, ranging in age from 104 to 302 days and body mass from 218 to 535 g. Monkeys were filmed with a high-speed digital video camera (MotionMeter 1000, Redlake MASD, San Diego, CA) at 250 Hz as they traversed a 2.75 m × 0.3 m × 0.53 m runway. The floor of the runway was constructed from vinyl-coated plywood (Omega Signboard, Laminators Incorporated, Hatfield, PA). The top and front walls of the runway were formed from a single piece of angled Plexiglas, allowing the subject to be easily lighted and filmed. Depending on experimental condition, e.g. floor versus pole, monkeys traversed either the flat runway floor or a 3.2 cm diameter PVC pipe elevated

10.7 cm above the surface of the runway. Both substrates were coated with a mixture of polyurethane and nonskid paint additive (Behr Process Corporation, Santa Ana, CA) in order to increase traction.

Before the beginning of each squirrel monkey experiment, individuals were weighed and the skin over the approximate centers of rotation of the shoulder and the hip were shaved and marked with retro-reflective tape, a procedure that did not require the use of anesthesia. Video files were imported into the MATLAB DLT Dataviewer 2 digitizing platform (Hedrick 2007) for coding of kinematic variables. More details about the experimental apparatus and procedure can be found in Young (2009).

Field Data

We collected field data from videotapes taken by A. Souther, of *Saimiri boliviensis* moving in its natural forest habitat at Cocha Cashu Biological Station, Manu National Park, Peru. Manu National Park sits on the bank of a large river (the Rio Manu) and consists of undisturbed primary forest encompassing several different vegetation types, from riparian successional vegetation, to dense lacustrine swamps, to high ground tropical forests. Because *Saimiri* is not a habitat specialist (Boinski et al. 2002), but rather ranges widely over several microhabitats during the course of a day, the variety of vegetation types ensures that individuals encounter a diversity of substrate sizes and inclinations during daily travel. We collected video data over a period of 2 months in September–October 1998. Individual subjects were not identified, so the number of individuals or their ages is unknown, although the sample does appear to include some juveniles with fully independent locomotion. Because the period of study corresponds to the beginning of the wet season at Manu, when most births take place (Terborgh 1983; Mitchell 1990), any juveniles filmed would have been no younger than ca. 10–12 months old. We filmed monkeys with a hand-held camcorder (Canon ES5000) at 30 Hz. Video fields were subsequently split, resulting in an effective frame rate of 60 Hz. We selected usable video clips, i.e., those in which the camera was close enough for good visibility of limbs and trunk, and then imported them into Peak Motus (v. 9.2, Vicon Motion Systems, Oxford, UK) for coding of kinematic variables.

Kinematic Variables

The kinematic variables used in this study represent a subset of those that have been shown to vary with substrate type in previous laboratory studies and that were also easily measured from the field videos.

Limb phase: Limb phase describes both footfall sequence and interlimb timing, i.e., couplets (Hildebrand 1966, 1976). Divisions between named gaits, e.g., DSDC, LSDC, are a slight modification of the divisions of Hildebrand (1966, 1976) and follow those of Cartmill et al. (2002), in which values between 50 and 75 are designated as diagonal sequence, diagonal couplets (DSDC) gaits; values between 25 and 50 are lateral sequence, diagonal couplets (LSDC) gaits; and values between 0 and 25 are lateral sequence, lateral couplets (LSLC) gaits. Although limb phase is usually calculated based on the time lag between ipsilateral limb touchdown events (Hildebrand 1967), theory and data indicate that when forelimb and hind limb duty factors are unequal, calculating limb phase from mid-support events provides a more accurate description of interlimb coordination (Hildebrand 1976; Griffin et al. 2004). Therefore, because forelimb and hind limb duty factors were highly divergent across substrate categories (paired *t*-tests: all $p < 0.001$), we calculated limb phase as the proportion of stride duration separating hind limb and ipsilateral forelimb mid-support events (where mid-support is defined as the midpoint between touchdown and lift-off).

Duty factor: Duty factor is the proportion of stride duration that a limb is in contact with the substrate. Because duty factors in hind and forelimbs often differ, we report the mean duty factor across all four limbs as an index of the overall response to variation in substrate type, size, and orientation.

Relative speed: Owing to the lack of absolute scale in our field videos, we measured relative speed as trunk lengths per second for both laboratory and field data. For field data, we measured trunk length as the distance between shoulder and hip joints. We calculated relative speed by scaling trunk length to the distance traversed by the individual on the substrate during a full stride, i.e., relative trunk length, and dividing by stride duration.

In the laboratory, absolute speed (in meters per second) was calculated from the displacement of either the hip or the shoulder, depending on marker visibility. After transforming raw pixel coordinates into meters using a standard calibration object, we used linear least-squares regressions of corrected displacement data on time to calculate overall speed across each stride. We then calculated trunk length as the mean distance between the hip and shoulder across six stride events, e.g., forelimb and hind limb touchdown, mid-support and lift-off. We calculated relative speed as the quotient of absolute speed divided by trunk length. The range of relative speeds sampled was similar across both conditions (field: 0.71–5.0; laboratory 2.4–5.0), allowing comparisons of other variables with respect to speed.

Symmetry: In a perfectly symmetrical walk or run, a fore- or hind limb contacts the ground at exactly 50% of the interval of time between footfalls of the contralateral fore or hind limb (Hildebrand 1966). For this study, we excluded asymmetrical strides that were obviously gallops, bounds, or half-bounds, i.e., with whole-body aerial phases. Because perfect symmetry is rare even in gaits normally classified as "symmetrical," we included walking or running gaits that deviated from perfect symmetry. The average fore-hind symmetry values in our dataset ranged predominantly from 40% to 60%, i.e., 97% of all strides in the data set (Fig. 17.1), matching previous boundaries used to define symmetry (Schmitt et al. 2006). Nevertheless, as long as there was no whole-body aerial phase, we did not exclude strides with symmetry values below or above this range, as we feel this captures the more naturalistic locomotor behavior of the animals.

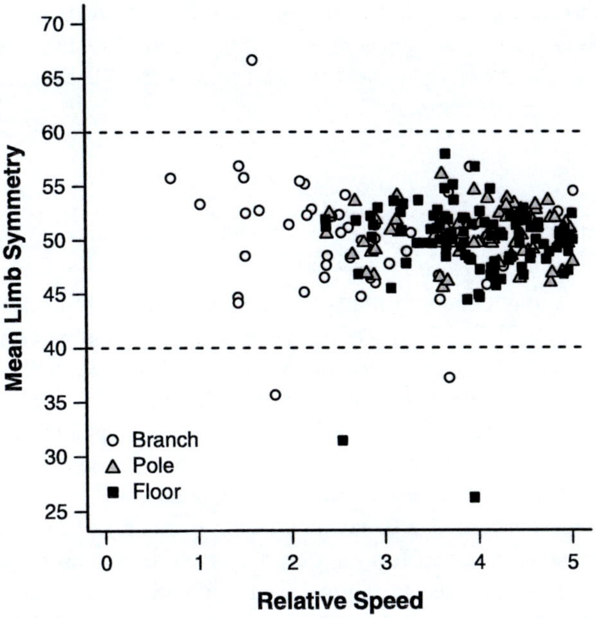

Fig. 17.1 Mean (fore-hind) limb symmetry *vs.* relative speed (trunk lengths/s)

Substrate Variables

In the field, we quantified substrates with respect to orientation and relative size. We determined substrate orientation by digitizing two endpoints of the substrate traversed by the individual for each stride included in the sample, and calculating its angular orientation relative to the horizontal plane. For categorical comparisons, "horizontal" included all substrates with orientations between –10 and +10 degrees. We categorized substrates with all other negative angles as declines and substrates with all other positive angles were categorized as inclines. Substrate orientations sampled ranged from –69° to +89°. Actual substrate sizes were unknown because the video images lacked an absolute scale. Therefore, we digitized substrate diameters, and calculated relative substrate size as substrate diameter divided by the animal's trunk length. Schultz (1963) reported that among squirrel monkeys, foot length is typically 41% of trunk length. Assuming the feet would need to grasp across at least half the circumference of a branch for it to be considered easily graspable (Cartmill 1974), we estimated that *Saimiri* would have more difficulty grasping branches with a circumference greater than 80% of trunk length. Since circumference is equal to π*diameter, we categorized relative substrate sizes (expressed as a proportion of trunk length) as "small" when below 0.25 and "large" when above 0.25. Overall, relative substrate sizes from the field data ranged from 0.04 to 0.73. In the lab, substrates consisted of an elevated 3.2 cm pole and the flat floor of the test runway, both of which were horizontal. So that "arboreal" laboratory and field data could be compared directly, the relative diameter of the pole was also expressed as a proportion of trunk length. Relative substrate size of the pole across all laboratory subjects ranged from 0.20 to 0.25 and was therefore comparable to a "small" branch.

Statistical Analyses

We used χ^2 tests of goodness of fit to examine proportional differences in categorical gait type, i.e., LSLC, LSDC, DSDC, attributable to substrate type, orientation, and relative size. Limb phase was non-normally distributed across most of our categorical subgroups. Categorical differences in limb phases were therefore examined using either nonparametric Kruskal-Wallis tests, supplemented by *post hoc* Wilcoxon rank-sum tests with a Sequential Bonferroni correction (Rice 1988), or rank-based analyses of covariance (ANCOVA: Conover and Iman 1981), specifying duty factor as the covariate. Mean duty factor, in contrast, was normally distributed across most subgroups. Because duty factor correlates strongly with relative speed across conditions (see later), we examined categorical differences in duty factor using ANCOVAs, specifying relative speed as the covariate. *Post hoc* analyses following significant ANCOVAs were examined using Tukey's T-method (Sokal and Rohlf 1995). Finally, we examined associations between continuous kinematic and substrate parameters using either Spearman rank-order (ρ) or Pearson product-moment correlations (r), depending on data normality.

Results

Comparison of laboratory- and field-based data on quadrupedal locomotion in Saimiri boliviensis

Limb Phase

In both laboratory and field, *Saimiri boliviensis* most frequently used DSDC gaits ($\chi^2_{[4]} = 35.7, p < 0.001$). Nevertheless, it is notable that squirrel monkeys also used LSDC or LSLC gaits in all three conditions, i.e., pole, floor, branches, albeit in lower frequencies (Fig. 17.2a, b). Limb phase correlates significantly negatively with duty factor on both branches and pole (branches: $\rho = -0.53, p < 0.001$; pole: $\rho = -0.50, p < 0.001$) but not on the floor ($\rho = 0.12, p = 0.24$). The lack of correlation between limb phase and duty factor during locomotion on the floor is due to the squirrel monkeys' flexible use of both DS and LS gaits at all duty factors. After controlling for the influence of duty factor when appropriate, we found *Saimiri boliviensis* to have used significantly higher average limb phases on the pole than

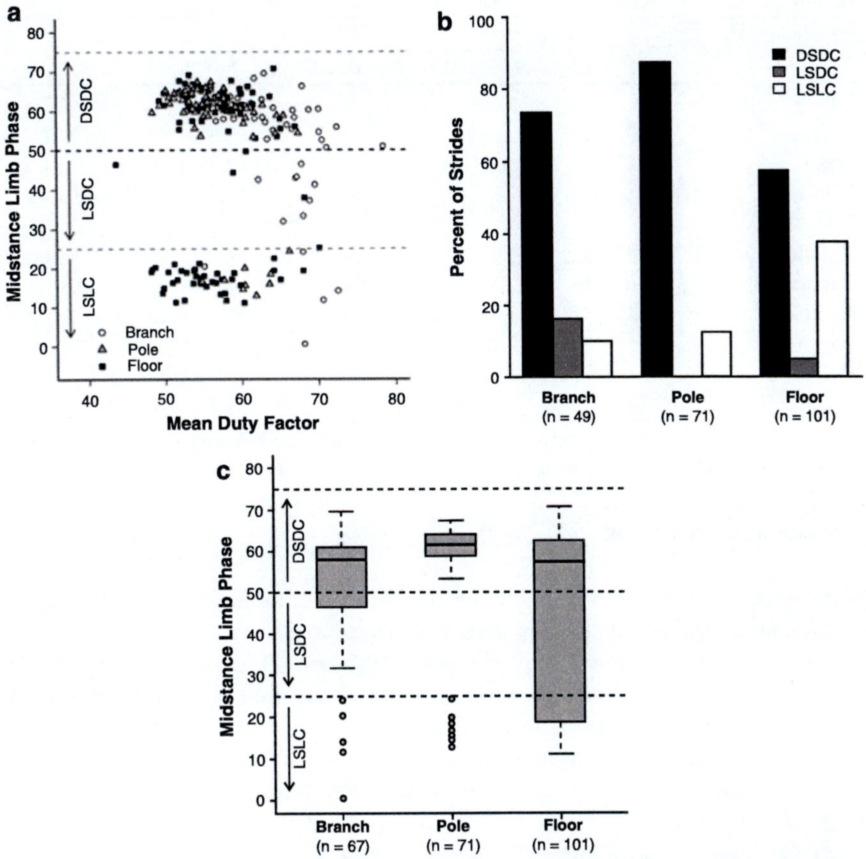

Fig. 17.2 Gait variation in *Saimiri boliviensis* across two laboratory substrates (pole, floor) and all natural substrates (branch). **(a)** Limb phase values for all strides plotted against mean duty factor. **(b)** Frequencies of each gait type. LS = lateral sequence, DS = diagonal sequence, DC = diagonal couplets, LC = lateral couplets. See text for definitions of limb phase values included in each gait category. **(c)** Box-and-whisker plots of limb phase, in which dark lines represent the median, gray boxes represent the interquartile range, dotted lines are 1.5 × interquartile range, and the circles are outliers

on the floor ($U_{S[101,71]} = 4627, p < 0.01$), but statistically similar average limb phases on the pole and on branches ($F_{[1,117]} = 1.0$, $p = 0.32$; Fig. 17.2c, Table 17.1). Limb phase did not differ between branches and the floor, perhaps due to increased variability in these conditions. On the floor, where balance issues are nonexistent, and hands and feet are not employed in grasping, any functional constraints on limb phase are likely alleviated, freeing the monkeys to utilize a wider variety of gaits (Vilensky and Larson 1989; Schmidt 2005). On the other hand, the variability of limb phase on branches cannot be interpreted without a more in-depth analysis of substrate variation encompassed by this category (see natural substrate section later).

Table 17.1 Summary statistics of the kinematic data set

		Gait type				
	n	LSLC (%)	LSDC (%)	DSDC (%)	Duty factor	Limb phase
Substrate type						
Branch	67	10	16	73	60.3 ± 0.05	51.5 ± 0.15
Pole	71	13	—	87	57.7 ± 0.04	56.1 ± 0.15
Floor	101	38	5	57	57.7 ± 0.04	43.9 ± 0.21
Branch orientation						
Decline	24	12	12	76	62.9 ± 0.05	51.0 ± 0.17
Horizontal	16	14	7	79	64.0 ± 0.05	50.7 ± 0.18
Incline	22	7	27	67	65.9 ± 0.05	51.7 ± 0.13
Branch size						
Small	34	11	7	87	60.2 ± 0.04	52.9 ± 0.15
Large	13	9	9	82	63.5 ± 0.04	54.7 ± 0.15

Means ± SE presented for duty factor and limb phase. Least-squares corrected means, evaluated at the overall mean of relative speed, are presented for duty factor.

Speed and Duty Factor

Consistent with the results of numerous laboratory studies, duty factor and relative speed had an inverse relationship across lab-based and field-based data (branch: $r = -0.62, p < 0.001$; pole: $r = -0.68, p < 0.001$; floor: $r = -0.53, p < 0.001$; Fig. 17.3). However, at a given relative speed, *Saimiri boliviensis* used higher duty factors on natural substrates than either the pole or floor ($F_{[2,219]} = 5.6$, $p < 0.01$). In other words, limb contact times were longer relative to stride duration on branches than on smooth poles or on a flat surface, even after controlling for the effects of speed (Fig. 17.3, Table 17.1).

Effects of Natural Substrate Variation on Quadrupedal Locomotion in Saimiri boliviensis

Substrate Orientation

Limb phase: Substrate orientation had no clear effect on limb phase. Squirrel monkeys used DSDC, LSDC, and LSLC gaits on inclining, declining, and horizontal branches but DSDC gaits were highly preferred on each type (Fig. 17.4a, b). There was no correlation between substrate angle and limb phase ($\rho = 0.006$, $p = 0.96$), and average limb phases did not differ across categories of substrate orientation ($H_{[2]} = 0.143$, $p = 0.931$; Fig. 17.4c, Table 17.1).

Speed and duty factor: Relative speed did not correlate with substrate orientation ($r = -0.15$, $p = 0.29$), indicating that squirrel monkeys used similar ranges of relative speed across declining, inclining, and horizontal branches. At a given relative speed, duty factors tended to be higher on inclines than on horizontal or declined branches (Fig. 17.5, Table 17.1), but this difference was not significant ($F_{[2,44]} = 1.2$, $p = 0.31$).

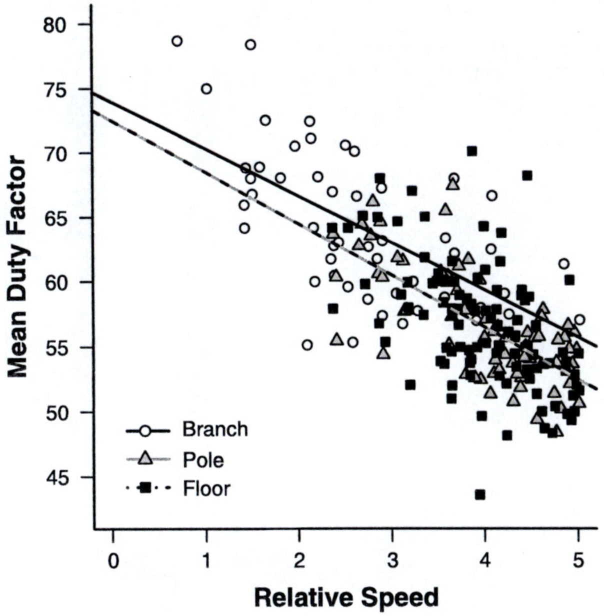

Fig. 17.3 The relationship between mean (fore-hind) duty factor and relative speed (trunk lengths/s) across two laboratory substrates (pole, floor) and all natural substrates (branch)

Relative Substrate Size

Limb phase: Relative substrate size had no clear effect on limb phase. Squirrel monkeys used DSDC, LSDC, and LSLC gaits on both small and large substrates (Fig. 17.6a, b). Similar to the results for substrate orientation, squirrel monkeys used predominantly DSDC gaits on both large and small substrates. There was no correlation between limb phase and relative substrate size ($\rho = 0.15$, $p = 0.36$), and average limb phases did not differ significantly between small and large branches ($U_{S[34,13]} = 124$, $p = 0.36$; Fig. 17.6c, Table 17.1).

Speed, duty factor: Squirrel monkeys did not modulate their speed in a consistent manner with respect to substrate size; relative speed and substrate size were uncorrelated ($r = 0.02$, $p = 0.90$). However, duty factor did vary with relative substrate size, even after controlling for the effects of speed ($F_{[2,114]} = 9.3$, $p < 0.001$). At a given relative speed, monkeys used significantly higher mean duty factors on large branches relative to small substrates, whether they are small branches or similarly sized poles (large branches – small branches: $t_{[55]} = 2.8$, $p < 0.05$; large branches – pole: $t_{[82]} = 4.3$, $p < 0.001$; Fig. 17.7, Table 17.1). Moreover, among all branches classified as "large," mean duty factor correlated significantly positively correlated with substrate size ($r = 0.73$, $p < 0.01$; Fig. 17.8). Although sample sizes were reduced when we examined limbs separately, both hind and forelimb duty factors increased with substrate size, and the correlation was stronger in the forelimb (hind: $r = 0.55$, $p = 0.051$; fore: $r = 0.71$, $p = 0.049$).

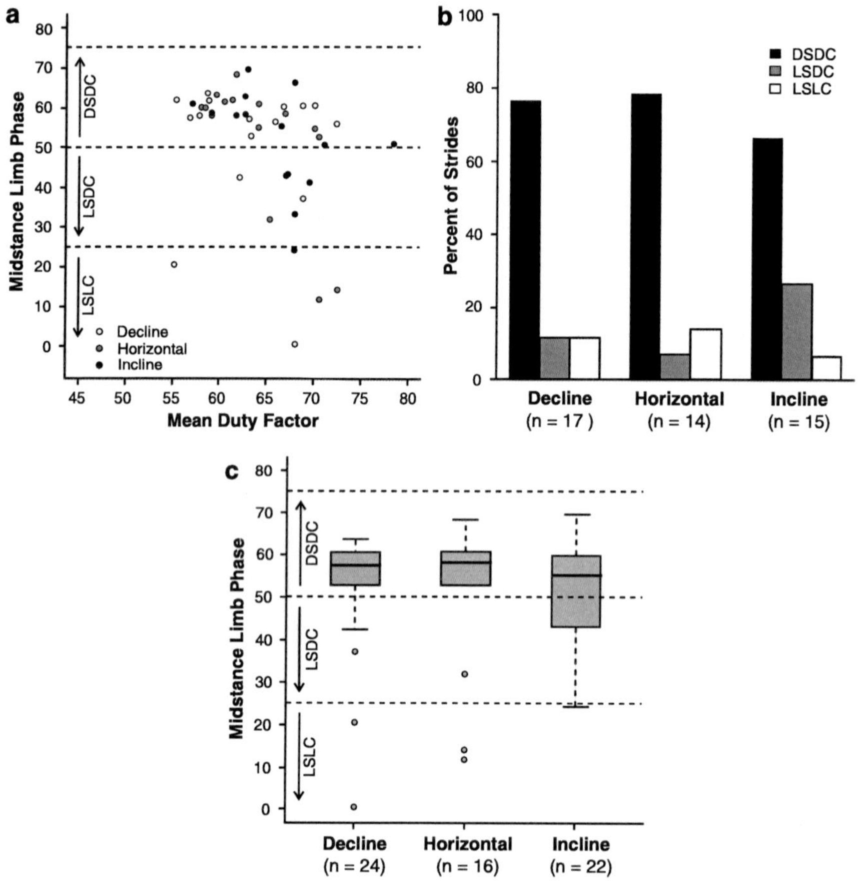

Fig. 17.4 Gait variation in *Saimiri boliviensis* across inclined, declined, and horizontal natural substrates. (**a**) Limb phase values for all strides plotted against mean duty factor. (**b**) Frequencies of each gait type. LS = lateral sequence, DS = diagonal sequence, DC = diagonal couplets, LC = lateral couplets. See text for definitions of limb phase values included in each gait category. (**c**) Box-and-whisker plots of limb phase, in which dark lines represent the median, gray boxes represent the interquartile range, dotted lines are 1.5 × interquartile range, and the circles are outliers

Summary of Results

1. Overall, limb phases did not differ with respect to substrate; DSDC gait was highly preferred by *Saimiri* on all substrates, in the laboratory as well as in its natural habitat.
2. LSDC and LSLC gaits were used occasionally by *Saimiri* on natural substrates of all three orientations (horizontal, incline, decline) and both sizes (small, large), and on a flat laboratory surface. LSLC, but not LSDC gaits, were used on the laboratory pole.
3. Variation in branch orientation or relative size did not affect relative speed.

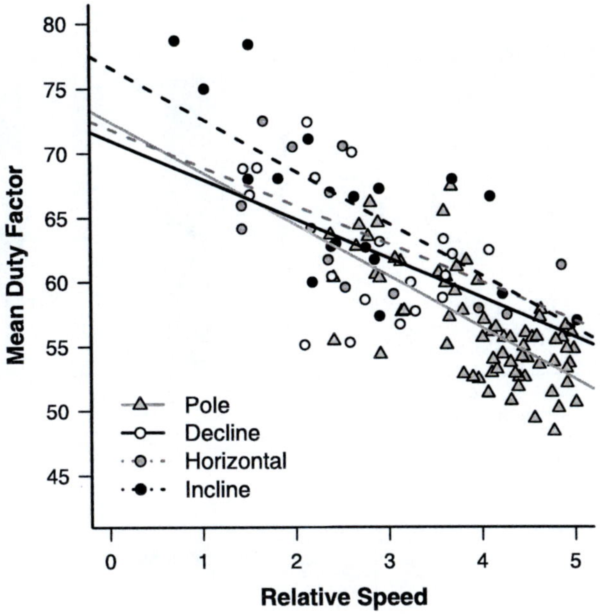

Fig. 17.5 The relationship between mean (fore-hind) duty factor and relative speed (trunk lengths/s) across inclined, declined and horizontal natural substrates, with horizontal laboratory pole included for comparison

4. Relative substrate size, but not substrate orientation, affected mean duty factor. At a given speed, *Saimiri* used significantly higher mean duty factors, i.e., relative contact times, on large branches than on small branches or poles of similar diameter to small branches, and mean duty factors increased as large branches increased in relative size.

Discussion

Previous laboratory studies on *Saimiri* assessed quadrupedal kinematics while subjects walked on flat horizontal or flat inclined "boards" (Prost and Sussman 1969); flat horizontal, inclined, or declined treadmills (Vilensky and Patrick 1985; Vilensky et al. 1994); horizontal wooden poles (Schmidt 2005); or a variety of substrates (Arms et al. 2002). Previous field studies of locomotion in *Saimiri* have provided ecological data on the relative frequency of quadrupedal locomotion on various substrates (Fleagle and Mittermeier 1980; Boinski 1989; Fontaine 1990; Mitchell 1990; Johnson and Shapiro 1998; Youlatos 1999). Our study expands on previous work by providing additional lab-based data, in conjunction with the first quantitative analysis of quadrupedal kinematics in *Saimiri* in a natural habitat. This allows us to assess 1) the consistency of

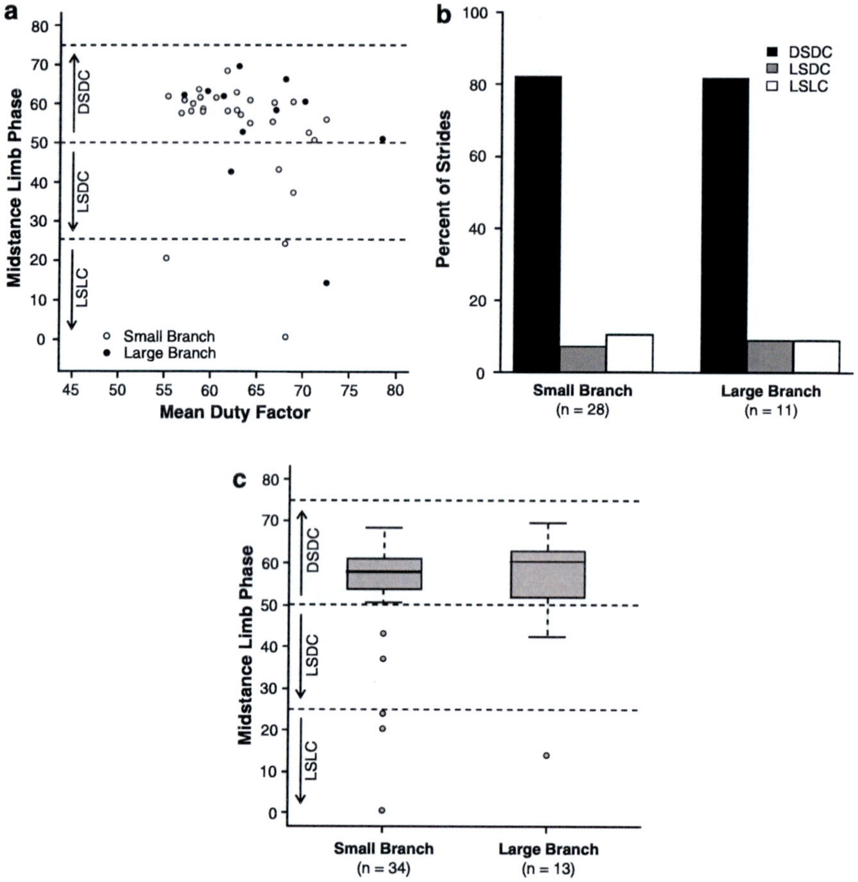

Fig. 17.6 Gait variation in *Saimiri boliviensis* across small and large natural substrates. (**a**) Limb phase values for all strides plotted against mean duty factor. (**b**) Frequencies of each gait type. LS = lateral sequence, DS = diagonal sequence, DC = diagonal couplets, LC = lateral couplets. See text for definitions of limb phase values included in each gait category. (**c**) Box and whiskers plots of limb phase, in which dark lines represent the median, gray boxes represent the interquartile range, dotted lines are 1.5 × interquartile range, and the circles are outliers

results across several laboratory-based studies, 2) the extent to which squirrel monkeys adjust their quadrupedal gait characteristics in response to substrate variation in size and angular orientation, and 3) the extent to which lab-based data represent natural locomotor behavior. Our data also provides insight on the benefits and limitations of both laboratory- and field-based kinematic analysis.

Our laboratory analysis tested horizontal substrates only. Prost and Sussman (1969) reported that *Saimiri* used LS gaits 63% of the time on level ground, and Vilensky et al. (1994) found exclusively LS gaits on a horizontal treadmill. Schmidt (2005) found that *Saimiri* used DS gaits exclusively on a 3 cm horizontal pole, and Arms et al. (2002) found that *Saimiri* used DS gaits nearly exclusively

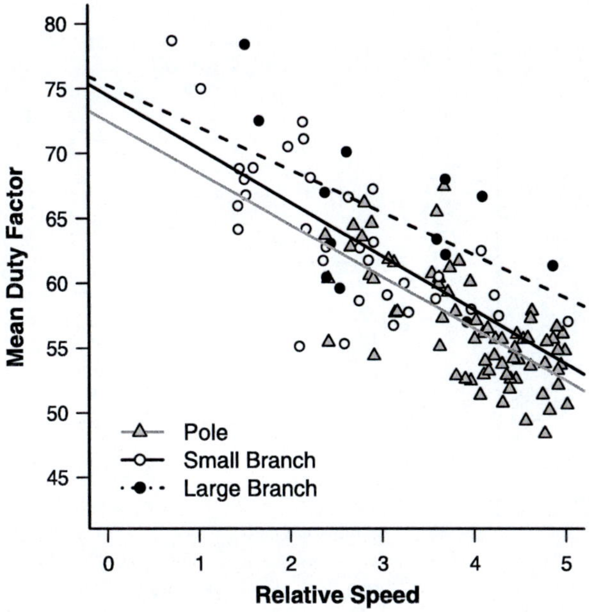

Fig. 17.7 The relationship between mean (fore-hind) duty factor and relative speed (trunk lengths/s) across small and large natural substrates with laboratory pole (similar in size to small branches) included for comparison

(95% of all strides) on a variety of laboratory-constructed substrates. Our laboratory data do not correspond to any of these previous studies. On the floor, rather than preferring LS gaits, our squirrel monkeys preferred DSDC gaits (57% of the time), even though they used LSLC fairly often (38%) and LSDC occasionally (5%). On our pole, rather than using DSDC gaits exclusively, squirrel monkeys used DSDC 87% of the time, and LSLC gaits the rest of the time. The variability of gait choice on horizontal flat surfaces both within our study and across other laboratory studies is consistent with Schmidt's (2005) observation that if DSDC gait evolved for an arboreal adaptive advantage, primates' limb phases should be less constrained in situations where grasping is not employed. Schmidt's (2005) explanation is also consistent with her laboratory animals' exclusive use of DS gaits on a pole, but begs the question as to why our squirrel monkeys used LSLC gaits in conjunction with DSDC gaits on a pole. It is possible that the use of LSLC gaits in our squirrel monkeys was attributable to the fact that they were infants and juveniles, as other primates have been shown to use this type of gait as a transitory ontogenetic phase (Hurov 1982; Nakano 1996; Shapiro and Raichlen 2005, 2006). We cannot exclude this explanation without further analysis, but at a minimum, our data do not reveal a strict correlation of age with limb phase. It is also possible that the difference in pole data between our study and that of Schmidt (2005) simply expresses the flexibility of gait choice in primates (Vilensky and Larson 1989; Vilensky and Moore 1992).

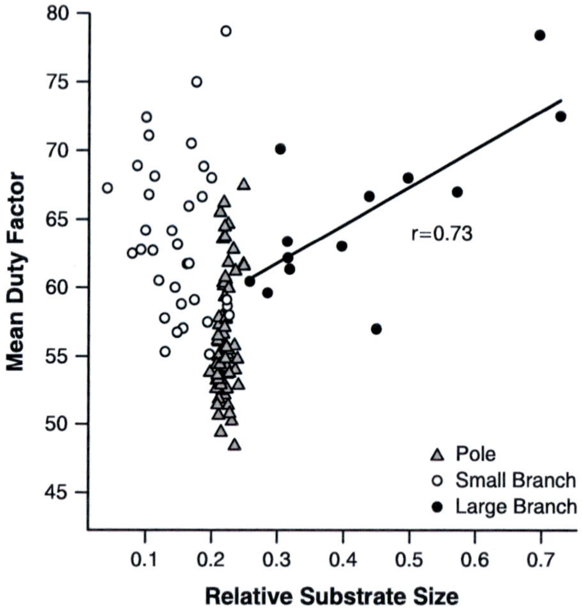

Fig. 17.8 The relationship between mean (fore-hind) duty factor and relative substrate size. Within branches categorized as "large," mean duty factor and substrate size correlate significantly positively

Our field analysis allowed us to test the effects of relative substrate size and orientation on quadrupedal kinematics. Current hypotheses emphasize that DS gait and other unusual aspects of primate quadrupedalism, e.g., accentuated forelimb protraction at touchdown, higher peak vertical forces on hind limbs than forelimbs, increased limb yield, long stride lengths, low stride frequencies, long limb contact times, most likely evolved because they provide a particular advantage for navigating "small" branches (Larson 1998; Schmitt and Lemelin 2002; Cartmill et al. 2002, 2007a,b). This view is supported by laboratory studies demonstrating that when primates switch from the floor to an artificial arboreal substrate such as a pole, or from larger to smaller poles, at least some aspects of their kinematics become more "primate-like" (Schmitt 1999, 2003b; Schmitt and Hanna 2004; Stevens 2007; Wallace and Demes 2008). Specific to the variables addressed in our study, laboratory studies have shown that DS gait increases in frequency on poles compared to floors (Wallace and Demes 2008; this study), and on inclines compared to declining or horizontal substrates (Prost and Sussman 1969; Vilensky et al. 1994; Stevens 2003; Nyakatura 2008). Duty factors (or limb contact times) have been shown to increase on poles compared to floors (Schmitt 1999), on relatively smaller compared to larger poles, and on declines (Stevens 2003). Therefore, previous laboratory studies combined with evolutionary hypotheses would lead to the prediction that limb phase should increase on relatively small and/or inclined substrates, and duty factor, i.e., relative limb contact time, should increase on relatively small and declined substrates.

To the contrary, at Manu, squirrel monkeys did not adjust limb phase in any consistent manner in response to changing substrate size or orientation. The influence of substrate size on limb phase has not been widely studied. However, contrary to the consensus view of primate quadrupedalism, Dunbar and Badam (2000) observed that juvenile macaques in a natural setting preferred DS on the large end of branches and LS on the smaller, distal stems. Our field results are more consistent with the only primate laboratory study directly assessing limb phase and relative substrate size (Stevens 2007) in which six strepsirrhines did not alter limb phase patterns on small versus large poles. In all other respects, however, our field study's results are not consistent with most primate laboratory studies or ecologically based evolutionary hypotheses. In *Saimiri boliviensis*, DS does not appear to be particularly (nor exclusively) functionally associated with smaller or inclined substrates; it is used just as frequently on horizontals, declines and relatively large substrates. In addition, *Saimiri boliviensis* occasionally uses LS gaits on substrates of both sizes and all orientations. This could represent random flexibility and lack of "fine-tuning," but it is also possible that instances of LS could be associated with substrate variables not measured here, such as branch surface continuity or branch compliance. In fact, Stevens (2006) showed that *Loris tardigradus* changed its limb phase when laboratory substrates were manipulated to challenge stability, i.e., rotated or displaced in different planes. Measuring substrate compliance and displacement is rare in field studies (e.g., Demes et al. 1996), but continued work in this area would enhance our understanding of variation in primate kinematics.

The fact that *Saimiri boliviensis* employed the longest limb relative contact times (duty factors) on the largest substrates is not consistent with adaptive hypotheses indicating that primates employ this kinematic adjustment to enhance stability and decrease branch oscillations on small, terminal branches. By comparison, Stevens (2003) found that strepsirrhines used higher duty factors on smaller substrates, but the substrate size effect was subtle and variable across species. Although it seems counterintuitive that large branches should present more of a functional challenge to squirrel monkeys than smaller ones, we hypothesize that larger duty factors may be a response to the increasing difficulty of grasping as branch circumference increases relative to hand or foot size. Certainly, we need more data from both laboratory and field to further test this hypothesis. Further study could also help reconcile the fact that we found no significant change in duty factors with substrate orientation, contra Stevens' (2003) observations of increased duty factors on declines in strepsirrhines and observations of Nyakatura et al. (2008) of increased hind/fore duty factor ratios on inclines in cotton-top tamarins.

Because we studied the same species in the laboratory and field, we can evaluate the extent to which laboratory data are representative of more natural behavior. As discussed earlier, although the *distribution* of gait types used on the floor and on branches (combined) are more similar than the distribution of gait types used on the pole, variability of gait selection on the floor has little to do with arboreality and more to do with freedom from constraints associated with balancing the body. With respect to gait variability on branches, our study did not find a significant influence of relative substrate size or orientation. However, it is possible that gait choice on branches is correlated with substrate variables we have not measured here, such as discontinuity or branch flexibility. After correcting for differences in duty factor, limb phases on the pole were similar on average to those observed on branches. However, the fact that gait choice was less variable on the pole than on branches, suggests that a single, stable, horizontal pole does not adequately capture the full extent of gait flexibility in *Saimiri boliviensis*. Nevertheless, some aspects of the arboreal environment are captured well by the use of horizontal poles in laboratory studies, as indicated by squirrel monkeys' use of similar mean duty factors (at a given relative speed) on the pole compared to branches of the same size range.

The benefits of laboratory studies are that animals are more easily filmed and kinematic variables are more easily and accurately measured than in the field. In addition, unique biomechanical hypotheses can be tested by coaxing subjects to move on substrates they might naturally avoid, or by artificially changing their biomechanical properties (e.g., Young et al. 2007). Primate laboratory studies are limited however, by the difficulty of housing or collecting data on more than a few individuals at a time, and by the difficulty of mimicking the complexity of an arboreal environment. Field studies allow one to collect data on many more individuals simultaneously and to test how "fine-tuned" primate locomotion is to the wide variety of substrate challenges found in the natural environment. In addition, field studies can be used to assess how kinematic characteristics might actually affect performance and therefore, evolutionary fitness (Arnold 1983). The disadvantages of locomotor field studies are that animals are difficult to film because they are either far away or obscured by foliage, and some variables cannot be measured without a fair amount of difficulty, e.g., substrate reaction forces.

The best solution is to use laboratory studies to test specific hypotheses in a controlled setting, while using field studies to evaluate the "messiness" of real locomotion and as a guide for selecting appropriate substrates for the lab. As a start, our comparative analysis has revealed that in order to capture the full range of quadrupedal kinematics employed by primates, laboratory studies would benefit from using a wider range of simulated arboreal substrates. Our field results suggest that varying substrate size may be even more critical than substrate orientation, while results from several laboratory studies imply that variation in substrate orientation is also very functionally informative. Of course, substrate variation is much more complex than simply size or orientation. Primates face other arboreal challenges such as discontinuous pathways and unstable branches. We were not able to assess these here, but such factors might have accounted for a portion of the kinematic variation we discovered. Although the true complexity of natural substrates and/or irregular locomotor movements are difficult to measure in the field, it is promising to see efforts to address some of these factors in a controlled laboratory setting (e.g., Stevens 2003, 2006; Demes et al. 2006; Higurashi et al. 2008).

Summary and Conclusions

To summarize, our field study of quadrupedal kinematics revealed that *Saimiri boliviensis* did not "fine-tune" its limb phases in a consistent manner with respect to relative substrate size or substrate orientation, but this species did significantly increase its duty factors in response to increased branch size. Our laboratory study revealed (not surprisingly) that quadrupedal kinematics on the floor is not a good representation of an arboreal primate's behavior in a natural habitat. Our laboratory monkeys used comparable duty factors and average limb phases when moving on the horizontal pole and similarly sized arboreal branches, but showed less variability in limb phase on the pole. Therefore, we conclude that laboratory studies can certainly be improved by incorporating more varied substrates, particularly with respect to size.

The flexibility exhibited by *Saimiri boliviensis* in limb phase across different natural substrates contradicts some previous laboratory studies on *Saimiri* and other primates that have found a clear effect of substrate orientation on limb phase, i.e., higher limb phases on inclines and lower limb phases on declines. It is also somewhat inconsistent with the view that DS gait in primates is functionally preferable to other gaits on relatively small branches, since both DS and LS gaits were used on small and large branches. Similarly, our results for duty factor are the opposite of what one would expect if primates were most challenged by stability on relatively small branches. There are several implications of the fact that our field results stand in distinction to laboratory studies. It is possible that our field study has revealed the need to examine aspects of substrate variation (in the laboratory or field) beyond size and orientation in order to determine what is driving kinematic variability in this species. Alternatively, *Saimiri* may happen to be a particularly flexible primate that does not require fine-tuning of its kinematic features to navigate complex and changing substrates. Either way, future studies of primate quadrupedalism, whether in the laboratory or the field, would benefit from a consideration of the unique biomechanical challenges presented by a complex natural environment as well as the distinctive approaches individual species may exhibit to those challenges.

Acknowledgments We thank the staff at the Center for Neotropical Primate Research and Resources (CNPRR, Mobile, AL) for assistance with data collection and animal care. We also thank staff and researchers at Manu National Park, Peru for permission to conduct this research at Cocha Cashu Biological Station, and for assistance with videotaping *Saimiri boliviensis*. Special thanks to Teresa Burgess, Abbas Haidry, and Parham Daghighi for assistance with video analysis. Laboratory research was supported by the L.S.B. Leakey Foundation (Grant 38648).

References

Alexander RM, Maloiy GMO (1984) Stride lengths and stride frequencies of primates. J Zool Lond 202:577–582.

Arms A, Voges D, Fischer MS, Preuschoft H (2002) Arboreal locomotion in small new-world primates. Z Morphol Anthropol 83:243–263.

Arnold SJ (1983) Morphology, performance and fitness. Am Zool 23:347–361.

Boinski S (1989) The positional behavior and substrate use of squirrel monkeys: ecological implications. J Hum Evol 18:659–677.

Boinski S, Sughrue K, Selvaggi L, Quatrone R, Henry M, Cropp S (2002) An expanded test of the ecological model of primate social evolution: competitive regimes and female bonding in three species of squirrel monkeys (*Saimiri oerstedii*, *S. boliviensis*, and *S. sciureus*). Behaviour 139:227–261.

Cartmill M (1972) Arboreal adaptations and the origin of the order Primates. In: Tuttle R (ed), The Functional and Evolutionary Biology of Primates. Chicago, Aldine 97–122.

Cartmill M (1974) Pads and claws in arboreal locomotion. In: Jenkins FA (ed), Primate Locomotion. New York, Academic Press 45–83.

Cartmill M, Lemelin P, Schmitt D (2002) Support polygons and symmetrical gaits in mammals. Zool J Linn Soc 136:401–420.

Cartmill M, Lemelin P, Schmitt D (2007a) Understanding the adaptive value of diagonal-sequence gaits in primates: a comment on Shapiro and Raichlen, 2005. Am J Phys Anthropol 133:822–825.

Cartmill M, Lemelin P, Schmitt D (2007b) Primate gaits and primate origins. In: Ravosa MJ, Dagosto M (eds), Primate Origins: Adaptations and Evolution. New York, Springer 403–435.

Conover WJ, Iman RI (1981) Rank transformations as a bridge between parametric and nonparametric statistics. Am Stat 35:124–133.

Demes B, Larson SG, Stern JT Jr, Jungers WL, Biknivicius AR, Schmitt D (1994) The kinetics of primate quadrupedalism: "hindlimb drive" reconsidered. J Hum Evol 26:353–374.

Demes B, Jungers WL, Fleagle JG, Wunderlich RE, Richmond BG, Lemelin P (1996) Body size and leaping kinematics in Malagasy vertical clingers and leapers. J Hum Evol 31:367–388.

Demes B, Carlson KJ, Franz TM (2006) Cutting corners: the dynamics of turning behaviors in two primate species. J Exp Biol 209:927–937.

Dunbar DC, Badam GL (2000) Locomotion and posture during terminal branch feeding. Int J Primatol 21:649–669.

Fleagle JG, Mittermeier RA (1980) Locomotor behavior, body size, and comparative ecology of seven Surinam monkeys. Am J Phys Anthropol 52:301–314.

Fontaine R (1990) Positional behavior in *Saimiri boliviensis* and *Ateles geoffroyi*. Am J Phys Anthropol 82:485–508.

Franz TM, Demes B, Carlson KJ (2005) Gait mechanics of lemurid primates on terrestrial and arboreal substrates. J Hum Evol 48:199–217.

Griffin TM, Main RP, Farley CT (2004) Biomechanics of quadrupedal walking: how do four-legged animals achieve inverted pendulum-like movements? J Exp Biol 207:3545–3558.

Hedrick T (2007) "DLT Data Viewer 2", Digitizing and DLT in MATLAB. http://www.unc.edu/~thedrick/software1.html

Higurashi Y, Hirasaki E, Kumakura H (2008) Gaits of Japanese macaques (*Macaca fuscata*) on a horizontal ladder and arboreal stability. Am J Phys Anthropol 138(4): 448–457.

Hildebrand M (1966) Analysis of the symmetrical gaits of tetrapods. Folia Biotheoretica 1–22.

Hildebrand M (1967) Symmetrical gaits of primates. Am J Phys Anthropol 26:119–130.

Hildebrand M (1976) Analysis of tetrapod gaits: general considerations and symmetrical gaits. In: Herman RM, Grillner S, Stein PSG et al (eds), Neural Control of Locomotion. New York, Plenum Press 203–236.

Hurov JR (1982) Diagonal walking in captive infant vervet monkeys. Am J Primatol 2:211–213.

Johnson SE, Shapiro LJ (1998) Positional behavior and vertebral morphology in atelines and cebines. Am J Phys Anthropol 105:333–354.

Kimura T, Okada M, Ishida H (1979) Kinesiological characteristics of primate walking: its significance in human walking. In: Morbeck ME, Preuschoft H and Gomberg N (eds), Environment, Behavior and Morphology: Dynamic Interactions in Primates. New York, G Fischer 297–311.

Larson SG (1998) Unique aspects of quadrupedal locomotion in nonhuman primates. In: Strasser E, Fleagle J, Rosenberger A, McHenry H (eds), Primate Locomotion. New York, Plenum Press 157–173.

Larson SG, Schmitt D, Lemelin P, Hamrick M (2000) Uniqueness of primate forelimb posture during quadrupedal locomotion. Am J Phys Anthropol 112:87–101.

Larson SG, Schmitt D, Lemelin P, Hamrick M (2001) Limb excursion during quadrupedal walking: how do primates compare to other mammals? J Zool Lond 255:353–365.

Lemelin P, Schmitt D, Cartmill M (2003) Footfall patterns and interlimb coordination in opossums (Family Didelphidae): evidence for the evolution of diagonal sequence gaits in primates. J Zool Lond 260:423–429.

Li Y (2000) Arboreal primates and the origin of diagonal gait. Acta Anthropological Sinica (Suppl) 19:83–89.

Mitchell CL (1990) The ecological basis for female social dominance: a behavioral study of the squirrel monkey (*Saimiri sciureus*) in the wild. PhD Dissertation, Princeton University, NJ.

Nakano Y (1996) Footfall patterns in the early development of the quadrupedal walking of Japanese macaques. Folia Primatol 66:113–125.

Nyakatura J, Fischer MS, Schmidt M (2008) Gait parameter adjustments of cotton-top tamarins (*Saguinus oedipus*, Callitrichidae) to locomotion on inclined arboreal substrates. Am J Phys Anthropol 135:13–26.

Pridmore PA (1994) Locomotion in *Dromiciops australis* (Marsupialia: Microbiotheriidae). Austral J Zool 42:679–699.

Prost JH, Sussman RW (1969) Monkey locomotion on inclined surfaces. Am J Phys Anthropol 31:53–58.

Reynolds TR (1985) Mechanics of increased support of weight by the hindlimbs in primates. Am J Phys Anthropol 67:335–349.

Rice WR (1988) Analyzing tables of statitical tests. Evolution 43:223–225.

Rollinson J, Martin RD (1981) Comparative aspects of primate locomotion, with special reference to arboreal cercopithecines. Symposia of the Zoological Society of London 48:377–427.

Rothman J, Chapman C, Twinomugisha D, Wasserman MD, Lambert JE, Goldberg TL (2008) Measuring physical traits of primates remotely: the use of parallel lasers. Am J Primatol 70:1191–1195.

Schmidt M (2005) Quadrupedal locomotion in squirrel monkeys (Cebidae: *Saimiri sciureus*): a cineradiographic study of limb kinematics and related substrate reaction forces. Am J Phys Anthropol 128:359–370.

Schmitt D (1994) Forelimb mechanics as a function of substrate type during quadrupedalism in two anthropoid primates. J Hum Evol 26:441–457.

Schmitt D (1998) Forelimb mechanics during arboreal and terrestrial quadrupedalism in Old World monkeys. In: Strasser E, Fleagle J, Rosenberger A, McHenry H (eds), Primate Locomotion: Recent Advances. New York, Plenum Press 175–200.

Schmitt D (1999) Compliant walking in primates. J Zool Lond 248:149–160.

Schmitt D (2003a) Evolutionary implications of the unusual walking mechanics of the common marmoset (*C. jacchus*). Am J Phys Anthropol 122:28–37.

Schmitt D (2003b) Mediolateral reaction forces and forelimb anatomy in quadrupedal primates: implications for interpreting locomotor behavior in fossil primates. J Hum Evol 44:47–58.

Schmitt D (2003c) Substrate size and primate forelimb mechanics: implications for understanding the evolution of primate locomotion. Int J Primatol 24:1023–1036.

Schmitt D, Lemelin P (2002) Origins of primate locomotion: gait mechanics of the woolly opossum. Am J Phys Anthropol 118:231–238.

Schmitt D, Hanna JB (2004) Substrate alters forelimb to hindlimb peak force ratios in primates. J Hum Evol 46:239–254.

Schmitt D, Cartmill M, Griffin TM, Hannah JB, Lemelin P (2006) Adaptive value of ambling gaits in primates and other mammals. J Exp Biol 209:2042–2049.

Schultz AH (1963) Relations between the lengths of the main parts of the foot skeleton in primates. Folia Primatol 1:150–171.

Sellers W, Crompton R (2004) Automatic monitoring of primate locomotor behaviour using accelerometers. Folia Primatol 75:279–293.

Shapiro LJ, Raichlen DA (2005) Lateral sequence walking in infant *Papio cynocephalus*: implications for the evolution of diagonal sequence walking in primates. Am J Phys Anthropol 126:205–213.

Shapiro LJ, Raichlen DA (2006) Limb proportions and the ontogeny of quadrupedal walking in infant baboons (*Papio cynocephalus*). J Zool 269:191–203.

Sokal RR, Rohlf FJ (1995) Biometry. New York, WH Freeman.

Stevens NJ (2003) The influence of substrate size, orientation and compliance upon prosimian arboreal quadrupedalism. PhD Dissertation, Stony Brook University, New York.

Stevens NJ (2006) Stability, limb coordination and substrate type: the ecorelevance of gait sequence pattern in primates. J Exp Zool 305A:953–963.
Stevens NJ (2007) The effect of branch diameter on primate gait sequence pattern. Am J Primatol 70:1–7.
Terborgh J (1983) Five New World Primates: a Study in Comparative Ecology. Princeton, NJ, Princeton University Press.
Vilensky JA, Patrick MC (1985) Gait characteristics of two squirrel monkeys, with emphasis on relationships with speed and neural control. Am J Phys Anthropol 68:429–444.
Vilensky JA, Larson SG (1989) Primate locomotion: utilization and control of symmetrical gaits. Annu Rev Anthrop 18:17–35.
Vilensky JA, Moore AM (1992) Utilization of lateral- and diagonal-sequence gaits at identical speeds by individual vervet monkeys. In: Matano S, Tuttle RH, Ishida H, Goodman M (eds), Topics in Primatology, Vol. 3: Evolutionary Biology, Reproductive Endocrinology and Virology. Tokyo, University of Tokyo Press 129–137.
Vilensky JA, Moore AM, Libii JN (1994) Squirrel monkey locomotion on an inclined treadmill: implications for the evolution of gaits. J Hum Evol 26:375–386.
Wallace IJ, Demes B (2008) Symmetrical gaits of *Cebus apella*: implications for the functional significance of diagonal sequence gait in primates. J Hum Evol 54:783–794.
White TD (1990) Gait selection in the brush-tail possum (*Trichosurus vulpecula*), the northern quoll (*Dasyurus hallucatus*), and the virginia opossum (*Didelphis virginiana*). J Mammal 71:79–84.
Youlatos D (1999) Comparative locomotion of six sympatric primates in Ecuador. Ann Sci Nat Zool Biol Anim 20:161–168.
Young JW, Patel BA, Stevens NJ (2007) Body mass distribution and gait mechanics in fat-tailed dwarf lemurs (*Cheirogaleus medius*) and patas monkeys (*Erythrocebus patas*). J Hum Evol 53:26–40.
Young JW (2008) Ontogeny of locomotion in *Saimiri boliviensis* and *Callithrix jacchus*: implications for primate locomotor ecology and evolution. PhD dissertation, Stony Brook University, New York.
Young JW (2009) Substrate determines asymmetrical gait dynamics in marmosets (*Callithrix jacchus*) and squirrel monkeys (*Saimiri boliviensis*). Am J Phys Anthropol 138: 403–420.

Index

K. D'Août and E.E. Vereecke (eds.), *Primate Locomotion: Linking Field and Laboratory Research*, Developments in Primatology: Progress and Prospects,
DOI 10.1007/978-1-4419-1420-0,

Q

R

S